Organic Thin Films

ACS SYMPOSIUM SERIES **695**

Organic Thin Films
Structure and Applications

Curtis W. Frank, EDITOR
Stanford University

Developed from a symposium sponsored by the Division
of Polymer Chemistry
at the 213th National Meeting
of the American Chemical Society,
San Francisco, California,
April 13–17, 1997

American Chemical Society, Washington, DC

Chemistry Library

Library of Congress Cataloging-in-Publication Data

Organic thin films : structure and applications / Curtis W. Frank, editor

    p.   cm.—(ACS symposium series, ISSN 0097–6156; 695)

    "Developed from a symposium sponsored by the Division of Polymer Chemistry at the 213th National Meeting of the American Chemical Society, San Francisco, CA, April 13–17, 1997."

    Includes bibliographical references and indexes.

    ISBN 0–8412–3564–3

    1. Thin films. 2. Organic compounds. 3. Monomolecular films. 4. Photonics—Materials

    I. Frank, C. W. II. American Chemical Society. Division of Polymer Chemistry. III. American Chemical Society. Meeting (213th : 1997 : San Francisco, Calif.) IV. Series.

TA418.9.T45074 1998
621.3815′2—dc21
                                          98–6368
                                              CIP

This book is printed on acid-free paper.

PRINTED IN THE UNITED STATES OF AMERICA

Foreword

THE ACS SYMPOSIUM SERIES was first published in 1974 to provide a mechanism for publishing symposia quickly in book form. The purpose of the series is to publish timely, comprehensive books developed from ACS sponsored symposia based on current scientific research. Occasionally, books are developed from symposia sponsored by other organizations when the topic is of keen interest to the chemistry audience.

Before agreeing to publish a book, the proposed table of contents is reviewed for appropriate and comprehensive coverage and for interest to the audience. Some papers may be excluded in order to better focus the book; others may be added to provide comprehensiveness. When appropriate, overview or introductory chapters are added. Drafts of chapters are peer-reviewed prior to final acceptance or rejection, and manuscripts are prepared in camera-ready format.

As a rule, only original research papers and original review papers are included in the volumes. Verbatim reproductions of previously published papers are not accepted.

ACS BOOKS DEPARTMENT

Contents

ix

MICROLITHOGRAPHY AND PACKAGING

INDEXES

Preface

This book is an outgrowth of the Symposium on Organic Thin Films sponsored by the Division of Polymer Chemistry and held at the San Francisco American Chemical Society Meeting. This meeting occurred on the 10th anniversary of the publication in Langmuir of the Department of Energy panel report on the science and applications of organic thin films. In the Overview chapter, the chairman of that panel, J. D. Swalen, provides an updated perspective on the extent to which these materials have fulfilled the expectations of a decade ago. Contributions selected for this monograph are divided into sections with initial emphasis on structures (monolayers, supramolecular assemblies, and nanostructures) followed by applications (photonics, microlithography, and microelectronics packaging). However, these divisions are somewhat arbitrary, and there is a general coupling between structure and applications throughout many of the chapters. Readers will find sufficient background on Langmuir monolayers, Langmuir–Blodgett–Kuhn multilayer films, self-assembled monolayers, layer-by-layer deposition, and ultrathin spin-cast films to appreciate the relationship between structure and properties in constrained geometries. They will also be introduced to the assembly strategies of supramolecular and nanostructured materials that contain "buried" interfaces. Finally, they will find a number of actual and potential applications of organic thin films in microlithography, integrated circuit packaging, liquid crystal and electroluminescent displays, nonlinear optical devices, and biosensors. Taken as a whole, the chapters in these proceedings demonstrate the vitality of the materials chemistry of organic thin films. Built upon a solid foundation, the new materials, new fabrication protocols, and new applications suggest that the next decade will provide additional fundamental and technological advances.

Although research into the structure of Langmuir films of amphiphilic compounds at the air–water interface extends back almost a half century, there has been a resurgence of activity because of the existence of new surface-active materials and novel approaches to the characterization of the structure and dynamics of these pseudo-two-dimensional systems, as described in the Monolayers section. For example, Goedel uses monolayers of polyisoprenes having sulfonate head groups that will anchor to the air-water interface to model polymer brushes; a thermodynamic analysis yields a plausible chain conformation in the "melt" state. In addition, Kampf et al. describe a two-dimensional stress-relaxation experiment to monitor the monolayer collapse of a pentaerythritol-based polyether dendrimer, which undergoes dramatic conformational changes during the compression process. Finally, Yim et al. describe rheologically well-defined extensional flow and simple shear flow experiments on dipalmitoylphosphatidylcholine monolayers and propose a surface capillary number to explain the flow-induced deformation process.

Whereas Langmuir films are of fundamental interest in the study of low-dimensionality systems, self-assembled monolayers (SAMs) formed by spontaneous adsorption on a solid substrate of hydrocarbons having surface-active functional groups may be more suited to technological applications. Because silicon technology is so advanced, there is interest in understanding how to marry organic thin films with the silicon substrate, the ultimate objective being to prepare molecular electronic devices. Zenou et al. address this through adsorption of various quinolinium-based chromophores, which leads to the ability to modulate the silicon work function. In one phase of a study on preparation of model surfaces for liquid crystal alignment, Yang and Frank explore how the hydrophilic/hydrophobic character of co-functionalized alkyl trichorosilanes influences the composition of binary SAMs. Liquid crystal anchoring is addressed directly by Abbott et al. in their study of alkanethiols on gold in which they exploit the manner of substrate gold fabrication, the alkanethiol structure, and the procedure for alkanethiol adsorption to monitor the orientation of nematic liquid crystals on these surfaces.

The second section of the book, on Supramolecular Assemblies, comprises chapters in which, ideally, the materials exhibit a higher-order state of organization. Moreover, the materials are either of biological origin or are inspired by biological structural motifs. For example, Heibel et al. describe their progress toward formation of a polymer-supported lipid membrane consisting of hydrophilic polymer to which is attached a lipid monolayer (or bilayer), with this whole supramolecular assembly adsorbed on a functionalized substrate. If such a system can be fabricated and stabilized with biologically-active enzymes, it has the potential for application as a biosensor. One route to the stabilization could come from polymerization of monomeric lipids, such as is described by Sisson et al. in their study of cross-linking efficiency of bis-substituted lipids. Polypeptides constitute another class of biomaterials of interest for substrate modification, as shown by Menzel et al. who investigate the thickness of polyglutamate layers grafted from binary SAMs with varying surface initiation density and with varying N-carboxyanhydride (NCA) monomer reactivity. Finally, Chang and Frank also examine polyglutamates grafted from silicon substrates but do so in a novel approach involving vapor-phase deposition of the NCA followed by polymerization to yield α-helical homopolypeptides as well as random and block peptide structures.

The third section on Nanostructures deals with materials that form heterostructured systems either because of thermodynamic driving forces or layered fabrication methods. Typically, an important aspect is the effect of confinement to a constrained geometry on the properties of the resulting system. This is demonstrated by Chen and Jenekhe in their study of quantum confinement effects in thin films of block conjugated copolymers, where they observe stimulated emission in blends and suggest that there may be potential applications in solid-state diode lasers. Block copolymers have also been of considerable interest in solution studies of swollen polymer brushes because adsorption from a selective solvent can lead to anchoring by the insoluble block with the soluble block free to expand into the solution. Liu uses this approach to prepare poly-(styrene-b-2-cinnamoylethyl methacrylate) brushes on silica particles. Layered assembly methods coupled with thermodynamic considerations are the focus of the chapter by Advincula et al. in which they prepare Langmuir–Blodgett–Kuhn (LBK) films of poly(vinylidene fluoride-co-trifluoroethylene) and poly(alkyl methacrylates). Clark et al. also use a multilayer fabrication approach, but instead of the LBK technique, they use the layer-by-layer deposition method developed by Decher and examine the effect of salt counterions on the efficiency of SAM surface templating. Coulombic interactions are also the basis of a related multilayer approach by Bliznyuk et al., but the materials are latex particles having positive or negative surface charges. Finally, Prucker et al. use surface plasmon spectroscopy to examine poly(methyl methacrylate) films and show that the glass transition temperature is substantially depressed in ultrathin films less than 300 Å thick.

The unifying feature of the Photonics section is that all systems exhibit nonlinear optical (NLO) behavior. This field has seen considerable effort directed toward development of new NLO materials having large electro-optic coefficients and long-term stability; both issues are addressed in this section. For example, Wang et al. describe a new perfluoroalkyldicyanovinyl acceptor chromophore with promising electro-optic characteristics. In addition, He et al. present results involving chromophore modification to reduce electrostatic intermolecular interactions, which typically lead to low poling efficiencies. Maintenance of the poled order is a critical issue for device performance. To address this, Roberts et al. use the LBK deposition method to assemble a polymeric salt at the air–water interface from a water-insoluble polycation and a water-soluble polyanion; this approach allows maintenance of noncentrosymmetric order through ionic and hydrogen bonding. In an approach directed at improved high-temperature stability, Ra et al. describe a new trifunctionalized Disperse-Red type of chromophore incorporated in a thermosetting polyurethane. By contrast, Sekkat et al. focus on the effect of molecular structure on sub-glass relaxation dynamics in azobenzene-containing polyimides and describe a combined photo and thermal isomerization protocol for achieving molecular orientation substantially below the glass transition temperature. Trollsås et al. take yet another approach and examine ferroelectric liquid crystal polymers, which have intrinsic, thermodynamically stable polar order. Finally, Prêtre et al. study a series of tricyanovinylaniline polymer thin films and demonstrate the use of these materials for ultrashort optical pulse diagnostics.

The final section on Microlithography and Packaging includes chapters describing the use of organic thin films in microelectronics, probably the most technologically significant application to date. First, Hinsberg et al. describe current understanding of the chemical and physical processes that influence the performance of advanced photoresist materials based on the acid-catalyzed, chemical amplification method. Related work on the effect of macromolecular architecture on the dissolution of phenolic polymers, which are typically the resins of choice for the chemically-amplified systems, is presented by Barclay et al. Although the organic thin film that is essential for the lithographic patterning does not appear in the final integrated circuit assembly, there are situations in which a polymer is incorporated. One such example is the interlayer dielectric that separates lines of metallization. Hedrick et al. point out that future generation devices will be limited by increased signal delays unless capacitive coupling and crosstalk between metal interconnects is reduced. They describe an approach based on polyimide–silsesquioxane organic–inorganic nanocomposites to address this problem.

Acknowledgments

The Editor wishes to acknowledge the Center on Polymer Interfaces and Macromolecular Assemblies (CPIMA), an NSF Materials Research Science and Engineering Center formed between Stanford University, IBM Almaden Research Center, and the University of California, Davis for the financial support of a significant fraction of the work presented herein. In addition, thanks are due to the Max Planck Institute for Polymer Research in Mainz, Germany and Prof. Wolfgang Knoll, who graciously hosted the Editor's sabbatical during which much of the organizational work was performed, and to Andreas Offenhäuser and Chistopher E. D. Chidsey, who handled the review process for manuscripts coming from the Editor's laboratory. Finally, sincere thanks are due to Jeannette Cosby, who kept the whole administrative process on track.

CURTIS W. FRANK
Center on Polymer Interfaces and
    Macromolecular Assemblies (CPIMA)
Department of Chemical Engineering
Stanford University
Stanford, CA. 94305–5025

OVERVIEW

Chapter 1

Some Emerging Organic-Thin-Films Technologies

J. D. Swalen

Department of Physics, University of California at Santa Cruz, Santa Cruz, CA 95064

Ordered thin organic films in the thickness range of a few to several hundred nanometers currently hold considerable technological promise. Electronic and optical devices incorporate structures that are in this thickness range, and organic thin films have been proposed, and in some cases applied, as passive or active components traditionally fabricated with other materials. Some proposed, but not all, applications of organic thin films will be discussed.

Organic thin films, which have some designed molecular order, can exhibit different material properties from the collective behavior of the molecules in a restricted geometry. This is materials science from the molecular point of view. Molecular solids are now being designed into organized films to perform new and special functions. This has stimulated many intensive scientific investigations in the preparation of new films and their characterizations. Many of the newer surface science techniques, designed and developed for semiconductors, metals, and dielectrics, are addressing specific details about structure and morphology of these organic films. In the past, organic films were considered to be too fragile and not of sufficient purity to give reliable and consistent properties to make them of much use. This is changing with new materials. We are seeing many new compounds and polymers being synthesized and made into thin films by a variety of techniques. These films are carefully constructed to avoid the common problems. Extensive scientific studies have revealing a wealth of knowledge useful in the development of these new thin film materials, exhibiting specifically desired behaviors. A number of books, reviews, and general articles have been published about organic thin films, both Langmuir-Blodgett-Kuhn films (LBK) and self-assembled films (SAM). *(1-9)* Other related topics cover polymer surfaces and interfaces *(10)* and the optical properties of organic thin films*(11-14)*. The reader is directed to these references for more details. Although the preparation and study of LBK films are still very active areas of research, *(15)* self-assembled films and polymeric films, either deposited as a polymeric film or polymerized in situ, are rapidly overtaking the field. Fluorinated surfaces are being applied to lower the surface energy.

Much of the scientific research on organic thin films has been on determining the orientation of the molecules and their packing. Is there order in the films? What are the domain sizes, if any? Does the substrate or superstrate induce any structure? Molecular motion is also being studied. This includes lateral diffusion, gas or ionic diffusion through the film, phase transitions, melting, and other relaxation processes. Interface properties between two layers of different materials are important. This is studied often between two different polymeric layers. Surface functionalization for specific interactions is an active area. This leads to wetting phenomena, adsorption, adhesion, and lubrication. Much work is being done in the characterization of the optical, spectroscopic, and electrical properties, including energy transfer between molecules in the film and between layers. Clearly, one needs to know the morphology of the substrate and its influence. The alignment of molecules and the order parameter are needed. Cooperative effects can change the behavior of any film significantly, and the extent of interaction, both laterally and vertically between layers or substrate is important. With this base of knowledge it is hoped that this will lead to our understanding of intermolecular interactions, energy transfer, and dynamic behavior for best optimizing a film for a specific research study or application.

The following articles in this ACS monograph report research along these topics and for the most part will not be covered in this review. These topics include new self-assembled layers, new Langmuir-Blodgett-Kuhn films, polymeric thin films, fluorinated surfaces, the alignment of liquid crystals by surface treatment, nonlinear optical films, light emitting diodes, photoresists, low dielectric films for faster electronic circuit speeds, sensors, and nano-particles. The reader is referred to these current scientific studies.

Current Applications of Organic Thin Films

There has been an increased utilization of organic thin films in many new electronic, optical, and mechanical devices; for example, organic photoconductors are being use in copiers and printers. The first organic photoconductors were charge transfer polymers which performed both the charge generation and conduction processes. Later versions separated these roles into different layers, each of which could be optimized. Liquid crystal displays are now common in watches and laptops. Not only have the liquid crystals been improved, but the surface treatment and manufacturability have been significantly advanced. Related to this is surface modification by organic materials for other specific applications. Photoresists and e-beam resists are the keys to the success of very large scale integrated (VLSI) electronic circuits. Without these resists, most electronic equipment we know today would not exist. These polymers are spun onto the semiconductor and the circuit pattern is exposed, leading to main chain scission or cross linking; with either a wet process or a dry one, sections are removed. Further treatment can include either diffusion of various semiconductor elements or metalization for conduction lines. Layer by layer, the total package is developed. Current research is directed towards finer features in the patterns and changes in the surface characteristics for subsequent layers. Polyimide and other related high temperature polymers are being used extensively in VLSI as insulating layers or for packaging. Previously, sputtered SiO_2 films were used. This process, however, required a large sputtering chamber, the films did not planarize well, and they were subject to cracks. Polyimide was, however, not without problems. Imidization gives water as a product which could be detrimental to semiconductor devices. Also, the adhesion of metal lines to polyimide was at times not good. Both of these problems now seem to be under control. To increase the storage density on magnetic disks, efforts to reduce the bit size by flying the read-write head closer to the disk surface have generated new problems called "stiction." Here, the head would stick or have high friction with the disk. To overcome these problems, lubricants were added to the disk surface. The most popular have been fluorinated polymeric ethers, usually on top of an amorphous carbon coating, allowing the read-write head to fly much closer to the magnetic storage disk surface. A number of groups are studying friction, lubrication and adhesion from a fundamental point of view *(16-22)*. On another topic, electrolytic capacitors were introduced by Sanyo in 1983 based on TCNQ *(23)*. Since that time, a number of other capacitors have been introduced into the marketplace. Polypyrrole, polyaniline and polythiophene have all been used. These capacitors range in values from 0.1 µF to 200 µF and have low equivalent series resistance and high frequency impedance with good reliability and lifetimes.

Some ten years ago, the Department of Energy realized that, in their materials research efforts, they were supporting very little on organic thin films. As a consequence of all this new activity in the applications of organic thin films, they commissioned a review panel, and the results of our discussions and concurrences were published in the journal *Langmuir(24)*. The applied topics, which were chosen because of their perceived significance, were thin film optics, sensors and transducers, protective layers for packaging and insulating, patterning of surfaces of electronic circuit components, functionalized surfaces, and coatings for electrodes. In addition, the scientific research on intermolecular forces, intermolecular order, and analytical methods for the characterizing thin films were addressed. This paper will be an update of these applications of organic thin films and add some new ones that have come into vogue.

New Applications of Organic Thin Films

Light Emitting Diodes. Some ten years ago Tang and VanSlyke *(25)* reported a two layer organic light emitting diode, consisting of holes ejected from an anode of indium-tin-oxide (ITO), a hole-transport layer, an electron-transport and light-emitting layer and a metal, e. g., magnesium with 10% silver, having a low work function. Electrons and holes are produced at opposite electrodes and diffuse to the middle of the film where they recombine to produce light. The light emitter was a quinoline complex of aluminum and the hole transport layer was a triphenyldiamine. Shortly thereafter, Friend and collaborators *(26)* made a single layer diode with the polymer, poly(p-phenylenevinylene). Although the light emission was rather low and the diodes had only a short term stability, many research groups entered the field and the progress has been significant *(27-29)*. Organic light emitting diodes are being considered for display application because with recent advances the intensity has been increased approximately to that of a fluorescent lamp, and most colors have been obtained with a great variety of fluorescing dyes. The life-time has also been increased to several thousands of hours; however, degradation is still a major problem. Oxygen, probably coming from the ITO anode, is believed to be one of the culprits *(30,31);* water is another *(29)*. Charges tend to be trapped. In the two-layer diode the aluminum complex appears to diffuse into the amine layer causing dark spots *(30)*. Research is continuing to increase intensity and lifetime*(28)*. New compounds are being tried and additional layers are being added to enhance performance.

Nonlinear Optical Materials. The research on nonlinear optical polymers has made significant advances over the last ten years *(13)*. These polymers contain a dye with an electron donor at one end and a charge acceptor at the other end. This chromophore dye can be a guest in the polymer host or be chemically attached. The "drosophila" of these NLO chromophores is the dye, disperse red No. 1. It is an azo dye with two phenyl groups, one on each side of the -N=N- group. In the para position on one benzene ring is the donor, an aliphatic substituted amine, and on the other benzene ring, also in the para position, is a nitro group. It has the chemical formula: $(C_2H_5-)HOC_2H_4-N-\varphi-N=N-\varphi-NO_2$, where the symbol φ represents a phenyl ring. A majority of the NLO experiments have been done with this dye, but recently many more chromophores have been synthesized. The optical nonlinearity has been increased by changing the commonly used charge transfer donor, the nitro group, to a tricyanovinyl group. Surprisingly, the increased unsaturated bonds in the tricyanovinyl group do not significantly degrade the thermal stability. In fact, they perform well. In addition, it was found *(32)* that by changing the aliphatic amine to an aromatic amine improved the thermal stability of the dye by 50 to 100 °C. This is important for fabricating any NLO device exhibiting long lifetimes. Further, by substituting a heterocyclic ring for one of the phenyl groups, it has been found that the nonlinearity increases, but the thermal stability is slighted degraded. Hence, with these improvements the technology is here for the production of modulators and switches made from NLO polymers. As yet, to the best of my knowledge, none have appeared in the market. In fact, a number of research laboratories have stopped or reduced research activities in NLO polymers. Hopefully, a viable product will be forthcoming to spur continued activity in this area.

Photorefractives. Although interest in nonlinear optical polymers for electrooptic modulators and deflectors seems to be waning, the activity on organic photorefractives for optical storage is vigorous *(33-36)*. In this material both photoconductivity and electrooptic activity must be present. Charges are generated by light and they become mobile and separate (An electric field has been found to enhance the effect.). Commonly only one charge moves and is trapped, and the other remains rather stationary. An internal electric field is thus generated by these separated charges which produces a difference in the index of refraction. Two mechanisms cause this effect: the electro-optic coefficient and a molecular rotational term from the anisotropy; both produce an index change. These indices of refraction patterns are then observable, usually as a holographic image. Recently, an organic glass was developed as a photorefractive material which had a higher concentration of active chromophores (36). This material has a higher refractive index change but is still somewhat slow.

Liquid Crystals. Ferroelectric liquid crystals can be switched much faster than the conventional nematic liquid crystal display. They consist of a Smectic C mesogen with a chiral center which causes each layer to be twisted with respect to neighboring layers. Clark and Lagerwall (37) were able to unwind the helical structure and stabilize it with a surface treatment of the cell walls. Although they have a fast response time and low power requirements, they suffer from poor mechanical stability to shock(38). During the same period of time, Ringsdorf and his associates (39) were studying polymers with liquid crystals attached as a side-chain. They found that for the systems to exhibit liquid crystal behavior a flexible chemical link was needed between the polymer backbone and mesogen. These systems have a slow response time but are mechanically stable. Therefore, the current efforts are to make a polymer with flexible spacers to which ferroelectric mesogens can be attached as side-chains (38). In this way, the polymer can give mechanical stability to the fast switching ferroelectric mesogens for displays and sensors.

Nematic liquid crystals dispersed in poly (vinyl alcohol) are being developed for paper-like displays (40). Normally, the liquid crystal droplets scatter light such that a film will appear translucent and white. Under an electric field, ordering leads to a clearer film. The scattering of these nematic droplets depends on the difference in the indices of refraction between the polymer and the liquid crystal and on the anisotropy in the index of the liquid crystal. This technology promises to become an inexpensive display material.

Organic Transistors. In the DOE report (24) on electronic circuit components, the panel concurred that "the use of molecular monolayers for circuit elements seems to be too speculative." Further, "the members of this panel have some reservations concerning the applicability of organic films to high speed switching...." This point of view has to be corrected. Recent advances (41) reported at the Device Research Conference by Garnier at CNRS in Thiais, France, Phillips at Lucent and Jackson at Penn State indicate that films of pentacene or α-hexathiophene can exhibit mobilities approximately the same as amorphous silicon. Jackson reported that with a slow vacuum evaporation better quality films with fewer charge-trapping defects could be made. Also, these organic materials show low loss characteristics. Hence, in spite of a negative prediction 10 years ago, significant progress has been made in adapting thin organic films to electronic devices.

Conclusions

Not too many years ago, polymers were identified with cheap, molded plastics. These materials were not very pure and, as such, were not consistent from batch to batch. Although inexpensive, they were soft and rather unstable in heat and sunlight. Their electrical and optical properties were not good. On the positive side, nylon, polyethylene, polystyrene, poly(methyl methacrylate), and polycarbonate all were active in commerce; they had a use. The major breakthrough in the electronic industry came with polyimide as an insulating layer in electronic chips, organic photoconductors in copiers and printers, photoresists in layer-by-layer circuit deposition, and liquid crystals for displays. Except for liquid crystals, most application, even the new applications discussed above, were with amorphous films. Many other applications not covered in the discussion are also under development. Many are proprietary and not published in the open literature, and others are optimized configurations of known applications. As time progresses, these additional technologies will become apparent by the announcement of new products or new patents.

With ordered films produced by, for example, self assembly, we can expect to discover new phenomena and effects which can be utilized both scientifically and technically. It is clear that the future is bright for organic thin films. New materials are needed and an interdisciplinary approach is required. Chemists, chemical engineers, applied physicists, materials scientists, and electrical engineers all must work together in a concerted effort to solve these problems in thin film materials science research and development at the molecular level of these organic and polymeric systems.

Literature Cited

1. Roberts, G.; Eds.; *Langmuir-Blodgett Films;* Plenum: New York, NY, 1990.
2. Ulman, A., *An Introduction to Ultrathin Organic Films from Langmuir-Blodgett to Self-Assembly;* Academic Press: Boston, MA, 1991.
3. Swalen, J. D.; *Annu. Rev. Mater. Sci.* **1991**, *21*, 373.
4. Dubois, L. H.; Nuzzo, R. G. *Annu. Rev. Phys. Chem.* **1992**, 437.
5. Whitesides, G. M.; Ferguson, G. S.; Allara, D.; Scherson, D.; Speaker, L; Ulman, A. *Crit. Rev. Surf. Chem.* **1993**, *3*, 49.
6. Bohn, P. W.; *Annu. Rev. Phys. Chem.* **1993**, *44*, 37.
7. Whitesides, G. M. *Sci. Am.* **1995**, *273*, 146.
8. Ulman, A., Ed.; *Organic Thin Films and Surfaces: Directions for the Nineties. in Thin Films;* Academic: San Diego, CA, 1995.
9. Allara, D. L. *Biosensors & Bioelectronics;* Elsevier: Amsterdam, 1995; Vol. 10, pp. 771-783.
10. Sanchez, I. C., Ed.; *Physics of Polymer Surfaces and Interfaces;* Butterworth-Heinemann: Boston, MA, 1992.
11. Swalen, J. D. *J. of Molecular Electronics* **1986**, *2*, 155.
12. Knoll, W. *Encyclopedia of Applied Physics;* VCH Publisher: Brooklyn, NY, 1996; Vol. 14, pp. 569-605.

8

13. *Organic Thin Films for Waveguiding Nonlinear Optics;* F. Kajzar, F.; Swalen J. D.; Eds.; Advances in Nonlinear Optics; Gordon and Breach Publisher: Amsterdam, 1996; Vol. 3.

14. Knoll, W. In *Handbook of Optical Properties, Optics of Small Particles, Interfaces, and Surfaces;* Hummel, R. E. and Wißmann, P., Eds.; CRC Press: Boca Raton, FL, 1997; Vol. 2, pp. 373-400.

15. For recent results on organic thin films see the proceedings of the conference "Organized Molecular Films" Gabrielli, G. and Rustichelli, F., Guest Eds.; in *Thin Solid Films* **1996**, 284.

16. Kendall, K., *Science* **1994**, *263*, 1720.

17. Yoshizawa, H.; Israelachvili, J.; *Thin Solid Films* **1994**, 246.

18. Yoshizawa, H., Chen, Y. L.; Israelachvili, J. *Polym. Prepr. (Am. Chem. Soc., Div. Polym. Chem.)* **1993**, *34*, 282.

19. Yoshizawa, H.; Chen, Y.L.; Israelachvili, J. *Wear*: **1993**, *168*, 161.

20. Reiter, G.; Demirel, A.; Levent, A.; Peanasky, J.; Cai, L. L.; Granick, S. *J. Chem. Phys.* **1994**, *101*, 2606.

21. Peanasky, J.; Cai, L. L.; Granick, S.; Kessel, C. R. *Langmuir* **1994**, *10*, 3874.

22. Thomas, R..C.; Houston, J. E.; Crooks, R. M.; Kim, T. M.; Terry A. *Am. Chem. Soc.* **1995**, *117*, 3830.

23. Miller, J. S. *Adv. Mater.* **1993**, *5*, 671.

24. Swalen, J. D.; Allara, D. L.; Andrade, J. D.; Chandross, E. A.; Garoff, S. J.; Israelachvili, J.; McCarthy, T. J.; Murray, R.; Pease, R. F.; Rabolt, J. F.; Wynne, K. J.; Yu, H. *Langmuir* **1987**, *3*, 932.

25. Tang, C. W.; VanSlyke, S. A. *Appl. Phys. Lett.* **1987**, *51*, 913.

26. Halls, J. J. M.; Baigent, D. R.; Cacialli, F.; Greenham, N. C.; Friend, R. H.; Moratti, S. C.; Holmes, A. B. *Thin Solid Films* **1996**, *276*, 13.

27. Kido, J. *Bull. Electrochem.* **1994**, *10*, 1.

28. Sheats, J. R.; Antoniadis, H.; Hueschen, M.; Leonard, W.; Miller, J.; Moon, R.; Roitman, D.; Stocking, A *Science* **1996**, *273*, 884.

29. Service, R. F. *Science* **1996**, *273*, 884.

30. Fujihira, M; Do, L.-M.; Koike, A.; Han, E.-M. *Appl. Phys. Lett.* **1996**, *68*, 1787.

31. Scott, J. C.; Karg, S.; Carter, S. A. *Appl. Phys. Lett.* **1997**, *70*.

32. Miller, R. D. In *Reference 13*, pp. 329-456.

33. Donckers, M. C. J. M.; Silence, S. M.; Walsh, C. A.; Hache, F.; Burland, D. M.; Moerner, W. E.; Twieg, R. J. *Opt. Lett.* **1993**, *18*, 1044.

34. Moerner, W. E.; Silence, S. M.; Hache, F.; Bjorklund, G. C. *J. Opt. Soc. Amer.* **1994**, *B11*, 320.

35. Meerholz, K.; Volodin B. L.; Sandalphon; Kippelen, B.; Peyghambarian N. *Nature* **1994**, *371*, 497.

36. Lundquist, P. M.; Wortmann, R.; Gelentneky, C.; Twieg, R. J.; Jurich, M.; Lee, V. Y.; Moylan C. R.; Burland, D. M. *Science* **1996**, *274*, 1182.

37. Clark, N. A.; Lagerwall, S. T.; *Appl Phys. Lett.* **1980**, *36*, 899.

38. Blackwood, K. M., *Science* **1996**, *273*, 909.

39. Finkelmann, H.; Happ, M.; Portugall, M.; Ringsdorf, H. *Makromol. Chem.* **1978**, *179*, 2541.

40. Drzaic, S.; Gonzales A. M. *Appl. Phys. Lett.* **1993**, *62*, 1332.

41. Service, R. F. *Science* **1996**, *273*, 879.

MONOLAYERS

Chapter 2

From Monolayers of a Tethered Polymer Melt to Freely Suspended Elastic Membranes

Werner A. Goedel

Max-Planck Institut für Kolloid- and Grenzflächenforschung, Haus 9.9, Rudower Chaussee 5, 12489 Berlin, Germany

Monolayers at the air/water interface of hydrophobic non glassy polymers with ionic head groups have been used as a model system to investigate thermodynamics of "polymer melt brushes". The polymer chains are tethered to the planar surface but are free to assume a distorted 3-dimensional conformation. The thermodynamic properties are dominated by the loss in entropy due to a stretching of the polymer coils away from the interface upon increasing tethering densities. Monolayers of polyisoprenes wit ionic head groups bearing photoreactive side groups (anthracene) have been cross-linked on the water surface via irradiation with UV light. These 40 nanometer thick stabilized films can be transferred to cover holes in solid substrates of 0.3 mm diameter and form freely suspended elastic membranes.

Block copolymers or (semi) telechelic polymers, in which one of the blocks or a small head group drastically differs from the main polymer chain, can be regarded as large amphiphiles. The head groups can aggregate to form (inverted) micelles (*1- 3*) or can strongly adsorb to interfaces in contact with the bulk or a solution of the polymer (*4,5*). If polymer chains are bound with a surface active head group to an interface with a lateral spacing considerably less than the dimensions of the undisturbed polymer coil, they form a so called "polymer brush". In this brush, the neighboring chains interact with each other and form a continuous film. In general, one has to distinguish two cases: a solvent free-brush of a polymer melt ("melt brush" or "dense brush", (*6-9*) Figure 1a) and a polymer brush in contact with a good solvent ("swollen brush", (*10*) Figure 1b). In the first case there is no solvent to fill voids between polymer segments. Therefore, the balance between strong attraction and hard core repulsion essentially fixes the concentration of segments at a value

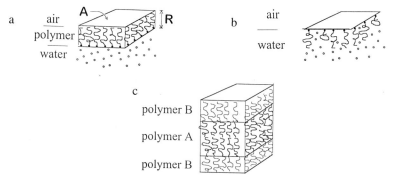

Figure 1: Schematic comparison of (a) monolayers of tethered polymers that are free of solvent, (b) monolayers of tethered polymers swollen with solvent, and (c) one lamella of phase separated block copolymers (adapted from ref. *24*)

close to the bulk density. In the latter case the interaction is primarily osmotic and the concentration of segments is variable.

In both cases, the interaction between neighboring chains leads to a distortion of the coils, which are forced away from the interface. The "brush" is considerably thicker than monolayers of polymers which adsorb with each repeat unit.

The "melt brush" has its main importance in the context of phase separated block copolymers and the often complex morphologies of these systems (e.g. lamellar [see Figure 1c], hexagonal, cubic...) are a result of the balance between the interfacial tension and the deformation of the closely packed polymer chains (7, 9, 11).

While there is considerable interest in these bulk systems, it has been very advantageous to study melt polymer brushes as monolayers at flat surfaces. In such a monolayer it is relatively easy to determine and tune the surface concentration of the head groups and to give the system a preferred orientation in space. In order to study equilibrium thermodynamics of these Monolayers it is necessary, however, to use polymers with low glass transition temperatures - e.g. polydienes (12), polydimethylsiloxanes (13 14) or perfluoropolyethers (15). If polymers like polystyrenes are investigated at temperatures below the glass transition, they form hard films or hard particles (16, 17) and the properties of these systems are not significantly changed by the presence of a surface active head group (18, 19).

Monolayers of suitable polymers with surface active head groups are convenient model systems to study the thermodynamics of melt polymer brushes; in addition they have a special advantage:

Crosslinking should lead to rubber elastic membranes. If the crosslinking is made on the water surface, transfer via the Langmuir-Blodgett technique (LB-technique) (20) offers the unique opportunity to cover holes in a solid substrate and thus create freely suspended membranes. While Langmuir-Blodgett films of low molecular weight substances and liquid polymers easily rupture during transfer, stable suspended membranes have been achieved using glassy polymers often stabilized in addition through crosslinking (21, 22). For various applications, like membrane separation processes or micromechanics, it will be advantageous to have a choice between rubbery and glassy membranes.

Freely suspended membranes can be generated as well via spreading a melt or a viscous solution of a polymer across the opening of a substrate followed by crosslinking. Using this process, however, it usually is difficult to achieve films of uniform thickness. Especially at the rim of the opening, one usually obtains a meniscus of excess material (23). By transferring preformed crosslinked membranes one might overcome this difficulty.

The paper presented here is composed of two parts: The first part focuses on using monolayers of polyisoprenes with ionic head groups as model systems for a "melt polymer brush"; the second part investigates the potential to form freely suspended elastic membranes via crosslinking.

Monolayers as Model Systems for a "Melt Polymer Brush"

Linear polyisoprenes with a sulphonate head group have been prepared via anionic polymerization followed by reaction with propanesultone. The characterization of the polymers is summarized in Table 1 (The polymers are copolymers of 71% 1,4-cis- 22% 1,4-trans- and 7% 3,4-isoprenyl repeat units, a simplified structure is included in Figure 2. The synthesis has been reported in ref. 24).

The polyisoprenes with sulfonate head groups can be spread onto the water surface. The isotherms of the five polymers investigated are shown in Figure 2. The isotherms significantly expand with increasing molecular weight.

Table 1 : Characterization of the polyisoprenes with sulfonate head groups.

number of repeat units, N	140	307	538	675	810
M_n (GPC) [g/mol]	9638	20820	36540	45900	55020
M_w/M_n	1.14	1.04	1.02	1.02	1.02

Thermodynamics. If one takes into account the conformational changes of the polymer chain, the isotherms can be interpreted quantitatively (a more detailed description of the analysis is given in ref. 24) In a simple picture one can assume that all polymer chains are bound to the aqueous phase by the head group, while the end group is at the polymer-air interface (see Figure 3a).

The surface pressure is given by the difference between the surface tension of the pure water and the film tension of the covered surface; surface and film tension are given by the first derivative of the Gibbs free energy, G, with respect to the area, A. If the volume of the system does not change, changes in the Gibbs free energy, G, are equal to changes in the Helmholtz free energy F:

$$\Pi = \left(\frac{\partial F}{\partial A} \right)_{water/air-interface} - \left(\frac{\partial F}{\partial A} \right)_{polymer film} \tag{1.}$$

The chain ends are separated by the film thickness, R. Therefore, the chains act like springs and store an elastic free energy (25-27):

$$F_{elastic} = nk_B T \cdot \frac{3}{2} \frac{1}{\langle r^2 \rangle_0} \cdot R^2 \tag{2.}$$

n = number of polymer chains in the monolayer, $\langle r^2 \rangle_0$ = the unperturbed mean square end-to-end distance, $\langle r^2 \rangle_0$ is proportional to the chain length. k_B = Boltzmann constant, T = temperature

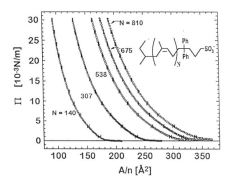

Figure 2 Isotherms of polyisoprenes with sulphonate headgroups and different chain lengths , N = number of repeat units. (averages of at least 5 measurements). (reproduced from ref. *24*)

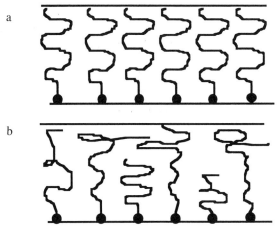

Figure 3
Schematic comparison between (a) an uniformly stretched brush, and (b) a non uniformly stretched polymer brush according to the Semenov scenario. (reproduced from ref. *24*)

If one assumes that the hydrophobic region of the monolayer is free of solvent and has a constant density (close to the bulk value of the polymer), the total volume of the film is given by the product of film thickness and area covered by the film, A. Thus the film thickness is inverse proportional to area per head group:

$$R \cdot A = n \cdot N \cdot v \quad \Leftrightarrow \quad R = N \cdot v \cdot \left(\frac{A}{n}\right)^{-1} \tag{3.}$$

N = number of repeat units per chain, v = volume of a repeat unit.
And the elastic free energy is inversely proportional to the square of the area.

$$F_{elastic} = k_B T \cdot \frac{3}{2} \frac{n^3 \cdot N^2 \cdot v^2}{\langle r^2 \rangle_0} \cdot A^{-2} \tag{4.}$$

The first derivative of the elastic energy is given by:

$$\frac{\partial F_{elastic}}{\partial A} = - 3 \cdot k_B T \cdot \alpha^2 \cdot N \cdot \left(\frac{A}{n}\right)^{-3} \tag{5.}$$

$$\text{with} \quad \alpha = v \Big/ \sqrt{\langle r^2 \rangle_0 / N}$$

In the case of a freely jointed chain, the constant α can be interpreted as the cross-sectional area of a chain segment

All other contributions to the surface pressure that are not due to the stretching of the polymer chain are assumed to be independent of the chain length, N. These contributions here are pointed out only in a general form as a function of the area per molecule f(A/n, not N) :

$$\Pi = f\left(\frac{A}{n}, \text{not } N\right) + c \cdot k_B T \cdot \alpha^2 \cdot N \cdot \left(\frac{A}{n}\right)^{-3} \tag{6.}$$

with : c = 3

Two features of this description are important in the following data treatment:
(i) at a given area per head group, A/n, the "elastic part" of the surface pressure depends linearly on the chain length; (ii) the "elastic part" of the surface pressure is proportional to the third power of the area per head group.

In order to test the linear dependency of the surface pressure on the chain length, Figure 4 shows the surface pressure data as a function of chain length, the third parameter being now the area per head group. The theoretically predicted linear dependency can only be tested in a regime of A/n where more than two isotherms can be measured. In this regime between 170 Å^2 and 250 Å^2 , all points fall onto a straight line in accordance with the predictions.

In order to test the second prediction, one might take the slope of the above-mentioned representation and thus eliminate the first two terms in eq. (6.):

$$\frac{\Pi_1 - \Pi_2}{N_1 - N_2} = c \cdot k_B T \cdot \alpha^2 \left(\frac{A}{n}\right)^{-3} \tag{7.}$$

If the above equation is rescaled by the "cross sectional area of the chain", α, one obtains the dimensionless representation:

$$\frac{\Delta \Pi}{\Delta N} \frac{\alpha}{k_B T} = c \left(\frac{A}{n} \cdot \frac{1}{\alpha}\right)^{-3} \Leftrightarrow \log\left\{\frac{\Delta \Pi}{\Delta N} \cdot \frac{\alpha}{k_B T}\right\} = \log c - 3 \cdot \log\left\{\frac{A}{n} \cdot \frac{1}{\alpha}\right\} \tag{8.}$$

The double logarithmic representation of eq. (8.) predicts a straight line, the slope of that line is -3 and the intercept is given by the prefactor c. In this rescaled plot all pairs of polymers should fall on a single line. The plot of the experimental data according to eq. (8.) is shown in Figure 5. The straight lines represent the theoretical predictions; the data are compared to the theoretical prediction without any fitting procedure.

In this double logarithmic plot the pressure differences between pairs of polymers actually superimpose into a single line; the slope is close to the predicted value of -3.

The assumption that the free ends are located at the "upper" surface of the film seems to be quite artificial. A scenario as depicted in Figure 3b) is more likely. It can be shown, however, that any type of affine deformation will lead to the same power law. The prefactor c, however, may differ from the uniformly stretched brush. For example, the more elaborate Semenov theory (7,8) yields the same power law, but with the prefactor c = $\pi^2/4 \approx$ 2.47. This result reflects the fact that in the Semenov brush the chains are allowed to assume thermodynamic equilibrium, while in simple scenario they are fixed at thermodynamically unfavorable positions. As can be seen from Figure 5, the result of the Semenov theory fits the experiment better than uniformly stretched brush.

Beside this general agreement between experiment and theory, the experimental data in Figure 5 lie upon a slightly bent curve, rather than on a straight line, the deviation being mostly pronounced at lower areas per head group or for polymer chains shorter than 300 repeat units. Earlier experiments indicated that there is no significant elastic contribution for chains shorter than 100 atoms (28) . Regardless of the details of the model used in the analysis, any affine deformation of the polymer film should give rise to a straight line with the slope -3 in the double logarithmic rep-

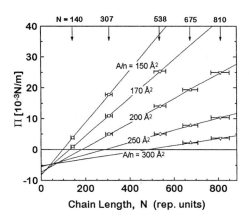

Figure 4
Surface pressure of Polyisopren-SO$_3$ films as a function of chain length at constant area per head group (same data as in Figure 2 for clarity only the data for a limited number of areas are included).(reproduced from ref. 24)

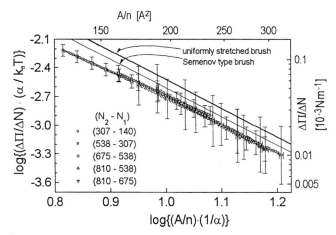

Figure 5
Double logarithmic plot of pressure difference over chain length difference, versus area per head group. The left and bottom axis are rescaled by the "cross-sectional area of a segment" $\alpha = v/(<r^2>_o/N)^{1/2}$. The experimental data are compared to the theoretical predictions (straight lines) without any fitting procedures. (reproduced from ref. 24)

resentation. Therefore, deviations from that slope might indicate deviations from affine deformation.

Lateral Structure and Film Thickness. The thermodynamic analysis of the previous section is based on two assumptions: (i) The film has no lateral structure and a (ii) the film thickness is proportional to the chain length and surface concentration.

   Lateral structure. The expanded isotherms of the tethered polymers do not show any phase transitions. It is therefore unlikely that films of a single polymer of uniform chain length laterally segregate to form domains. Indeed, microscopic images of films on the water surface (Brewster angle microscopy) and of films transferred to glass slides do not reveal any sign of lateral structure as long as the film is imaged or transferred at surface pressures above zero mN m^{-1} and below the collapse pressure (see Figure 6) (*24*). At zero pressure (areas per molecule larger than the onset of the isotherm), however, the film breaks up into patches of two-dimensional foam, which is in coexistence with the "bare" water surface. This observation is a general feature of any insoluble monolayer (see for example ref. *29*) and can be interpreted as a two dimensional liquid/gas coexistence.

Film Thickness. Films on the water surface and transferred films have been investigated by X-ray reflection techniques (*30,24*). The reflection data have been analyzed assuming a homogeneous film of constant height. As can be seen in Figure 7, the measured height is in agreement with the predictions derived from the assumptions of constant density (eq.(3.)).

   Thus, in principle one can fine tune the film thickness of a polymer monolayer just by selecting the appropriate area per molecule at transfer. This dependency of the film thickness on the surface concentration might be of great value for the application of transferred films.

Crosslinking and Transfer to form Suspended Membranes

The films considered until now are liquids at room temperature; the transferred films can easily be washed away by water and may creep away from the substrate with time.

   In order to create rubbery membranes, polymers with photoreactive anthracene side chains have been synthesized and the corresponding monolayers have been cross-linked via irradiation. The experiments were aimed towards the generation of freely suspended membranes because the elastic properties will be most useful if the film is not attached to a hard substrate.

   The properties of the polymer synthesized for this purpose and the chemical structure are given in Table 2 (for synthesis and purification of the anthracene-tagged polymer and preliminary crosslinking reactions see ref. (*31*))

a

Figure 6
Monolayers of polyisoprene with a sulfonate head group and N=140 repeat units imaged with Brewster angle microscopy at various areas per head group. The pictures cover an area of approximately. 0.75 mm x 0.75 mm. a) A/n = 220 Å^2 b): A/n = 130 Å^2, c: A/n = 85 Å^2. The corresponding isotherm is included in Figure 2 left side. (adapted from ref. *24*)

Continued on next page.

b

Figure 6. *Continued.*

c

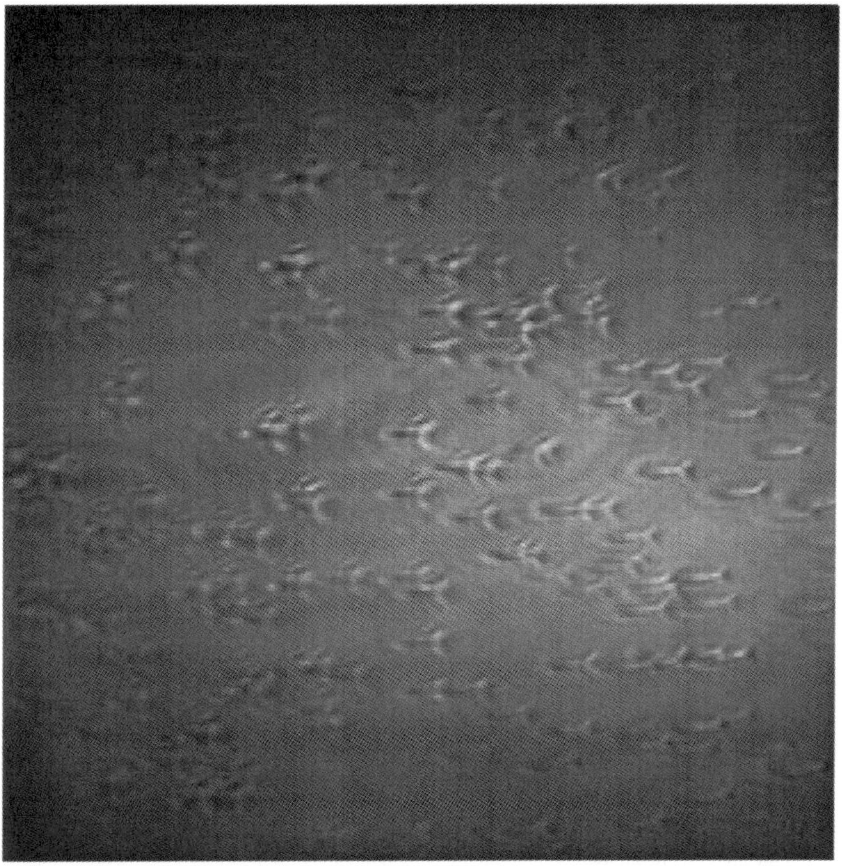

Figure 6. *Continued.*

22

Table 2: Characterization of the anthracene-tagged polyisoprene

M_w/M_n (calculated from GPC of parent polymer)	1.02
M_n (calculated from GPC of parent polymer)	39000 g/mol
number of repeat units, N	538
average number of anthracene groups per chain	8

Characterization of the monolayers at the water surface before crosslinking.
Compared to the parent polymer, the isotherm of the anthracene-tagged polymer is
shifted to higher values of area per molecule and surface pressure. In addition, the
isotherm shows hysteresis, which is more pronounced at higher compression / expansion rate (see Figure 8).

The shift to larger areas per molecule can at least partially be attributed to the
fact that the molecule increases in size due to the anthracene side groups, which
change the molecular weight by approximately 5%. The hysteresis might be due to
two effects: (i) The anthracene side groups might reversibly associate and introduce
additional friction (ii) the side groups might be attracted by the polymer water interface and thus slightly increase the surface pressure; on fast compression the concentration of the side groups at the polymer/water interface might be slightly enriched
(and thus the surface pressure increases), while at compression it is slightly de-
pleted (surface pressure decreases).

Crosslinking of films transferred to solid substrates. The non cross linked
monolayers can be transferred to glass and silicon substrates at a surface pressure of
25×10^{-3} N/m with a transfer ratio of $r_{tr} = 0.85$.

Illumination of the anthracene-tagged polymer should lead to crosslinking via
the $(4 + 4)$-cycloaddition of anthracene side groups according to Figure 9 (*32-34*).
Figure 10 shows the UV-vis spectra of a transferred monolayer before illumination
and after increasing illumination times. The intensity of the absorption bands in the
spectral region of $\lambda = 330$ to 430 nm, which are characteristic for the anthracene
side group, decrease monotonically with illumination time. After the illumination
with an integral intensity of 3 J/cm^2 the adsorption characteristics for anthracene are
nearly gone.

The spectra indicate that after illumination nearly all anthacene groups have
reacted. If all these reactions give rise to crosslinking, the resulting network has an
average chain length between crosslinking sites of approximately 70 repeat units or
4650 g/mol. However, anthracene groups might be lost due to side reactions, e.g.
addition of oxygen (*35*).

**Crosslinking on the water surface and transfer to cover a hole in a solid sub-
strate.** (A detailed paper on this subject has been submitted to Langmuir) Mono-

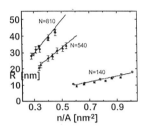

Figure 7
Film thickness, R, of films at the water surface and transferred films of polyisoprenes with hydrophilic head groups and different chain lengths, N (= number of repeat units) as a function of the surface concentration, n/A (= inverse
area per head group). The straight lines represent the theoretical film thickness
derived from eq.(3.) Filled symbols: films on the water surface, (30) open
symbols : Transferred films (24)

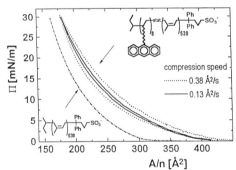

Figure 8
The isotherm of the anthracene-tagged polymer compared to the non-tagged
parent polymer. (compression speed is given in area/molecule/time) (adopted
from ref.31)

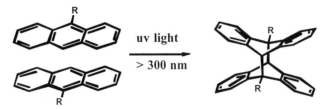

Figure 9

Dimerisation of anthracene side chains upon illumination.

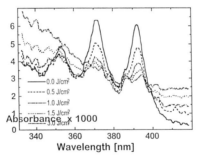

Figure 10

UV-vis spectra of a monolayer of the anthracene-tagged polymer transferred to a glass slide at 25×10^{-3} N/m. The spectra are recorded of the same film after increasing illumination times. The adsorption characteristics for the anthracene side groups vanish if the film is irradiated with an integrated intensity of 3 J/cm^2 or more.(adopted from ref.*31*)

layers of the anthracene-tagged polymer, compressed to an area of approximately 200 $Å^2$ at a surface pressure of 25×10^{-3} N/m, were illuminated on the water surface with an integrated intensity of 5 J/cm^2.

The monolayers were transferred to a thin brass plate with a 0.3 mm diameter hole, which was attached flat to a glass slide. In approximately half of the trials the opening is covered with a thin membrane. On the other hand, the hole is never covered if the transfer is done with the non illuminated polymer. At the location of the hole the film presumably is separated from the supporting glass slide by a millimeter thin layer of water. The film seems to have enough mechanical stability to withstand the mechanical stress occuring when the brass plate is separated from the glass slide and the water film ruptures. Figure 11 shows the top view of the hole in the brass plate which is covered by the cross-linked polymer membrane.

The elastic properties of the film are qualitatively demonstrated in Figure 12: The film is deformed by applying a gas overpressure from below. The deformation captured in Figure 12b is completely reversible upon release of the overpressure and the procedure can be repeated back and forth.

Conclusions

Monolayers of hydrophobic polymers, tethered to the water surface, are suitable model systems for the investigation of polymer "melt brushes". Especially, they offer a convenient way to vary the area per tethered head group and to give the brush a well-defined orientation and make it accessible to experimental techniques not available in the case of bulk phases. Simple measurements of isotherms already yield valuable information on the thermodynamics of these brushes.

From the applied point of view, they offer a convenient method to prepare nanometer thick coatings and thus might be an alternative to other coating techniques. Especially promising is the possibility to generate approximately 40 nanometer thick rubber elastic membranes, which can span holes of the size of one third of a millimeter.

Acknowledgments

The hospitality and great support of Prof. Antonietti and Prof. Möhwald is gratefully acknowledged. The work has been supported by the Deutsche Forschungsgemeinschaft (Go693/1-1, Go693/1-2) and the Max-Planck-Gesellschaft

Figure 11

Freely suspended film, generated *via* crosslinking of a monolayer of tethered polyisoprene on the water surface followed by transfer to a solid substrate with an opening of 0.3mm diameter (Top view) (submitted to *Langmuir*)

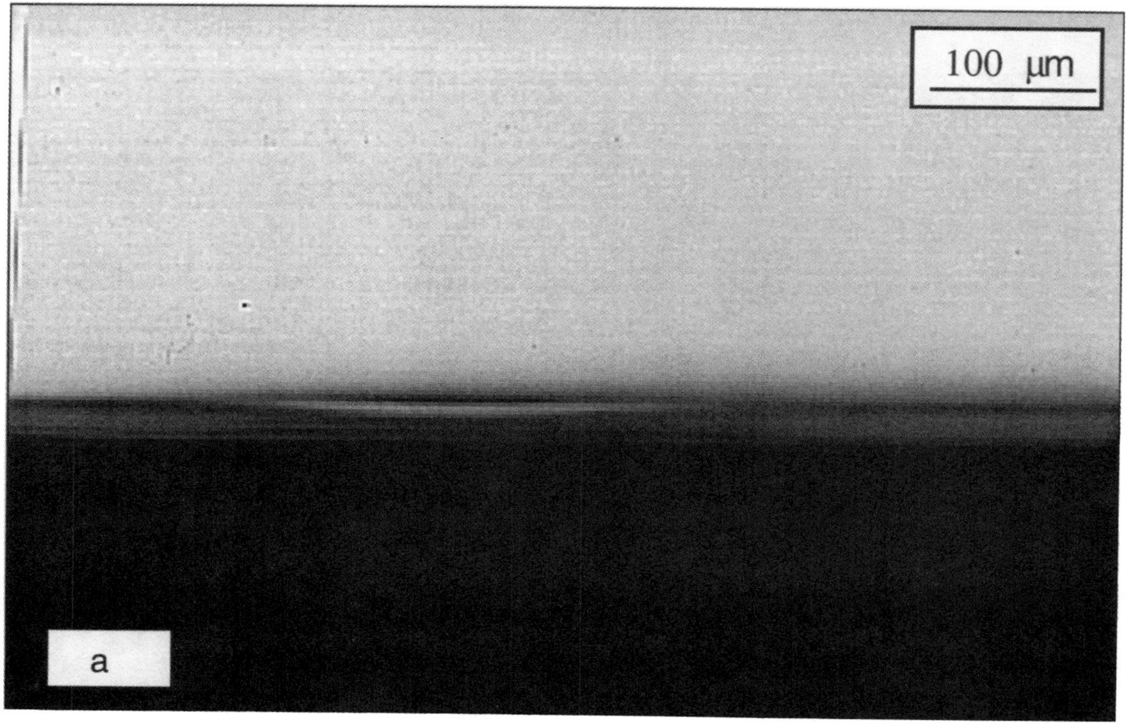

Figure 12
Side view of the same film as in Figure 11. The film is elastically deformed via
a slight overpressure from below. (submitted to *Langmuir*)

Continued on next page.

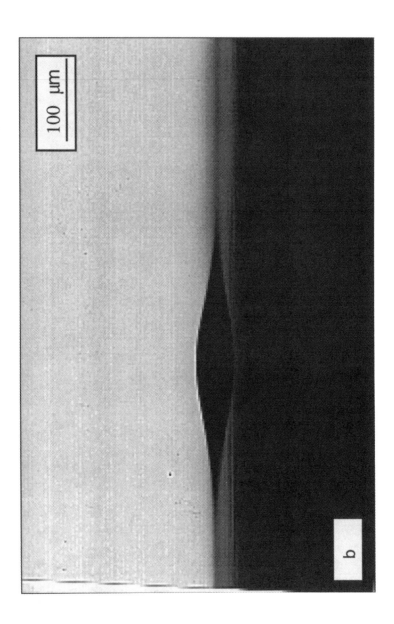

Figure 12. *Continued.*

Literature Cited

1 Jerome, R. In *Telechelic Polymers*; Goethals E. J., Ed; CRC Press, Boca Raton, Florida, 1989, pp. 261-287

2 Broze, G.; Jerome, R.; Teyssié, Ph.; Marco, C. *Polym. Bull* **1981**, *4*, 241

3 Möller, M.; Mühleisen, E.; Omeis, J.; In *Physical Networks-Polymers and Gels*; Burchard, W.; Ross-Murphy, S. B., Eds; Elsevier Applied Science, London, UK, 1990, pp. 45-64

4 Watanabe, H.; Tirell, M. *Macromolecules* **1993**, *26*, 6455

5 Tripp, C.P.; Boils, D.; Hair, M. L. *Macromolecular Reports* **1992**, *A29/2*, 131

6 Helfand, E. *Macromolecules* **1975**, *8*, 552

7 Semenov, A. N. *Sov. Phys. JETP* **1985**, *61*, 733

8 Semenov, A. N. *Macromolecules* **1993**, *26*, 6617

9 Ohta, T.; Kawasaki, K. *Macromolecules* **1986**, *19*, 2621

10 Millner, S. T. *J. Polym. Sci. B* **1994**, *32*, 2743

11 Bates, F. S.; Fredrickson, G. H. *Ann. Rev. Phys. Chem* **1990**, *41*, 525

12 Christie P.; Petty M. C.; Roberts G. G. *Thin Solid Films* **1985**, *134*, 75

13 Elman J. F.; Lee D. H. T.; Koberstein J. T.; Thompson, P. M. *Langmuir* **1995**, *11*, 2761

14 Lenk T. J.; Lee D. H. T.; Koberstein J. T. *Langmuir* **1994**, *10*, 1857

15 Goedel, W. A.; Xu, C.; Frank, C. W. *Langmuir* **1993**, *9*, 1184

16 Kumaki, J. Macromolecules 1986, 19, 2258

17 Kumaki, J. Macromolecules 1988, 21, 749

18 Niwa, M.; Hayashi, T.; Higashi, N.; Langmuir 1990, 6, 263

19 Yoshikawa, M.; Worsfold, D. J.; Matsuura, T.; Kimura, A.; Shimidzu, T. Polymer Communications 1990, 31, 414

20 For an introduction to the characterisation and transfer of insoluble films see: Gaines Jr, G. L. *Insoluble Monolayers at Liquid-Gas Interfaces*; Interscience: NY, 1966 or Roberts, G. *Langmuir-Blodgett Films*; Plenum Press: NY, 1990

21 Seufert, M.; Fakirov, C.; Wegner, G. *Advanced Materials* **1995** *7* 52

22 Kunitake M.; Nishi T.; Yamamoto H.; Nasu K.; Manabr O.; Nakashima N. *Langmuir* **1994**, *10*, 3207

23 Reibel, J.; Brehmer, M.; Zentel, R.; Decher, G. *Adv. Mater.*, **1995**, *7*, 849

24 Heger R.; W. A. Goedel *Macromolecules*, **1996**, *29*, 8912

25 Treolar L. R. G. *The Physics of Rubber Elasticity*, Clarendon Press, Oxford, 1975, p. 56/57

26 deGennes, P. G. *Scaling Concepts in Polymer Physics*; Cornell University Press: Ithaca & London, 1979, p. 31

27 Mark J. E.; Erman B. *Rubberlike Elasticity a Molecular Primer*, Wiley & Sons; NY, Chichester, Singapore, 1988, p. 30

28 Goedel W. A.; Wu H.; Friedenberg M. C.; Fuller G. G.; Foster M.; Frank C. W. *Langmuir* **1994**, *10*, 4209

29 Mann, E. K.; Henon, S.; Langevin, D.; Meunier, J. *J. Phys. II France*, **1992**, *2*, 1683

30 Baltes, H.; Schwendler, M.; Helm, C. Heger, R.; Goedel, W. A.; *Macromolecules* (accepted May 1997)

31 Heger R., Goedel W.A., *Progr. in Colloid & Polymer Science* **1997**, *105*, 167

32 Trecker D. J.; in: *Photodimerisations in Organic Photochemistry*, Chapman O. L. Ed., 2. Edit., Marcel Dekker Inc., New York, 1969, pp. 63-116

33 Fritzsche J. *J. prakt. Chem.* **1967**, *101*, 337

34 Tazuke S.; Hayashi N. J.; *Polymer Science* **1978**, *16*, 2729

35 Foote G. S. *Acc.Chem. Res.*, **1968** *1*, 104

Chapter 3

Pentaerythritol-Based Polyether Dendrimers at the Air–Water Interface

J. P. Kampf[1], T. L. Einloth[1,3], Curtis W. Frank[1,4], Yufei Li[2], Jae-Min Oh[2], and J. A. Moore[2]

[1]Center on Polymer Interfaces and Macromolecular Assemblies (CPIMA) and Department of Chemical Engineering, Stanford University, Stanford, CA 94305–5025
[2]Department of Chemistry, Rensselaer Polytechnic Institute, Troy, NY 12180–3590

Surface pressure versus area isotherms and film relaxation experiments of Langmuir films of second generation, pentaerythritol-based polyether dendrimers were recorded to study the behavior of these molecules at the air-water interface. These dendrimers form highly compressible films at molecular areas between 200 and 700 $Å^2$ and undergo a collapse to a multilayer film at a surface pressure of approximately 17 mN/m and surface area of 200 $Å^2$. Empirically, the collapse is found to involve three relaxation times when fit with a series of exponentials. Increasing the temperature from 5 to 20 °C increases the collapse rate and the extent of collapse during compression and slightly decreases the collapse pressure from 17.4 to 16.5 mN/m.

The highly branched, well-defined macromolecules known as dendrimers have been studied by several investigators in the area of interfacial science (*1-3*). Dendrimers are of interest in this field not only because of their high degree of branching and three-dimensional structure, but also because it is possible to change the functionality of the end groups through various synthetic techniques. More specifically, pentaerythritol-based polyether dendrimers with functionalized end groups have been studied for possible uses as catalysts and as displacers for ion-exchange displacement chromatography (*4,5*) and may prove to have interfacial applications as well. An understanding of the shape of dendrimers when confined to an interface and the interaction between the dendrimers and an adjacent surface is of utmost importance when examining interfacial applications. A useful technique for studying molecules at interfaces is to make Langmuir films of the molecules by spreading them on a water surface, where they form a monolayer and where the packing of this monolayer can be controlled by changing the available water surface area. Several investigators have

---

[3]Current address: Johnson & Johnson, 199 Grandview Road, Skillman, NJ 08558.
[4]Corresponding author.

recently studied the behavior of poly(benzyl ether) dendritic molecules at the air-water interface (*6-10*). In this study, we use surface pressure versus area isotherms and film relaxation measurements to investigate the interactions between pentaerythritol-based polyether dendrimers and the water surface. We then compare experimental results with a molecular model of the polyether dendrimer studied to gain a better understanding of the shape that the molecule takes at the air-water interface.

Experimental

The polyether dendrimers were synthesized by a divergent growth method based on the work of Hall and Padias (*11*) as modified by Moore and Lee (*12*). Starting from a pentaerythritol (PE) initiator core, successive generations of dendrimers were produced by a sequence of reiterative reactions (*4,5*). A two-dimensional representation of the dendrimer used in this study, a second generation polyether dendrimer (PE-MBO12) formed by successive reactions of 1-methyl-4-(hydroxymethyl)-2,6,7-trioxabicyclo-[2.2.2]-octane with the PE core, is given in Figure 1.

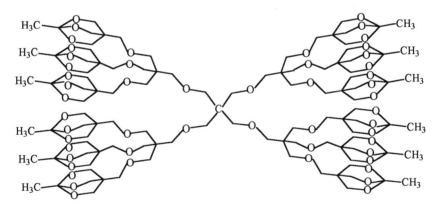

Figure 1. Sketch of PE-MBO12. Note that the two-dimensional figure does not represent the anticipated three-dimensional structure.

A 50 cm x 15 cm symmetric-compression KSV-5000 Langmuir-Blodgett trough (KSV, Helsinki) was used for the isotherm measurements. The subphase was an aqueous potassium hydroxide solution with an approximate pH of 10, held at a specified constant temperature to within ± 0.2°C. Deionized water, purified with the Milli-Q system (Millipore Corp.), and semiconductor grade potassium hydroxide pellets were used to prepare the subphase. A basic environment was needed to prevent the subphase from hydrolyzing the dendrimers. Solutions of approximately 0.5 mg/ml of dendrimer in HPLC grade chloroform were prepared for spreading. The spreading solution was added dropwise onto the water surface using a microsyringe.

After a 20 minute waiting period for solvent evaporation, the barriers were compressed at a constant rate. The surface pressure was measured using the Wilhelmy plate method (*13*) with an experimental error of approximately ± 0.1 mN/m.

For the hysteresis experiments, the barriers were immediately expanded at the same speed used for compression after the desired surface pressure had been reached. The barriers were allowed to reach their starting position before further compression/ expansion cycles were performed. For the surface pressure relaxation experiments, the film was compressed to the desired molecular area with a barrier speed of 10 mm/min, and the change in surface pressure was monitored at a constant surface area for one hour or more. During the area relaxation experiments, the surface pressure was held constant after the film had been compressed to the target value, and the change in film area was recorded.

The molecular modeling was conducted with software provided by Molecular Simulations, Incorporated. After constructing the polyether dendrimer, we minimized the potential energy of the structure using the steepest descent and conjugate gradient methods to a maximum derivative less than 0.001 kcal mol^{-1} Å^{-1}. The molecule was then allowed to relax via a 100 ps molecular dynamics simulation at 298 K in which Newton's equation of motion was solved using the Verlet velocity integrator. After completion of the molecular dynamics run, the molecular structure was again minimized in the manner described above.

Results

Figure 2 shows the surface pressure versus area isotherms of PE-MBO12 for compression speeds of 3, 10, and 100 mm/min and a subphase temperature of 5 °C. Higher compression speeds produce isotherms with slightly greater molecular areas. All isotherms have similar qualitative behavior, however. The monolayer films are highly compressible below 18 mN/m and have a measurable surface pressure beyond a molecular area of 700 Å^2. Depending on the compression speed, a plateau region appears at about an area of 200 Å^2 and a surface pressure of 18 mN/m, which we attribute to a collapse from a monolayer to a multilayer. After the plateau, the film experiences a more condensed phase with a much lower compressibility than that seen prior to the plateau region. The final, irreversible collapse of the film occurs at surface pressures greater than 60 mN/m.

Temperature effects are illustrated by the isotherms at 5, 12, 20, and 27 °C shown in Figure 3 for a compression speed of 10 mm/min. Slight differences can be seen between the four isotherms throughout the compression range. Although the changes seem small compared to the temperature differences, increasing the temperature at which the isotherm is taken increases the compressibility in the expanded region and decreases the surface pressure of the monolayer-to-multilayer collapse. There also exists a more obvious variation, however, between the length of the plateau regions of the four isotherms. The higher temperature isotherms have a more extended plateau region than the lower temperature isotherms, suggesting a more complete collapse at higher temperatures. All isotherms display a small peak at the beginning of the plateau, as is shown in the inset of Figure 3.

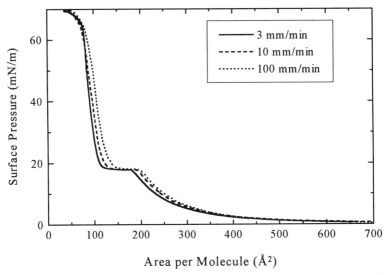

Figure 2. Surface pressure versus area isotherms for varying compression rates and a temperature of 5 °C.

Two compression/ expansion cycles at a subphase temperature of 20 °C and a compression speed of 10 mm/min are shown in Figure 4. For this experiment, the film is compressed to a surface pressure of 40 mN/m and immediately expanded, and then the cycle is repeated. Upon expansion after the first compression, the surface pressure drops rapidly to approximately 17 mN/m, where it begins to rise slightly and forms a plateau similar to that seen during compression but at a lower surface pressure. The surface pressure then decreases slowly and smoothly to zero in a manner much the same as that observed during compression. The second compression/ expansion cycle displays a slightly smaller molecular area than the initial cycle but otherwise retains a qualitatively similar behavior, demonstrating that the monolayer-to-multilayer collapse is repeatable.

To probe the hysteresis behavior further, surface pressure relaxation experiments at constant area have been performed at the start of the monolayer-to-multilayer transition as well as at molecular areas of 180, 240, and 300 Å², which correspond to surface pressures of 17.9, 11.1, and 6.1 mN/m at 5 °C and 17.1, 10.6, and 5.4 mN/m at 20 °C. These experiments can be considered two-dimensional analogs of stress relaxation experiments. The start of the monolayer collapse is taken as the inflection point in the isotherms that occurs just prior to the plateau. For a subphase temperature of 5 °C the start of collapse is determined to be at 198 Å² and 17.4 mN/m; while for 20 °C the start of collapse is found to be at 203 Å² and 16.5 mN/m. Relaxation experiments at 240 and 300 Å² show only a small decrease in surface pressure after one hour. A substantial decrease in the surface pressure can be observed in the collapse region of the isotherm, however. The pressure relaxation in the collapse region can be fit with the following empirical equation, which assumes

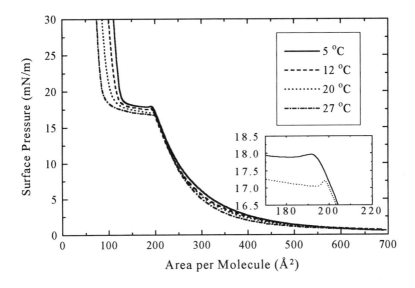

Figure 3. Surface pressure versus area isotherms at 5 and 20 °C and a barrier speed of 10 mm/min.

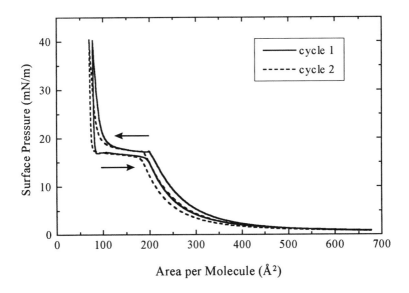

Figure 4. Compression/ expansion cycles at 20 °C and a compression rate of 10 mm/min

that there are three relaxation processes involved in the monolayer collapse to a multilayer:

$$\Pi(t) = \Pi_{eq} + \Pi_1 e^{-t/\tau 1} + \Pi_2 e^{-t/\tau 2} + \Pi_3 e^{-t/\tau 3}$$

where $\Pi(t)$ is the surface pressure at any time t; Π_{eq} is the equilibrium surface pressure or the pressure at infinite time; Π_1, Π_2 and Π_3 are the amounts that each relaxation process contributes to the surface pressure at the start of relaxation; and $\tau 1$, $\tau 2$ and $\tau 3$ are the time constants for each relaxation process. An example of a typical relaxation curve is given in Figure 5 for an area of 198 Å^2 and a subphase temperature of 5 °C. The longest relaxation time constants along with the total number of relaxation modes for each experiment are given in Table I. The films show slower relaxation times at 5 °C than at 20 °C; while the relaxation times are much faster at 180 Å^2 than at the start of collapse. In fact, PE-MBO12 films at 20 °C have only two measurable relaxation modes instead of the three present at 5 °C.

Table I. Pressure relaxation parameters.

Temperature (°C)	Area (Å^2)	Relaxation Modes	Longest τ (min)
5	198	3	289
5	180	3	35.1
20	203	2	260
20	180	2	33.0

In an effort to better understand the collapse mechanism, the change in film area has been monitored over time at the surface pressure denoting the start of collapse and at 20 mN/m for temperatures of 5 and 20 °C. Measuring the area change as a function of time at constant surface pressure can be thought of as a two-dimensional creep experiment. As previously stated, the monolayer-to-multilayer collapse is found to start at 17.4 and 16.5 mN/m for subphase temperatures of 5 and 20 °C, respectively. As can be seen in Figure 6, the film experiences a greater relative area loss at 20 °C than at 5 °C both at the start of collapse and at a surface pressure of 18 mN/m. The films also show a greater relative area loss at higher surface pressures. None of the films appear to have reached a steady area even after 14 hours of relaxation.

Discussion

From the surface pressure versus area isotherms, one can gain some insight into the behavior of the dendritic molecules at the water surface. First, the isotherms demonstrate that PE-MBO12 is surface-active, which is not necessarily apparent *a priori*. The highly compressible part of the isotherm that extends to molecular areas greater than 700 Å^2 implies that the dendrimer molecules lie somewhat flat on the water surface in an attempt to maximize the number of ether linkages in contact with the water, possibly forming hydrogen bonds. The suggestion that the molecules are in an expanded state at the start of the isotherm is supported by the molecular structure

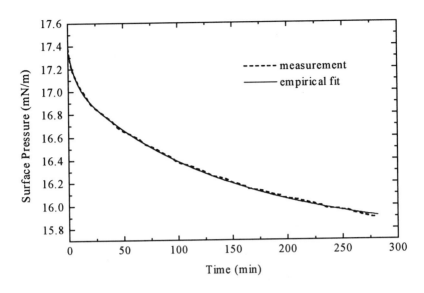

Figure 5. Surface pressure versus time at 5 °C and a constant area of 198 Å^2 and a three time constant empirical fit to the data.

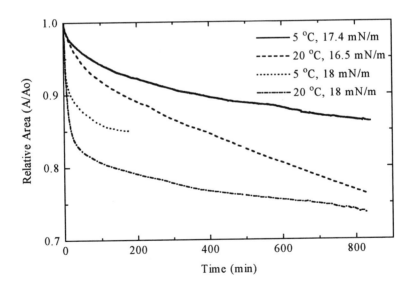

Figure 6. Relative area change over time at constant surface pressures of 17.4 and 18 mN/m for 5 °C and 16.5 and 18 mN/m for 20 °C.

obtained through our molecular modeling calculations. Although we do not intend to suggest that the model represents an actual conformation of a polyether dendrimer molecule at the water surface, one can gain some insight by comparing our model with the isotherm data. Figure 7 shows a top and a side view of an energy-minimized structure for PE-MBO12. In the state shown, the molecule is rather flat and extended. As a crude estimate of the largest molecular area this specific structure could have, one can assume the top view of the molecule to be a circle with a diameter of 28 Å. Even with this overestimation, the calculated molecular area is only 616 Å². The isotherms show a measurable surface pressure beyond 700 Å² suggesting that the molecules indeed take a rather flat and extended shape at the air-water interface.

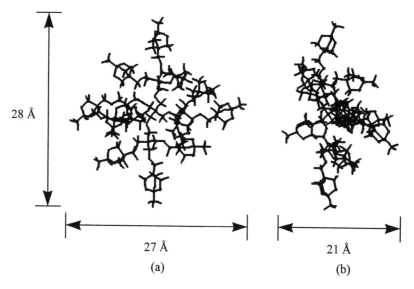

28 Å

27 Å

21 Å

(a)

(b)

Figure 7. Expanded structure of PE-MBO12 obtained from molecular modeling calculations: (a) top view and (b) side view.

As the film is compressed, the molecules become more globular, which most likely leads to a decline in the number of ether linkages per molecule in contact with the water surface. When the surface pressure becomes great enough, the attractive interactions between the dendrimer molecules and the water surface can no longer hold the individual molecules to the surface, at which point the molecules are expelled from the water surface to sit on top of the original monolayer. In this way, the monolayer collapses to a bilayer or some other multilayer film. The small peak and plateau region seen in the isotherms correspond to this collapse. The peak seen in the PE-MBO12 isotherms seems to be representative of a nucleation-and-growth process for the collapse, and similar peaks have been observed by other investigators (*6, 7, 9, 14, 15*). After some number of successive layers has been produced, the upper layers prevent the molecules from the original monolayer from leaving the water surface and the surface pressure begins to rise significantly again resulting in the condensed phase that appears after the plateau region in the isotherm.

The compression rate dependence of the isotherms shown in Figure 2 shows that, at least for the higher speeds, the Langmuir films are not at equilibrium during the isotherm measurements. Further experiments at even slower barrier speeds would be required to determine the equilibrium behavior. The relaxation data suggest that below a surface pressure of 10 mN/m the film is relatively stable, but the small drop in the surface pressure while holding at constant area indicates that there is still some conformational relaxation that occurs in the dendrimer molecules. On the other hand, the compression/ expansion cycles as well as the relaxation experiments demonstrate that the dendrimer film is not stable above the monolayer-to-multilayer collapse pressure. The hysteresis between compression and expansion seen in Figure 3 shows that the film has not fully relaxed at high surface pressures. In fact, the change in molecular area after recompression suggests that the film has still not reached equilibrium even after expansion to the original surface area. This conclusion is supported by the long relaxation times (10^3 - 10^4 sec) observed within the monolayer-to-multilayer transition region. One would expect to see an elimination of the compression/ expansion hysteresis at sufficiently slow compression rates.

If one accepts the collapse mechanism as described above, then one might expect the temperature to have a strong effect on the surface pressure at which the collapse occurs. It is possible to imagine an energy barrier, possibly related to the breakage of hydrogen bonds, to the transition from a monolayer to a multilayer, so that an increase in temperature would increase the rate of the collapse and lower the surface pressure at which the collapse occurs. Figure 4 shows that increasing the temperature of the subphase does lower the collapse pressure, even though the difference may appear to be relatively small, especially considering the large difference in temperature of 15 °C. The relaxation times at the start of the monolayer to multilayer transition also suggest that the films collapse faster at 20 °C than at 5 °C. In fact, the number of measurable relaxation processes involved in the collapse drops from three to two when the temperature is increased, implying that one of the relaxation modes occurs so fast as to become negligible compared to the other two relaxation processes.

Additionally, subphase temperature has a strong effect on the extent of collapse. The plateau region extends to much smaller molecular areas at 20 °C than at 5 °C, suggesting that the films collapse further at higher temperatures. This observation could be a result of faster collapse rates for higher temperatures, in which case one would expect the dependence of the extent of collapse on the temperature to disappear near equilibrium; or the above observation could be caused by the formation of different equilibrium structures for films at the same surface pressure but different temperatures. In an attempt to clarify this matter, we have monitored the area change over time when the film is held at a constant surface pressure. Because of evaporation and acidification of the subphase due to carbon dioxide absorption, data collection is limited to approximately 14 hours per experiment. During this time we see that the area change is greater at 20 °C than at 5 °C, but the films have not reached equilibrium by the end of the experiments. No firm conclusions can be made about the number of layers formed during the collapse process given the current data. Attempts to fit our data with nucleation-and-growth kinetic models found in the literature (*16, 17*) have thus far been unsuccessful. Further work is needed in this area.

The relaxation data show that film relaxation times are faster for smaller molecular areas (higher surface pressures) within the collapse region. The observation that the collapse is faster at smaller molecular areas, or higher surface pressures, is not surprising, since the surface pressure is the driving force behind the collapse. The relaxation data also show that there are three relaxation processes involved in the collapse of the monolayer to a multilayer. With the data available it is difficult to predict what these three processes are. The transition from a monolayer to a multilayer is complicated and involves the expulsion of the dendrimer molecules from the monolayer and the subsequent relaxation of the molecules remaining at the water surface as well as those in the newly formed layers.

Previous investigators have studied film relaxation behavior in the reversible collapse region for other types of macromolecules, and it is therefore tempting to try to draw comparisons between various types of molecules to gain a better understanding of the film relaxation process. For example, Duran and coworkers (14) have shown that, when spread at the air-water interface, side-chain polysiloxanes, which form smectic liquid-crystalline phases in the bulk, undergo a reversible collapse to a bilayer during compression. From stress relaxation studies similar to those discussed earlier, it has been shown that the film collapse can be modeled as consisting of two relaxation processes, which the authors suggest can be related to the difference between the mobility of the side chains and the polymer. The fast relaxation is associated with the movement of the side chains, while the slow relaxation is attributed to the rearrangement of the polymer backbone. Similarly, Ringsdorf and coworkers (15) have studied the behavior of side-chain polyethyleneimines at the air-water interface. This polymer is highly crystalline in the bulk and also experiences a reversible collapse to a bilayer at the water surface. The collapse behavior is similar to that of the polysiloxanes and can be described in terms of the same two relaxation processes mentioned above.

Einloth and Frank (18) have shown that a polyglutamate film will collapse to a bilayer upon sufficient compression and that, unlike the before-mentioned systems, the collapse is found to have three relaxation modes. The polyglutamate molecules studied have side chains with two different lengths, and the backbone takes an α-helical structure at the air-water interface. These molecules are often referred to as "hairy rod" polymers because of the rigid backbone and flexible side chains. Einloth and Frank attribute the longest relaxation time to the rearrangement of the backbone, with the other two relaxation processes being related to the movement of the different length side chains. These observations are in accord with the film relaxation processes hypothesized for side-chain polymers discussed above.

Studies of polyvinylacetate at the air-water interface performed by Wang et al. (19) suggest that the film relaxation process might not be so easily generalized. Although a plateau region is not observed in the isotherm, the authors claim that polyvinylacetate monolayers collapse to stable multilayers when confined to sufficiently small molecular areas. Wang and coworkers have studied the surface pressure relaxation behavior in the collapse region and have found that the collapse consists of two relaxation processes. The fast relaxation is associated with the growth and collapse of chain loops that form when the film is at small molecular areas. The slow relaxation, on the other hand, is thought to be related to the repulsion effects between the polar head groups of the polymer molecules at the water surface.

It is difficult to make generalizations about the manner in which these macromolecular films relax in the collapse region. The process by which the films collapse depends on the polymer-polymer intermolecular interactions, polymer-subphase interactions, and intramolecular interactions and structure. Nevertheless, it seems as though the side-chain polymers behave similarly in that it is believed that the individual molecules are squeezed out of the monolayer, and it is the mobility of the polymer backbone that is associated with the slow relaxation process. The dendrimers studied in this work may be related to the side-chain polymers in that it is proposed that individual dendrimer molecules are expelled from the monolayer. The slow relaxation may then be related to the mobility of the entire dendrimer, while the fast relaxation modes may be associated with some movement of the branches. Further investigations are needed, however, before a more conclusive argument can be made.

Summary

We have shown that the polyether dendrimer discussed in this study is surface-active. The dendrimer molecules most likely lie somewhat flat on the water surface at large molecular areas and then form more globular structures when they are compressed. At even smaller molecular areas the monolayer collapses to a multilayer film, possibly a bilayer. At low surface pressure the monolayer is stable; however, the film experiences substantial relaxation in the collapse region. The collapse pressure decreases and the collapse rate increases with increasing temperature implying that there exists an energy barrier to the transition to the multilayer phase. Future studies are needed in order to understand the film relaxation process more completely. The above results combined with future studies using higher generation dendrimers should provide valuable knowledge regarding the interactions between polyether dendrimers and surfaces.

Acknowledgment

This work was supported in part by the NSF-MRSEC Center on Polymer Interfaces and Macromolecular Assemblies (CPIMA) and by an NIH Biotechnology Training Grant.

Literature Cited

1. Wells, M.; Crooks, R. M. *J. Am. Chem. Soc.* **1996**, *118*, 3988.
2. Bar, G.; Rubin, S.; Cutts, R. W.; Taylor, T. N.; Zawodzinski, T. A., Jr. *Langmuir* **1996**, *12*, 1172.
3. Watanabe, S.; Regan, S. L. *J. Am. Chem. Soc.* **1994**, *116*, 8855.
4. Lee, J. J.; Ford, W. T.; Moore, J. A.; Li, Y. *Macromolecules* **1994**, *27*, 4632.
5. Jayaraman, G.; Li, Y. F.; Moore, J. A.; Cramer, S. M. *J. Chromatogr. A* **1995**, *702*, 143.

6. Saville, P. M.; White, J. W.; Hawker, C. J.; Wooley, K. L.; Frechet, J. M. J. *J. Phys. Chem.* **1993**, *97*, 293.
7. Saville, P. M.; Reynolds, P. A.; White, J. W.; Hawker, C. J.; Frechet, J. M. J.; Wooley, K. L.; Penfold, J.; Webster, J. R. P. *J. Phys. Chem.* **1995**, *99*, 8283.
8. Karthaus, O.; Ijiro, K.; Shimomura, M. *Langmuir* **1996**, *12*, 6714.
9. Kampf, J. P.; Frank, C. W.; Hawker, C. J.; *Polymer Prepr.* **1997**, *38*, 908.
10. Bo, Z.; Zhang, X.; Yi, X.; Yang, M.; Shen, J.; Rehn, Y.; Xi, S. *Polymer Bulletin* **1997**, *38*, 257.
11. Padias, A. B.; Hall, H. K., Jr. *J. Org. Chem.* **1987**, *52*, 5305.
12. Moore, J. A.; Lee, Y. Unpublished results.
13. Gaines, G. L., Jr. *Insoluble Monolayers at Liquid-Gas Interfaces*; Interscience Publishers: New York, NY, 1966.
14. Adams, J.; Buske, A.; Duran, R. S. *Macromolecules* **1993**, *26*, 2871.
15. Seitz, M.; Struth, B.; Preece, J. A.; Plesnivy, T.; Brezesinski, G.; Ringsdorf, H. *Thin Solid Films* **1996**, *284*, 304.
16. Smith, R. D.; Berg, J. C. *J. Colloid Interface Sci.* **1980**, *74*, 273.
17. Vollhardt, D.; Retter, U. *J. Phys. Chem.* **1991**, *95*, 3723.
18. Einloth, T. L.; Frank, C. W. Unpublished results.
19. Wang, L. F.; Kuo, J. F.; Chen, C. Y. *Colloid Poly. Sci.* **1995**, *273*, 426.

Chapter 4

Flow-Induced Deformation and Relaxation Processes of Polydomain Structures in Langmuir Monolayer

Kang Sub Yim, Carlton F. Brooks, Gerald G. Fuller[1], Curtis W. Frank, and Channing R. Robertson

Department of Chemical Engineering, Stanford University, Stanford, CA 94305–5025

Flow-induced deformation and relaxation processes of monolayer domains at the air-water interface have been studied by using simple shear flow and extensional flow. It is found from fluorescence microscopy that deformed domains of dipalmitoyl-phosphatidylcholine (DPPC) tend to relax to circular shapes due to the line tension effects. By examining this relaxation process at low surface pressure, the value of the line tension can be calculated. In the limit of low surface pressure, the line tension was found to be $\lambda = 1.11 \pm 0.20 \times 10^{-12}$ N, which is of the same order as values found for other monolayers. A surface Capillary number $Ca_s = \dfrac{\eta_b \dot{\gamma} R^2}{\lambda}$ is proposed to explain the two-dimensional deformation process in systems dominated by the subphase viscosity, η_b. R is the undeformed domain radius and $\dot{\gamma}$ is the velocity gradient. For slightly deformed domains, the deformation D and orientation angle φ are linear functions of Ca_s in two dimensions, and extensional flow is twice as effective in deforming domains as simple shear flow, which is similar to the case of three-dimensional droplets.

Chemical substances such as phospholipids, fatty acids, alcohols and polymers, which form insoluble monolayers at the air-water interface, have been studied extensively.[1-3] These Langmuir monolayers have been used as models of biological membranes[4] and have a wide range of possible applications.[5] A monolayer can be thought of as a separate phase with thermodynamic properties similar to those of three-dimensional systems.[6] Fluorescence microscopy, using a low concentration of fluorescent lipid probes, can be used to observe the structure of monolayers.[7-9] These experiments permit the visualization of domains of various thermodynamic phases in monolayers.

[1]Corresponding author.

The shape of domains in monolayers is assumed to be governed by two distinct intermolecular forces: the line tension and electrostatic forces.[10-11] The line tension arises from short-range interactions that favor circular domain shapes and is opposed by long-range interactions that lead to non-circular shapes, mainly due to electrostatic forces. For this reason, the line tension is a key parameter determining the stability and shape of two-dimensional monolayer domains. Several groups have studied the line tension using a variety of measurement techniques. One class of methods uses the rate of activated homogeneous nucleation[12] and the long-range interactions between domains.[13] Alternatively, the other investigators have determined the line tension by deforming the monolayer in the coexistence region using hydrodynamic forces and observing the relaxation dynamics of the phase separated domains. The deformation has been produced using techniques where the applied hydrodynamic forces were rather uncontrolled: a gas jet blowing on the interface[14,16] or dragging a needle through the interface.[15] The undefined nature of the resulting flow fields can lead to difficulties in the hydrodynamic analysis of two-dimensional monolayer systems. In the present study, simple shear flow using a parallel band device and extensional flow using a four-roll mill were used to deform a monolayer. These well-defined flow systems also provide the additional advantage of making it possible to analyze the fluid dynamics of the monolayers *in situ* during flow.

In three dimensions, as first noted by Taylor[17,18], droplet deformation and orientation are governed by two dimensionless variables: the viscosity ratio η_{rel} between the droplet phase and the continuous phase, and the Capillary number Ca, which is the ratio between viscous and interfacial forces. In the case of Newtonian droplets in a Newtonian matrix, Ca can be expressed as $Ca = \dfrac{\eta_b \dot{\gamma} R}{\sigma}$ where η_b is the viscosity of the bulk matrix, $\dot{\gamma}$ is the strain rate of the flow, R is the equilibrium radius of a drop, and σ is the interfacial tension. Taylor's analysis was performed in the limit that Ca $\ll 1$. Since Taylor's early work, a number of investigators have extended the deformation and orientation predictions with higher order solutions valid over a larger range of Ca[21-24], and have verified these analyses experimentally.[19,20] These theoretical and experimental studies of deformation in three dimensions have been mainly performed to ascertain how much distortion is produced by a given flow, how strong the flow must be to break the drop, and the number and size of the droplets that result from droplet breakup. On the other hand, only a few theoretical studies have considered the analogous deformation process in two-dimensions.[25,26] It has been found that the solutions obtained in two dimensions showed remarkable similarities with the observed behavior of the three-dimensional droplet, such as the development of a re-entrant cavity at its rear end and the tendency to migrate towards the axis in a circular tube. However, the relationships between two-dimensional drop deformation and the two-dimensional equivalents of a Capillary number and viscosity ratio have not been examined theoretically or experimentally. In this study, Langmuir monolayers of dipalmitoyl-phosphatidylcholine (DPPC) subject to both shear and extensional flow were studied in order to explore deformation phenomena in two-dimensional systems.

Background

Relaxation Processes. A deformed, two-dimensional droplet will relax towards its equilibrium shape in a process characterized by two regimes depending on its initial deformed shape: bola-shaped and ellipse-shaped. At high distortion, two-dimensional domains have a bola-shaped structure with two heads and a thin connecting strip. Benvegnu and McConnell[14] first measured the line tension by observing the relaxation rate of each bola. It is assumed that the two bola ends are circular with radius R, which is much larger than the width w of the connecting strip. The line tension force causes the two bola ends to coalesce and is opposed by the viscous drag of the subphase. By balancing these two forces and neglecting the electrostatic forces, the following equation can be obtained for the line tension in the low Reynolds number limit.

$$\lambda = \frac{8}{3}\eta_b RU \qquad (1)$$

where λ is the line tension, η_b is the bulk viscosity of the subphase (i.e. water) and U is the velocity of a single bola. Thus, the line tension at high distortion can be calculated by measuring the bola radius and its velocity in the relaxation process of highly deformed domains.

In the case of slightly deformed domains, it is useful to characterize the definition of the distortion as $\Theta = L/W - 1$, where L and W are the length and the width of an elliptically deformed domain. At small distortions ($\Theta \leq 1$), it is known that Θ relaxes exponentially with time ($\Theta = \Theta_0 \exp(-t/Tc)$) where Tc is a characteristic relaxation time. The characteristic time, Tc, of relaxation can be determined from dimensional analysis, and it is a function of the mean radius of the domain $R \equiv (Area/\pi)^{1/2}$, the line tension, and viscosity.[15,27]

$$T_C \propto \frac{\eta_S R}{\lambda} \qquad \text{if the surface viscosity is dominant}$$

$$T_C \propto \frac{\eta_b R^2}{\lambda} \qquad \text{if the bulk viscosity is dominant}$$

η_s, η_b are the surface viscosities of the monolayer and the bulk viscosity of the subphase, respectively. In the experimental conditions of this study (low surface pressure $\Pi \ll 1$ mN/m), it can be shown that dissipation by the bulk viscosity dominates the surface viscosity effects due to very low values of the latter property in the region of low Π.[28-30] The following equation is derived under the assumption of an incompressible monolayer.[28]

$$T_c = \frac{5\pi}{16} \frac{\eta_b R^2}{\lambda} \qquad (2)$$

Therefore, the line tension at small distortions can be obtained by measuring the characteristic time of the relaxation process. Interdomain electrostatic forces are also neglected in the derivation of this equation. This assumption is reasonable when individual domains are sufficiently far apart from each other; all experiments in this study were performed under such conditions.

Deformation Processes. In this paper, the deformation process under *in situ* flow as well as the relaxation process after flow cessation in two-dimensional monolayers has been studied. In a three-dimensional droplet, Taylor provided the first theoretical analysis of the steady deformation and orientation of a drop suspended in a viscous liquid. He solved the creeping motion equations for the drop surface for small deformations where the normal stress components were balanced by the interfacial tension through changes in the surface curvature. It was found that the deformation D and orientation angle of droplets, φ, relative to the velocity gradient direction in three dimensions were linear functions of the Capillary number for weak flows. The deformation parameter D and orientation angle φ are defined in Fig. 1 for both shear and extensional flows.

[1] In a simple shear flow given by $u=(\dot{\gamma} x_2, 0)$[18,24], the results are

$$D = \left(\frac{19\eta_{rel} + 16}{16\eta_{rel} + 16}\right) \cdot Ca \qquad (3)$$

$$\varphi = \frac{\pi}{4} + \frac{(19\eta_{rel} + 16)(2\eta_{rel} + 3)}{80 \ (\eta_{rel} + 1)} \cdot Ca \qquad (4)$$

[2] In an extensional flow given by $u=(\dot{\gamma} x_1, -\dot{\gamma} x_2)$[18] , we have

$$D = 2\left(\frac{19\eta_{rel} + 16}{16\eta_{rel} + 16}\right) \cdot Ca \qquad (5)$$

$$\varphi = \frac{\pi}{2} \qquad (6)$$

Note that in extensional flow the deformation increases with Ca twice as fast as in the case of shear flow.

In contrast, for 2D droplets, the interfacial tension σ in Ca should be replaced with the line tension λ. This proposed modification to the Capillary number will henceforth be called the surface Capillary number, Ca_s. Dimensional analysis suggests that the surface Capillary number is given by $Ca_s = \dfrac{\eta_b \dot{\gamma} R^2}{\lambda}$ if the bulk viscosity dominates the system. If the

surface viscosity is the dominant mode of dissipation, then $Ca_s = \dfrac{\eta_s \dot{\gamma} R}{\lambda}$.

Materials and Methods

Dipalmitoyl-phosphatidylcholine (DPPC) and the fluorescent probe Texas Red (TR) were purchased from Sigma and Molecular Probes, respectively. The phospholipid solution was prepared by dissolving 99.5 mol% DPPC and 0.5 mol% TR in chloroform to a final concentration of 0.36 mM. Before spreading a monolayer, the air-water interface was aspirated extensively with a polypropylene pipette to remove contaminants. A small amount of this solution was spread with a microsyringe on deionized water purified by a Milli-Q system (Millipore corp.)

Monolayers were formed on a 25.0 cm x 7.5 cm Langmuir trough made of Teflon with a Wilhelmy balance for surface pressure measurements. All experiments were performed in the two phase coexistence region of the gas and liquid expanded phases ($\Pi \ll 1$ mN/m) at 20 °C.

Monolayers were observed with a fluorescence microscope with a 10X objective. The images of monolayer were recorded on videotape and analyzed by the help of image analysis software (NIH-Image 156 ppc)

Simple shear flow and extensional flow were generated to deform domains by using a parallel band made of polypropylene and a four-roll mill made of Teflon, respectively. These devices are shown in Fig. 2. In each flow field, measurements were performed in the stagnation region, where material has a long residence time. Also this stagnation region is the farthest from the device, resulting in sufficient flatness of the surface, which means there is no meniscus effect.

In order to provide a qualitative description of the extensional flow kinematics, a particle tracking method was used. Monolayers were dusted with sulfur particles whose motion was analyzed by a CCD camera (SONY XC771). This experiment produced the relationship between the angular velocity of the rollers and the strain rate of the extensional flow field. The sulfur particle trajectories are shown in Fig 3. The symbols represent the center of mass of the sulfur particles, and the solid curves correspond to the best fit hyperbolic streamlines through these coordinates. The agreement between the experimental and theoretical streamlines is very good, indicating that the four-roll mill gives a good approximation to planar extensional flow in monolayers.[31] The relationship between the mean strain rate and roller velocity was obtained by the simple regression of particle coordinates and time. As expected, a good linear correlation was found and used to analyze the deformation process and is shown in Fig. 4.

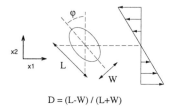

$$D = (L-W) / (L+W)$$

Figure 1. The definition of deformation parameter (D) and the orientation angle (φ) in a simple shear flow.

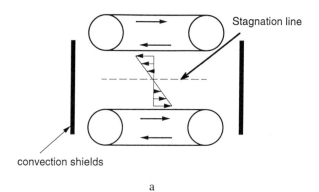

a

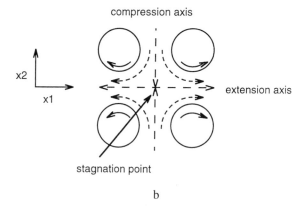

b

Figure 2. (a) Top view of the parallel band for a simple shear flow. The direction of band rotation and the resulting flow field are shown. A line of flow stagnation exists along the center line. (b) Top view of the four-row mill to produce extensional flow. The direction of roller rotation and the resulting flow field are shown. A stagnation point of the flow field exists at the center of the device.

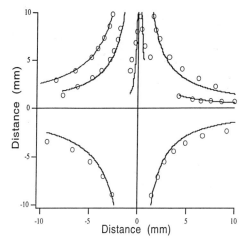

Figure 3. Sulfur particle trajectories and best fit hyperbolic streamlines for DPPC monolayers subject to flow in the four roll mill.

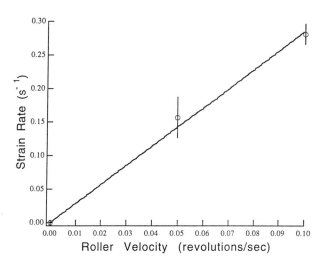

Figure 4. Relationship between roller velocity and the observed strain rate for monolayers of DPPC.

Results and Discussions

Relaxation Processes. As explained in the introduction, there are two relaxation regimes depending on the initial shape of deformed domains; (either a bola-shape or an elliptical shape). In each regime, the line tension can be obtained by measuring the relaxation of deformed domains towards to their original circular shape. Fig. 5 gives an example of a bola-shaped domain as it relaxes to a circle. It is observed that a domain initially deformed by shear (Fig. 5(a)) is restored to an equilibrium circular shape (Fig. 5(f)) when the shear is removed. Equation (1) was used to calculate the line tension from measured values of velocity U and bola radius R.

In the small distortion limit, a deformed domain of elliptical shape relaxes exponentially with time, and the line tension can be obtained by measuring the characteristic relaxation time Tc from equation (2). An example of such a relaxation process is shown in Fig. 6 for a monolayer domain. In both regimes, it was found that the relaxation of domains was hindered by the presence of adjacent domains, which reduced the apparent line tension value. This hindrance seems to be due to electrostatic and hydrodynamic forces between adjacent domains. Therefore, domains were selected that were sufficiently far apart from neighboring domains in order to minimize these effects.

Fig. 7 shows the mean line tension value by taking averages over all values in each regime. In the size range studied (R=10-80 μm), the obtained line tension, as expected, is independent of both the domain size and the relaxation regime. The independence of domain size indicates that the subphase is the dominant mode of dissipation in our case and that the proper scaling for Tc has been used. The line tension values obtained by various groups for different monolayer systems are compared in Table I. As can be seen, the measured line tension in this study was found to be $1.11 \pm 0.20 \times 10^{-12}$ N, and this value is of the same order as values found for different monolayer systems. All experiments were performed at low surface pressures, which leads to bulk viscosity dominant systems. However, it has been reported that different line tension values are obtained at higher surface pressures,[13,14] and the line tension of bilayers has been found to be 10 times higher than that of corresponding monolayers.[16] The result of the present study suggests that for systems for which the bulk viscosity dominates, monolayers of different chemical materials such as phospholipids, polymers and fatty acids all have similar line tension values if there is no particular conformational and/or aggregational change. The well-defined flow used in this study does not cause improvement on the magnitude of error of the measured line tension value because relaxation is the process after flow stops.

Deformation Processes. In this section, the *in situ* measurement of the deformation of monolayer domains subjected to flow is presented, providing information on the relationship between domain deformation and the applied flow strength and flow type in two dimensions. To our knowledge, no previous investigations of *in situ* deformation have been

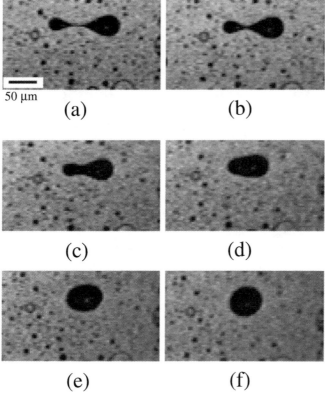

Figure 5. Relaxation of a bola-shaped domain at a surface area of A=110 Å²; (a) t=0 s, (b) 0.5 s, (c) 1 s, (d) 1.5 s, (e) 2 s, (f) 2.5 s.

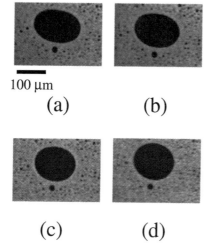

Figure 6. Relaxation of an elliptical shaped domain at A=140 Å²; (a) t=0 s, (b) 1 s, (c) 2 s, (d) 4 s.

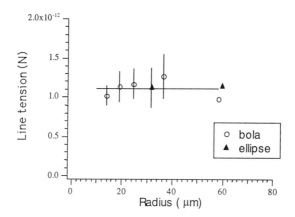

Figure 7. Line tension vs. radius for both relaxation regimes. The radius of the bola in the bola domain and the radius of the circular shape after relaxation in the ellipse domain were used, respectively: bola-shaped(○), elliptical-shaped(▲).

Table I. The line tension value obtained from different monolayer systems by several groups.

	Line tension	Monolayer	Surface pressure and Temperature
Benvegnu and McConnell[14]	$(1.12\pm0.28)\times10^{-12}$ N[a]	the mixture of DMPC (dimyristoyl phosphatidylcholine) and cholesterol	1 mN/m 20 °C
Mann et al.[15]	$(1.1\pm0.3)\times10^{-12}$ N	Poly(dimethyl)siloxane	≤ 1 mN/m 22 °C
Läuger et al.[16]	$(1.2\pm0.3)\times10^{-12}$ N	4-octyl[1,1-biphenyl]-4-carbonitride	≤ 1 mN/m 18 °C
This study	$(1.11\pm0.20)\times10^{-12}$ N	**DPPC**	**≤ 1 mN/m 20 °C**

[a] The value recalculated by excluding the electrostatic term.

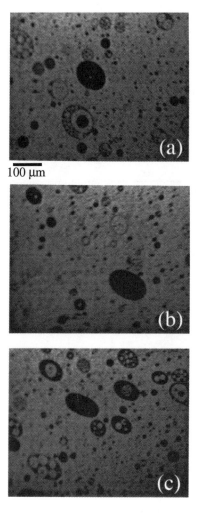

100 μm

Figure 8. Domain structures in response to simple shear flow at a variety of velocity gradients A=110 Å^2: (a) $\dot\gamma$=0.22 s^{-1}, (b) 0.44 s^{-1}, (c) 0.88 s^{-1}.

reported on the two-dimensional systems. The micrographs shown in Fig. 8 are examples of the deformed domains over different shear rates under *in situ* simple shear flow. As the shear rate increases, both the deformation D and the orientation angle φ increase. This result suggests the existence of a relationship between deformation and strain rate in two dimensions, as found in the case of three dimensions. In order to examine this relationship, D and φ are plotted against Ca$_s$ for simple shear flow in Fig. 9. This plot indicates that (D, φ) are well correlated against Ca$_s$ in two

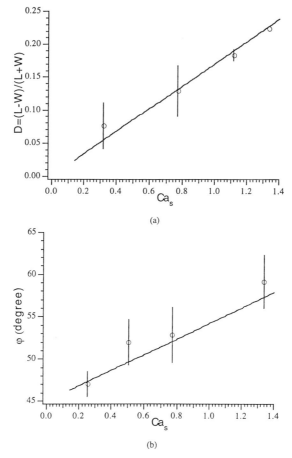

Figure 9. (a) Deformation, D, and (b) orientation angle, φ, as a function of Ca_s during simple shear flow.

dimensions and are linearly related at these shear rates. From this result, it was found that Ca_s is the appropriate dimensionless number to gauge the two-dimensional deformation process.

The deformation D in extensional flow is plotted against the rate of strain in Fig. 10. As expected, extensional flow is more effective in deforming domains than simple shear flow. The slope of D as a function of Ca_s in extensional flow, 0.334, is about twice as high as that in simple shear flow, 0.170. This two-to-one relationship is also found in three-dimensional droplet systems.[18,19] From these results, it is verified that the deformation and orientation angle in two dimensions show a linear correlation with the surface Capillary number for small value of Ca_s, analogous to three dimensions. Indeed, if the theoretical relationship between D and flow strength were known, the line tension could be obtained through this type of measurement.

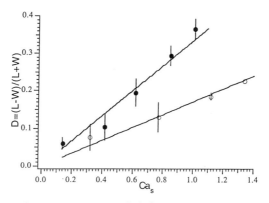

Figure 10. The comparison of deformation in a simple shear flow and an extensional flow: a simple shear flow(○), an extensional flow(●).

Conclusion

We studied the relaxation process after flow cessation and the deformation process under *in situ* flow in a two-dimensional DPPC monolayer system. Both the relaxation of both bola shaped and elliptical shaped domains yield the same line tension value of $1.11 \pm 0.20 \times 10^{-12}$ N, which is similar to values found for different monolayer systems investigated by other groups. In order to examine the deformation process, a surface Capillary number was proposed. It was found that the deformation and orientation angle are linear functions of the surface Capillary number for slightly deformed domains in two dimensions. These findings suggest that the line tension can be obtained from the correlation between deformation and surface Capillary number as well as from the relaxation of deformed domains.

Literature Cited

1. McConnell, H. M. *Annu. Rev. Phys. Chem.* **1991**, 42, 171.
2. Möhwald, H. *Annu. Rev. Phys. Chem.* **1990**, 41, 441.
3. Knobler, C. M. *Adv. Chem. Phys.* **1990**, 77, 397.
4. Phillips, M. C.; Chapman, D. *Biochim. Biophys. Acta.* **1968**, 163, 301.
5. Swalen, J. D.; Allara, D. L.; Andrade, J. D.; Chandross, E. A.; Garoff, S.; Israelachvili, J.; McCarty, T. J.; Pease, R. F.; Rabolt, J. F.; Wynne, K. J.; Yu, H. *Langmuir* **1987**, 3, 932.
6. Knobler, C. M. *Annu. Rev. Phys. Chem.* **1992**, 43, 207.
7. Peters, R; Beck, K. *Proc. Natl. Acad. Sci. USA* **1983**, 80, 7183.
8. von Tscharner, V.; McConnell, H. M. *Biophys. J.* **1981**, 36, 409.
9. Lösche, M.; Möhwald, H. *Rev. Sci. Instrum.* **1984**, 55, 1968.
10, Andelman, D.; Brochard, F.; Joanny, J. F. *J. Chem. Phy.* **1987**, 86, 3673.

11. Keller, D. J.; McConnell, H. M.; Moy, V. T. *J. Phys. Chem.* **1986**, 90, 2311.
12. Muller, P.; Gallet, F. *Phys. Rev. Lett.* **1991**, 67, 1106.
13. Riviere, S.; Henon, S.; Meunier, J.; Albrecht, G.; Boissonnade, M. M.; Baszkin, A. *Phys. Rev. Lett.* **1995**, 75, 2506.
14. Benvegnu, D. J.; McConnell, H. M. *J. Phys. Chem.* **1992**, 96, 6823.
15. Mann, E. K.; Henon, S.; Langevin, D.; Meunier, J.; Leger, L. *Phys. Rev. E* **1995**, 51, 5708.
16. Läuger, J.; Robertson, C. R.; Frank, C. W.; Fuller, G. G. *Langmuir*, **1996**, 12, 5630.
17. Taylor, G. I. *Proc. Roy. Soc. Ser. A* **1932**, 138, 41.
18. Taylor, G. I. *Proc. Roy. Soc. Ser. A* **1934**, 146, 501.
19. Rumscheidt, F. D.; Mason, S. G. *J. Colloid Sci.* **1961**, 16, 238.
20. Torza, S.; Cox, R. G.; Mason, S. G. *J. Colloid Interface Sci.* **1972**, 38, 395.
21. Barthes-Biesel, D.; Acrivos, A. *J. Fluid Mech.* **1973**, 61, 1.
22. Hinch, E. J.; Acrivos, A. *J. Fluid Mech.* **1980**, 98, 305.
23. Cox, R. G. *J. Fluid Mech.* **1969**, 37, 601.
24. Chaffey, C. E.; Brenner, H. *J. Colloid Interface Sci.* **1967**, 24, 258.
25. Richardson, S. *J. Fluid Mech.* **1968**, 33, 475.
26. Richardson, S. *J. Fluid Mech.* **1973**, 58, 115.
27. Mann, E. K.; Henon, S.; Langevin, D.; Meunier, J. *J. Phys. II* (France) **1992**, 2, 1683.
28. Stone, H. A.; McConnell, H. M. *Proc. R. Soc. London Ser. A* **1995**, 448, 97.
29. Gaub, H. E.; McConnell, H. M. *J. Phys. Chem.* **1986**, 90, 6830.
30. Krägel, J.; Li, J. B.; Miller, R.; Bree, M.; Kretzschmar, G.; Möhwald, H. *Colloid Polym. Sci.* **1996**, 274, 1183.
31. Higdon, J. J. L. *Phys. Fluids A* **1993**, 5, 274.

Chapter 5

Tuning the Electronic Properties of Silicon via Molecular Self-Assembly

Noemi Zenou[1], Alexander Zelichenok[1], Shlomo Yitzchaik[1,3], Rami Cohen[2], and David Cahen[2,3]

[1]Department of Inorganic and Analytical Chemistry, The Hebrew University of Jerusalem, Jerusalem 91904, Israel
[2]Department of Materials and Interfaces, Weizmann Institute of Science, Rehovot 76100, Israel

Control over the surface chemistry and physics of electronic and optical materials is essential for constructing devices and fine-tuning their performance. In the past few years we have started to explore the use of organic molecules for systematic modification of semiconductor surface electronic properties. In this paper, manipulation of silicon surfaces by self-assembly of various quinolinium-based chromophores is reported. The progress of the assembly process is monitored by XPS, UV-Vis, and FTIR spectroscopies as well as with surface wettability. The effect of the monolayer's dipole-moment on the Si surface potential and the interaction with surface states is monitored by CPD measurements. A pronounced effect of a sub-nanometer coupling-agent layer alone on the electron affinity and band-bending of Si was observed. We also show a way to modulate the Si work-function by tuning the dipole strength of the chromophore-containing organic, self-assembled monolayer and of its orientation with respect to the silicon surface.

Molecule-based, self-assembled, ultra-thin films have attracted considerable attention in recent years, because of potential technological applications owing to their inherent significance in surface modifications. These supramolecular assemblies[1-3] are attractive candidates for advanced photonics[4], such as nonlinear optics, and for molecular-electronics[5] applications. Assemblies of molecules can also be used for nano-scale modification of non-molecular semiconductors, such as the control over electron transfer efficiency through them.[6-9] Utilization of molecules, whose properties can be changed systematically and that chemisorb onto a semiconductor surface, presents a versatile and reproducible way to change electrical properties of the semiconductors in a controllable manner.[10-17]

Ideally, control over the surface electronic properties will be considered as the ability to alter the work function (f) of the semiconductor by either regulating the electron affinity (EA) or the band bending (BB) or both. These properties are illustrated schematically in Figure 1, by a simplified one electron energy diagram. Here the electron affinity, or the surface potential, is the difference between the energies of an electron in vacuum and in the deepest level of the conduction band, at the surface.

[3]Corresponding authors.

This property is influenced by the electric-field at the surface or interface and thus is highly sensitive to the presence of molecular dipoles on the surface. By controlling the magnitude and direction of the molecules' dipole we can change the semiconductor's electron affinity. Band bending (at a free surface) depends on the electric charge density on the surface, i.e., on the occupation of semiconductor surface states. The smaller the charge in the surface states, the smaller is the band bending. These surface states, which might originate from defect sites at the surface, can function as trapping or recombination centers for electrons and holes.

In general, surface modification can be achieved by binding of different molecules at the surface, using either inorganic or organic materials. The advantage in utilizing organic modifications as compared with inorganic treatments stems from the greater flexibility in molecular design for surface modification because of the possibility of incorporating several functional groups into a single molecule (*i.e.,* surface binding sites, molecular packing motifs, and electron susceptibility). These groups can be modified systematically and independently. Earlier studies have shown that exposure of semiconductors to organic ligands can change semiconductor luminescence [11-13,18,19] and, in solution, its flat band potential.[14,20]

In contrast to several attempts that have been made in the past to passivate and control the Si surface using inorganic treatments,[21,22] our approach uses organic manipulations and self-assembly of organo-silanes. These assemblies can be well-ordered, with a relatively dense structure[4, 23-25]. They exhibit a remarkable stability in common organic solvents and in acidic media. Control over the EA of the silicon surface is performed by incorporating a polar group within the monolayer, which can be modified systematically. For this purpose we selected polarizable molecules, quinolinium-based chromophores, that can easily be substituted with an electron-accepting or -donating group. In this way it is possible to control the dipole, direction and magnitude of the resulting monolayer, without altering the overall packing motif of this nano-structure. Moreover, this assembly technique produces interfacial polar alignment with a maximized chromophore number density and enhanced temporal stability of the molecular structure.

EXPERIMENTAL

All the synthetic operations were carried out using Schlenk techniques. Hexane, pentane, and tetrahydrofuran (T. J. Baker) were distilled from Na/K alloy immediately before use. (3-bromopropyl)trichlorosilane (Gelest Inc.) was purified by vacuum distillation. Quinoline and quinolines, substituted in the 6th position (Aldrich), were vacuum-distilled and the solid precursors (IV and V, *vida infra*) were recrystalized, prior to their use.

Quartz windows (ChemGlass) and silicon wafers (n-type, <100>, 0.1 $\Omega \cdot$cm, Virginia Semiconductor, Inc.) were cleaned in 0.5 vol. % aqueous detergent solution, rinsed copiously with triple-distilled (TD) water, cleaned for 1 h in 90°C H_2SO_4 : H_2O_2 (70:30 v/v) solution and then allowed to cool down to room temperature (over 30 min.). The clean substrates were then rinsed three times with TD water. Further cleaning was carried out by sonication in an $H_2O/H_2O_2/NH_3$ (5 : 1 : 1 by vol.) cleaning solution for 30 min. After subsequent washing with TD-water, the substrates were immersed for 5 min. in pure acetone (T. J. Baker, electronic grade) and dried in a clean convection oven for 20 min. at 110°C.

Monolayer Assembly: Step *i*: Freshly cleaned quartz and/or silicon single-crystal substrates were immersed in a 1 : 100 (v/v) solution of (3-bromopropyl) trichlorosilane in *n*-hexane for 20 min. and then washed with copious amounts of *n*-pentane, followed by sonication in acetone for 10 min. The samples were then air-cured for 10 min. at 110°C. Step *ii*: The silylated substrates (after step *i*) were immersed in 5 mM solution of the various quinoline precursors in toluene (for the nitro derivative 1:1 (v/v) THF/toluene solution was used) for 54 h at 100°C. After the substrates had reached room temperature, they were rinsed with toluene, toluene/methanol, and methanol, followed by further sonication in methanol for 10 min.

Surface wettabilities were measured with a Ramé-Hart (model 100-00) goniometer. UV-Vis absorption spectra were recorded in the transmission mode, on samples of quartz substrates with a Shimadzu UV3101PC spectrophotometer. X-ray photoelectron spectra were recorded using a monochromatic Al K_a X-ray source on a Kratos Axis-HS instrument. FTIR data were collected (Bruker FTIR IFS66) in the grazing (80°) angle mode on silicon substrates. Contact potential difference (CPD) measurements[26] were made with a commercial instrument (Delta Phi Besocke, Jülich, Germany) in ambient conditions, in a home-built Faraday box. BB was calculated by subtracting the dark value from the value measured under photosaturation condition (QTH) lamp, intensity at sample 130mW/cm^2), which was assumed to be that at flat band. The CPD measurements were taken after the signal stabilized. To eliminate molecular decomposition during photosaturation measurements, we used an optical filter that blocks absorption by the molecules (Schott, RG-780). For more details see reference.[27]

Model Chromophores Synthesis: All the 6-quinoline derived salts, see Figure 2, were prepared by stirring the appropriate quinoline with excess of methyl-iodide (MeI) in methylene chloride solution for 12 hours at room temperature. The precipitates were filtered, washed with ethyl acetate and recrystalized from ethanol.

(I), 6-methoxy (N-methylquinolinium-iodide): ^{1}H NMR (DMSOd$_6$, 300MHz) 3.99 (S, 3H [OCH$_3$]); 4.60 (S, 3H [CH$_3$-N]); 7.88 (m, 2H); 8.08 (t, 1H, J=6.5Hz); 8.43 (d, 1H, J=9.0 Hz); 9.2 (d, 1H, J=9.0Hz); 9.31 (d, 1H, J=6.5Hz). EA (for C, H, N) Calc'd: 43.88, 4.02, 4.65; Found: 44.54, 4.11, 4.72. λ_{max}(MeOH) = 350 nm. (II), 6-methyl (N-methylquinolinium-iodide): ^{1}H NMR (D$_2$O, 300MHz) 2.57 (S, 3H [CH$_3$]); 4.57 (S, 3H [CH$_3$-N]); 8.0 (m, 3H); 8.16 (d, 1H, J=9.0Hz); 8.89 (d, 1H, J=9.0Hz); 9.08 (d, 1H, J=6.5Hz). EA (for C, H, N) Calc'd: 46.34, 4.24, 4.91; Found: 45.99, 4.15, 4.74. λ_{max}(MeOH) = 321 nm. (III), N-methylquinolinium-iodide: ^{1}H NMR (D$_2$O, 300MHz) 4.66 (S, 3H; [CH$_3$-N]); 8.03 (m, 2H); 8.26 (t, 1H, J=6.5Hz); 8.34 (t+t, 2H, J=9.0Hz); 9.1(d, 1H, J=9.0Hz); 9.2 (d, 1H, J=6.5Hz). EA (for C, H, N) Calc'd: 44.30, 3.72, 5.17; Found: 44.56, 3.56, 4.66. λ_{max}(MeOH) = 297 nm. (IV), 6-chloro (N-methylquinolinium-iodide): ^{1}H NMR (DMSOd$_6$, 400MHz) 4.62 (S, 3H [OCH$_3$]); 8.23 (td, 1H J$_1$=6.5Hz, J$_2$=2.5Hz); 8.30 (dd, 1H, J$_1$=6.5Hz, J$_2$=2.5Hz); 8.55 (d, 1H, J=9.0Hz); 8.65 (d, 1H, J=2.5Hz); 9.20 (d, 1H, J=9.0Hz) 9.54 (d, 1H J=6.5Hz). EA (for C, H, N) Calc'd: 39.31, 2.97, 4.58; Found: 39.36, 2.91, 4.38. λ_{max}(MeOH) = 272 nm. (V), 6-nitro (N-methylquinolinium-iodide): ^{1}H NMR (D$_2$O, 300MHz) 4.76 (S, 3H [CH$_3$-N]); 8.24 (t, 1H, J=6.5 Hz); 8.66 (d, 1H, J=9.0 Hz); 8.97 (d, 1H, J=9.0 Hz); 9.37 (m, 2H); 9.47 (d, 1H, J=6.5 Hz). EA (for C, H, N) Calc'd: 37.97, 2.87, 8.86; Found: 38.21, 2.87, 8.31. λ_{max}(MeOH) = 267 nm.

RESULTS AND DISCUSSION

The progress of the assembly process is monitored by X-ray photoelectron spectroscopy (XPS), advancing aqueous contact angle (θ_a) measurements, grazing angle FT-IR and transmission optical spectroscopies. The effect of the self-assembled monolayer on the surface electrical properties was measured by contact potential difference (CPD), using a capacitance method, the vibrating Kelvin probe.[28]

Our synthetic approach to chromophoric assemblies utilizes a sequential process of two layer-building steps, as shown in Figure 2. Starting from a clean, hydroxylated surface (e.g. SiO_2 or the native-oxide layer of single-crystal silicon) step i introduces a bromoalkyl functionality. The presence of the bromoalkyl is evident from the Br 3d XPS peak (at 69.1 eV, characteristic of bromoalkane), an increase in θ_a (from 15° for SiO_2 to 75°) indicative of a more hydrophobic surface, and the appearance of a C-H stretching band at 1278 cm^{-1}. The cure component of this step is designed to achieve maximum intrafilm and film-substrate cross-linking. The resulting monolayer thickness was determined, in an earlier study,[29] to be about 10Å. Step ii anchors the quinoline chromophore precursor via quaternization to the surface-bound bromopropyl. This step forms the quinolinium chromophoric monolayer. The course of the quaternization /assembly was verified by the decrease in the Br 3d XPS energy (67.3 eV), indicative of Br$^-$ formation [30], the appearance of an N1s XPS signal (398.5 eV) and an IR peak at 1616 cm^{-1} (C-N$^+$ stretch). We ascribe the last two to quinolinium. Figure 3 demonstrates the formation of covalent C-N$^+$ bonds (formation of a quaternary amine[31]) between the quinoline precursor and the surface bound bromopropyl anchoring layer. The 1606 cm^{-1} band is not observed on the surface. Its absence and the presence of the 1616 cm^{-1} band suggest that the adsorbate is bound to the surface by the formation of the expected quaternary ammonium site. Functionalization of the surface was supported by the change in θ_a from 75° after step i to 84°[X=CH$_3$], 81°[X=H], 78°[X=Cl], 76°[X=OCH$_3$], and 66°[X=NO$_2$] after step ii.

The course of step ii can also be monitored by UV-Vis absorption spectroscopy. Figure 4 illustrates the anchoring reaction for the 6-methoxyquinoline substituent. The chromophore precursor absorption at $\lambda_{max} = 330$ nm is bathochromically shifted upon quaternization in solution to yield a charge-transfer (CT) band at $\lambda_{max} = 350$ nm ("model" in Figure 4). The surface-bound chromophore exhibits the same CT band as the model chromophore [I] in solution. Similar bathochromic shifts in the absorption spectra were observed for the other model chromophores, $viz.$: [II] 317 → 321 nm; [III] 286 → 297 nm; [IV] 264 → 272 nm; [V] 256 → 267 nm. Assuming a similar extinction coefficient of the chromophores in solution and in a surface bound monolayer, a rough estimate of surface coverage can be made. For all of the investigated chromophores we find a full monolayer coverage with a molecular "footprint" in the range of 40 - 60 Å^2/molecule. These chromophore surface number densities are comparable to the ones that are obtained for stilbazolium derived monolayers.[4] Tailing of the absorption spectra into the red can be attributed to a small concentration of aggregates with parallel dipole stacking. Due to their low concentration such aggregates will influence the measured electrical properties only marginally $(vide\ infra)$.

Figure 5 illustrates the change in WF (ϕ) and in the EA of the modified Si as a function of the Hammett parameter[32] (σ) of the quinoline substituent at the 6th position only. The changes are relative to a surface, treated with the coupling agent. The experimental points represent an average of at least five measurements. Recently we showed a similar linear relationship with the dipole moment of the substituted phenyl part of the quinolinium adsorbate.[33] The linear fit between σ and the change in both the WF and EA, agrees with the linear correlation between the surface potential and the over-all dipole moment of the molecular layer,[17,34,35], due to the quinolinium group. The relation between Hammett parameter and dipole can be understood by realizing that the electron withdrawing properties reflected by Hammett parameters also determine the dipole moment. According to this, a substituent group that has a

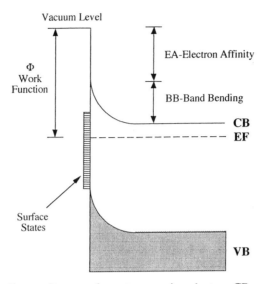

Figure 1: Energy diagram of an n-type semiconductor. CB and VB are the bottom of the conduction and top of the valence band, respectively, and EF is the Fermi level. The work function Φ is equal to EA + BB+ (CB-EF).

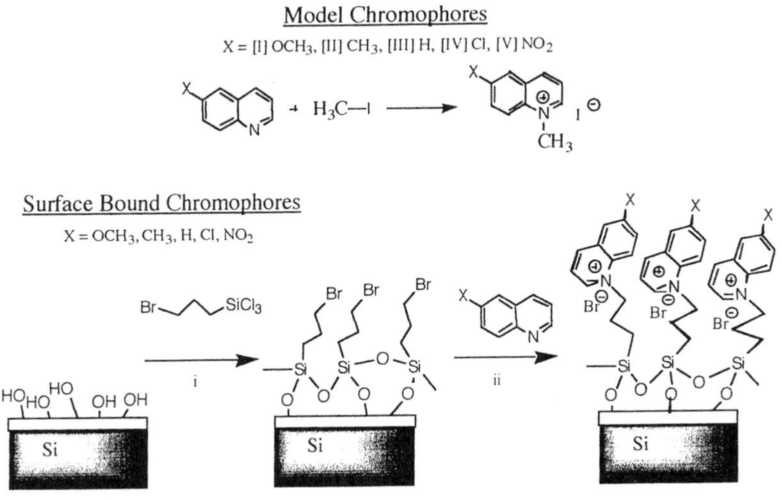

Model Chromophores

X = [I] OCH$_3$, [II] CH$_3$, [III] H, [IV] Cl, [V] NO$_2$

Surface Bound Chromophores

X = OCH$_3$, CH$_3$, H, Cl, NO$_2$

Figure 2: Schematic representation of self-assembled quino-linium-derived chromophore monolayers and of model chromophores, used in this study.

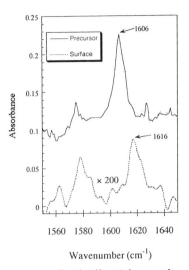

Wavenumber (cm^{-1})

Figure 3: FTIR spectrum of quinoline (chromophore precursor, KBr disk) and surface-FTIR spectrum of quinolinium-bromide self-assembled monolayer (surface).

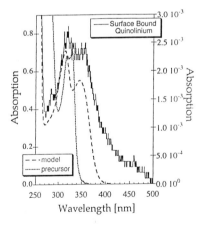

Figure 4: Absorption spectra of 6-methoxyquinoline (precursor, dotted line), of compound **I** (model, dashed line), both in 10 mM methanol solution and of the bound chromophore on quartz (surface bound quinolinium; solid line).

positive/negative Hammett value (electron accepting/donating properties, respectively) is expected to induce a net molecular dipole pointing away/towards the surface, and thus will increase/decrease the surface potential (i.e. the electron affinity), as is indeed observed in Figure 5. The differences in the slopes of the curve fits, for the change in f and EA is due to the change in BB which is different for the various substituted quinolinium groups. This change in BB is only reflected in φ as demonstrated schematically in Figure 1. The difference in φ between surfaces treated with quinoline units of methoxy and nitro groups demonstrates the high sensitivity of the electrical properties of Si to small modifications in the monolayer chemical structure and to surface modifications as a whole. Linear correlations, similar to the one shown above, between the change in φ or EA and the Hammet parameters or the dipole moment of the substituent groups of the adsorbed molecules were found for benzoic acids on CdTe[16] CdSe,[27] CuInSe$_2$,[16] and GaAs,[17], hydroxamic acids on CdTe and CuInSe$_2$,[16] and diphenyl-dicarboxylic acids on CdTe.[36] These data emphasize our ability to transfer a molecular property, the molecular dipole moment, to macroscopic electrical properties of the semiconductor.

The good linear fit of the data suggests that the different substituents on the quinolinium group have only minor effects on the self-assemblies' architecture in terms of tilt angle and surface coverage, and on the quinolinium-bromine bond in terms of its polar character. A molecular dipole, parallel to the surface will not affect the EA. We note, though, that in principle the observed changes might be attributed to simultaneous changes in orientation and magnitude of the dipole, because the Kelvin probe measurement is sensitive only to the dipole component that is perpendicular to the surface. This is also the component that scales with the Hammett parameter.

In addition to the change in EA, the chromophore assembly affected the surface BB (Figure 6). Bromopropyl coupling to Si (step i, Figure 1), via the chlorosilane condensation reaction, replaces the surface silanol groups with a siloxane-based network. Note the strong decrease in BB upon the reaction of the hydroxylated surface (Si-OH) with the coupling agent (Pr-Br), as shown in step i in Figure 2. This is evident also in the XPS spectra where a decrease in the ratio between Si° and oxidized Si was observed (i.e., an increase in the thickness of the SiO$_2$ layer). This step decreases the BB by about 300 mV. Such a decrease corresponds to a decrease in negative charge, localized on the surface of the Si wafer. This can come about because of elimination of states with energy levels between the Si Fermi level and its valence band edge at the surface, or because the surface states' levels are removed to outside this energy region, upon surface-molecule interaction. Thus, due to the elimination of negative surface charge (localized on silanols) by silylation, and formation of the neutral siloxane, the net surface charge and as a result the BB is reduced. We note that chemically we can look at the silylation as a Lewis acid/base reaction[12-13]

Our ability to further tune the BB of silicon by chromophore anchoring can be explained by taking into account the substituents' chemical nature and their subsequent effect on the electronegativity of the molecular layer. Figure 6 demonstrates this: unsubstituted quinolinium, as well as methyl- and methoxy- substituted chromophores have a negligible effect on BB, as compared to silicon with the coupling agent alone. However, adsorption of the chromophores, substituted with electron-withdrawing units (chloro and nitro) leads to an increase of 70 - 90 mV in BB, as compared to the silicon surfaces passivated by the coupling agent alone. To understand this, we note that these groups are stronger acceptors than the nitrogen of the quinolinium's quaternary amine. Therefore, their adsorption can add net surface charge to the molecular layers. Other ways to look at this are in terms of electronegativities (higher with assemblies of these acceptors than with those with electron-donating substituents) and in terms of new surface states, introduced due to molecule-surface state interaction.

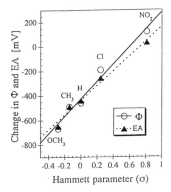

Figure 5: Change in the ϕ and in the EA of Si as a function of the Hammett parameter of the quinolinium substituents.

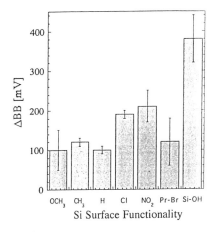

Figure 6: Band bending (BB) of clean silicon and after binding the coupling layer and the various quinolinium-derived chromophore monolayers.

If these states have energy levels within the band gap, lower than the fermi level, the states will be charged and increase the BB. Considering the main interacting orbital of the molecule as the lowest unoccupied molecular orbital (LUMO),[36] quinolines with electron-withdrawing substituents will have a lower lying LUMO state and thus are expected to have stronger interaction or, alternatively, stronger coupling[37] with the surface states than those with donor substituents.

SUMMARY

In conclusion, we have shown that self-assemblies of organo-silanes can be used to tune the electronic properties of silicon surfaces. The changes in EA correlate linearly with the Hammett parameters and the net molecular dipole moment of the quinoline substituents, thus illustrating the power of applying a systematic molecular modification approach. The BB was strongly changed after the formation of a two-dimensional siloxane-based monolayer and was further modified after the quinolinium chromophores were grafted subsequently. These results point to a simple yet very useful route in exploring new directions to fine-tune the electrical properties of rationally designed Si-based molecular electronic devices.

ACKNOWLEDGMENTS

S.Y. thanks the support for this research by the Israel Science Foundation, Jerusalem, Israel. D.C. acknowledges the US-Israel Binational Science Foundation, Jerusalem, Israel, and the Minerva Foundation, Munich, for partial support.

REFERENCES

(1)   Lehn, J. M. *Pure. Appl. Chem.* **1994**, *66*, 1961-1966.
(2)   Bissell, R. A.; Córdova, E.; Kaifer, A. E.; Stoddart, J. F. *Nature* **1994**, *369*, 133-137.
(3)   MacDonald, J. C.; Whitesides, G. M. *Chem. Rev.* **1994**, *94*, 2383-2420.
(4)   Yitzchaik, S.; Marks, T. J. *Acc. Chem. Res.* **1996**, *29*, 197-202 and references therein.
(5)   Petty, M. C.; Bryce, M. R.; Bloor, D. *An Introduction to Molecular Electronics*; Oxford University Press: NY, 1995.
(6)   Dharmadasa, I. M.; Roberts, G. G.; Petty, M. C. *Electronics Letters* **1980**, *16*, 201-202.
(7)   Haran, A.; Waldeck, D. H.; Naaman, R.; Moons, E.; Cahen, D. *Science* **1994**, *263*, 948-950.
(8)   Moons, E.; Bruening, M.; Libman, J.; Shanzer, A.; Beier, J.; Cahen, D. *Synthetic Metals* **1996**, *76*, 245-248.
(9)   Vuillaume, D. ; Boulas, C.; Collet, J.; Davidovits, J. V.; Rondelez, F. *Appl. Phys. Lett.* **1996**, *69*, 1646-1648.
(10) Sturzenegger, M.; Lewis, N. S. *J. Am. Chem. Soc.* **1996**, *118*, 3045-3046.
(11) Lunt, S. R.; Ryba, G. N.; Santangelo, P. G.; Lewis, N. S. *J. Appl. Phys.* **1991**, *70*, 7449-7467.
(12) Kepler, K. D.; Lisensky, G. C.; Patel, M.; Sigworth, L. A.; Ellis, A. B. *J. Phys. Chem.* **1995**, *99*, 16011-16017.
(13) Lisensky, G. C.; Penn, R. L.; Murphy, C. J.; Ellis, A. B. *Science* **1990**, *248*, 840-843.
(14) Thackeray, J. W.; Natan, M. J.; Ng, P.; Wrighton, M. S. *J. Am. Chem. Soc.* **1986**, *108*, 3570-3577.
(15) Natan, M. J.; Thackeray, J. W.; Wrighton, M. S. *J. Phys. Chem.* **1986**, *90*, 4089-4098.

66

(16) Bruening, M.; Moons, E.; Yaron-Marcovich, D.; Cahen, D.; Libman, J.; Shanzer, A. *J. Am. Chem. Soc.* **1994**, *116*, 2972-2977.

(17) Bastide, S.; Butruille, R.; Cahen, D.; Dutta, A.; Libman, J.; Shanzer, A.; Sun, L.; Vilan, A. *J. Phys. Chem. B* **1997**, *101*, 2678-2684.

(18) Neu, D. R.; Olson, J. A.; Ellis, A. B. *J. Phys. Chem.* **1993**, *97*, 5713-5716.

(19) Lunt, S. R.; Santangelo, P. G.; Lewis, N. S. *J. Vac. Sci. Technol. B* **1991**, *9*, 2333-2336.

(20) Uchihara, T.; Matsumura, M.; Ono, J.; Tsubomura, H. *J. Phys. Chem.* **1990**, *94*, 415-418.

(21) Mönch, W. *Semiconductor Surfaces and Interfaces;* Second ed.; Springer: Berlin, 1995, pp 274-284.

(22) Stutzmann, M.; Herrero, C. P. *Phys. Scripta* **1989**, *T25*, 276-282.

(23) Ulman, A. *Chem. Rev.* **1996**, *96*, 1533-1554.

(24) Maoz, R.; Sagiv, J. *J. Colloid Interface Sci.* **1984**, *100*, 465-496.

(25) Wasserman, S. R.; Tao, Y.; Whitesides, G. M. *Langmuir* **1989**, *5*, 1074-1087.

(26) Prutton, M. *Introduction to Surface Physics*; Oxford University Press: Oxford, 1994.

(27) Bruening, M.; Moons, E.; Cahen, D.; Shanzer, A. *J. Phys. Chem.* **1995**, *99*, 8368-8373.

(28) Lüth, H. *Surfaces and Interfaces of Solids;* Second ed.; Springer-Verlag: Berlin, 1993; Vol. 15.

(29) Lin, W.; Yitzchaik, S.; Lin, W.; Malik, A.; Durbin, M. K.; Richter, A. G.; Wong, G. K.; Dutta, P.; Marks, T. J. *Angew. Chem. Int. Ed. Engl.* **1995**, *34*, 1497-1498.

(30) Moulder, J. F.; Stickle, W. F.; Sobol, P. E.; Bomber, K. D. *Handbook of X-ray photoelectron spectroscopy* Ed. J.Chastain; Perkin-Elmer Corp.: Minesota, 1992.

(31) Li, D.; Swanson, B. I.; Robinson, J. M.; Hoffbauer, M. A. *J. Am. Chem. Soc.* **1993**, *115*, 6975-6980.

(32) March, J. *Advanced Organic Chemistry;* 3rd ed.; John Wiley & Sons: NY, 1985, pp 244.

(33) Cohen, R.; ZenouZenou, N.; Cahen, D.; Yitzchaik, S. *Submitted to Chem.Phys.Lett.*

(34) Evans, S. D.; Urankar, E.; Ulman, A.; Ferris, N. *J. Am. Chem. Soc.* **1991**, *113*, 4121-4131.

(35) Taylor, D. M.; Oliveira, O. N.; Morgan, H. *J. Coll. Int. Sci.* **1990**, *139*, 508-518.

(36) Cohen, R.; Bastide, S.; Cahen, D.; Libman, J.; Shanzer, A.; Rosenwaks,Y. *Advanced Materials* **1997**,*9*, 746-749.

(37) Kadyshevitch, A.; Naaman, R.; Cohen, R.; Cahen, D.; Libman, J.; Shanzer, A. *J. Phys. Chem.* **1997**,*101*, 4085.

Chapter 6

Composition of Binary Self-Assembled Monolayers of Alkyltrichlorosilanes

Jean Y. M. Yang and Curtis W. Frank

**Department of Chemical Engineering, Stanford University,
Stanford, CA 94305–5025**

Two binary self-assembled monolayer systems of alkyltrichlorosilane are examined for preferential deposition on a silicon oxide substrate. The surface compositions of these monolayers are monitored with contact angle, FTIR and X-ray photoelectron spectroscopy and compared to single component self-assembled monolayers. SAMs containing the small, relatively hydrophobic bromine end group show no detectable preferential deposition, but the more strongly hydrophilic methyl ester end group shows significant preferential deposition.

Self-assembled monolayers (SAMs) represent a class of well-defined surfaces that are especially suited to examine complex surface phenomena such as LC alignment *(1,2)*, wetting *(3-5)*, protein adsorption *(6-9)*, and other types of adsorption studies *(10-13)*   Generally, SAMs are formed through the thermodynamically-driven adsorption of a particular class of amphiphilic molecules on several types of solid substrates. Detailed discussions of different types of self-assembled monolayers and possible uses can be found in several reviews *(14,15)*.

The formation of covalently-bound silane monolayers on surfaces with free OH groups was postulated as early as 1968 *(16)*. However, Sagiv was the first to demonstrate more than a decade later that long chain alkyltrichlorosilanes will form ordered monolayers of fairly uniform orientation and packing on hydroxylated surfaces *(17)*. The order in the monolayer stems partially from the van der Waals interactions between adjacent chains, which contribute a few kcal/mol of energy to the monolayer formation process *(15)*. Although these alkyltrichlorosilane SAMs are not as ordered as the well-known SAMs made from alkanethiol or disulfide on gold, they are nonetheless organized and excellent model surfaces for the study of complex surface interactions.

We have chosen to study binary systems of alkyltrichlorosilane SAMs because of their unique set of physical and chemical properties. First, the densely-packed structure of these SAMs allows good control of surface topology. Secondly, alkyltrichlorosilane SAMs are believed to be covalently bound to the substrate, resulting in films that are more stable thermally and chemically than those obtained by Langmuir-Blodgett *(18)* deposition or alkanethiol SAMs on gold *(19-21)*. For example, contact angle, SFM, and FTIR studies have shown octadecyltrichlorosilane SAMs to be stable at least up to 125°C, whereas alkane thiol SAMs begin to desorb from gold at 70°C *(19-21)*. Lastly, the alkyltrichlorosilanes allow for the convenience of using transparent substrates, which are necessary for optical studies of liquid crystal anchoring, our ultimate objective. In addition to the properties mentioned above, the formation of binary self-assembled monolayers allows tailored access to a whole range of surface properties. In spite of these advantages, the majority of SAM studies so far have been on alkanethiol/gold systems with significantly fewer studies on binary alkyltrichlorosilane SAMs

Formation of hydroxyl/methyl-terminated binary alkanethiol SAMs on gold has been studied by Bain and Whitesides *(22)*. They found that, although the two components do not phase separate on the surface, the hydroxyl-terminated molecule seems to bond at a higher percentage in the monolayer than in the solution when coadsorbed from alkane solution. This behavior is not observed when an ethanol solution is used. Subsequently, the hydroxyl/methyl binary system has been subject to many studies involving wetting *(4,23,24)*, composition *(25,26)* and phase separation of the two species *(27-29)*. XPS data suggest that the two components deposit at similar composition as the solution when ethanol is used as the solvent *(26)*. Theoretical and experimental (XPS, FTIR, ellipsometry and wetting) approaches have been used to examine whether phase separation of the two species takes place *(26-29)*. Although individual results point both to mixing *(26,27,29)* and phase separation *(27)*, the results usually indicate a single phase of the molecular mixture, at least when the two chain lengths are similar and the system is in equilibrium *(26,28,29)*. On the other hand, contrary experimental results have been shown for methyl ester/methyl- terminated alkanethiol SAMs where scanning tunneling microscopy reveals 20-25Å diameter size domains *(30)*. The actual composition of the domains is unknown, however.

Long and short alkanethiols and disulfides have also been co-deposited, and longer chains have been shown to preferentially adsorb *(2,31-34)*. Experimental results suggest good mixing, but like the other binary systems, the possibility of phase separation cannot be completely ruled out *(34)*. Additional binary systems of alkanethiol and disulfides that have been studied also include mixed alkanethiol SAMs of cyano/methyl *(22)*, bromo/methyl *(22)*, caboxylic acid/methyl *(5)*, azobenzene/methyl *(35)*, anthraquinone/methyl *(36)* and mixed disulfide SAMs of fluoroalkyl with amide, ester or methyl *(37,38)* and hydroxyl/methyl *(35)*. Much has been learned about alkanethiol and disulfide on gold binary SAMs, especially in the past few years, but there remain areas such as phase separation and mixing where definitive answers are lacking.

Some of the observations from binary alkanethiol SAMs may be applicable to alkyltrichlorosilane SAMs, which has added cross-linking complexities due to the tri-functional head group. We note here that the formation mechanism of alkyltrichlorosilane SAMs is not simple, possibly involving an adsorption step, followed by some reorganization, then finally reaction *(39,40)*. Since it is unclear exactly how one component may be preferentially incorporated into the SAM over the other component, we will refer to any deviation of the monolayer composition from the solution composition as preferential deposition rather than adsorption to acknowledge that the overall deposition phenomenon is far more complex than a simple adsorption.

Binary alkyltrichlorosilane SAMs of long and short chains display no preferential deposition *(41)* unlike alkanethiol SAMs which show a preference for long chains *(2,31-34)*. The two types of binary SAMs are alike in that no detectable phase separation was found in either system *(2,31-34,41)*. Vinyl/methyl binary trichlorosilane SAMs have also been studied, with both components shown to deposit in the same composition as the solution and no phase separation detected *(3,42)*. Post-deposition treatment of binary vinyl/methyl SAMs was conducted by Wasserman et al. *(43)*. who used $KMnO_4/NaIO_4$ to convert surface vinyl groups to carboxylic acid.

However, the conversion was incomplete. No detailed composition data were given for the binary, post-treated SAMs, but the contact angle trends show that the composition of the vinyl/methyl surface is near or the same as that of the solution. Rieke et al. post-treated similar vinyl/methyl binary SAMs with SO_3 gas and achieved high conversion (85-95%) in conditions which cause little degradation of pure methyl SAMs (water contact angle decreased from 109° to 103°) *(13)*.

A bulkier non-polar end group has been incorporated in binary SAMs by Mathauer and Frank using a naphthyl-terminated undecyltrichlorosilane *(44)* which co-deposited with octadecyltrichlorosilane (OTS) with no preferential deposition. The fluorescence behavior of the naphthalene chromophore in the naphthyl/methyl monolayers was examined using an energy migration model, and the fitted results showed no detectable phase separation down to the nanometer range. Additional UV spectroscopy of the naphthyl-terminated silane mixed with long or short methyl-terminated chains also supports good mixing of the naphthyl/methyl components at the molecular level *(1)*.

Based on these previous results, we are interested in exploring binary trichlorosilane SAM formation with additional functional groups to better understand these systems and because they are useful in model studies of surface interactions. We have, therefore, synthesized two bromo-terminated silanes and a methyl ester-terminated silane to examine the question of whether preferential deposition will occur for polar terminal groups.

Experimental

Bromo/Methyl-Terminated Binary SAMs. 1-bromo-16-(hexadecyl) trichlorosilane was synthesized by the method reported by Balanchander and Sukenik *(45)* with some modifications. The Grignard reagent of ω-undecenyl bromide (Pfaltz and Bauer) was coupled to 1,5 dibromopentane (Aldrich) in anhydrous THF with a LiCl /CuCl$_2$ catalyst (0.1M in THF) at -10°C for 12 hours. The reaction was stopped with a saturated aqueous solution of NH$_4$Cl, then washed with NaCl (sat. aq.) and dried with Na$_2$SO$_4$. The product was purified by vacuum distillation (135-148°C, 1.1mm Hg) after rotovapping. Finally, ω -hexadecyl bromide was purified by silica gel (Aldrich, 230-400 mesh) flash chromatography with hexane(Baker) (R$_f$=0.43).

Hydrosilation was conducted by first dissolving ω-hexadecenyl bromide in trichlorosilane with H$_2$PtCl$_6$·6H$_2$O catalyst in a pressure tube under nitrogen atmosphere, and then heating to 90°C for 15 hours. The product 1-bromo-16-(hexadecyl)trichlorosilane was separated by trichlorosilane distillation (N$_2$, 37°C) and vacuum distillation (195-205°C, 2.2mmHg). 1-bromo-11-(undecyl)trichlorosilane was synthesized by direct hydrosilation of ω -undecenyl bromide and then purified by vacuum distillation (120-150°C, 0.8mm Hg). Both products and the ω -hexadecenyl bromide intermediate were verified by ^{1}H NMR to be at least 95% pure, and complete reaction of the double bonds was verified in the products.

Binary SAMs were formed from mixed silane solution totaling 0.1 volume% in 5:1 anhydrous isooctane:carbon tetrachloride. The solutions were made in oven-dried glassware in a nitrogen-purged glovebag. Undoped silicon [100] wafers (Virginia Semiconductor), glass microscope slides (VWR) and fused silica microscope slides (ESCO) were used as substrates. These substrates were first cleaned by piranha solution (3:1 concentrated sulfuric acid/ 30% hydrogen peroxide) for 30 minutes, then rinsed with DI water followed by high purity (18MΩ) water, and dried under a stream of nitrogen. The cleaned substrates were then immediately taken into the nitrogen glovebag and placed in the freshly-made silane solution. The deposition takes place overnight (15-20 hours) at room temperature (20-25°C) in the nitrogen glovebag.

Finally, the deposition was terminated by removing the substrates from the solution and dipping in dichloromethane before they were exposed to the atmosphere. The substrates were cleaned by lightly brushing with a camel hair brush while rinsing with copious amounts of isopropanol to remove any solution-deposited materials *(44)*. This was followed by chloroform and ethanol rinses and then drying with a stream of nitrogen.

Two systems of binary bromo/methyl SAMs were chosen to examine the question of preferential deposition and monolayer homogeneity using different experimental techniques. The first system consists of binary SAMs produced from the coadsorption of 1-bromo-11-(undecyl)trichlorosilane (C11Br) and a longer *n*-eicosyltrichlorosilane (C20). The *n*-eicosyltrichlorosilane and the *n*-octadecyltrichlorosilane were used as received from Hüls with care taken to not expose contents to moisture. By co-adsorbing two components that are different in length, preferential deposition can be easily detected with ellipsometry measurements and verified with XPS. This is a method that has been successfully employed to study hydroxyl/methyl alkanethiol SAMs *(27)*.

The second binary SAM system was made from 1-bromo-16-(hexadecyl)-trichlorosilane (C16Br) and *n*-octadecyltrichlorosilane (C18) coadsorbed from solution. The fully-extended molecular lengths of the two species are within 2 Å of each other, with C18 being slightly longer. Therefore, the amount of bromine deposited can be quantified with XPS without normalizing for different layer thicknesses.

Methyl Ester/Methyl-Terminated Binary SAMs. Methyl 17-trichlorosilyl hepta-decanoate (C16ES) was synthesized by a four-step reaction from undecenoyl chloride to methyl 16-heptadecanoate, as described by Chen *(46)*, and then hydrosilated according to the method described in the previous section. The hydrosilation was followed by purification via distillation in N_2 to remove excess trichlorosilane reactant (37°C). Again, the product was verified by ^{1}H NMR to be at least 95% pure and to exhibit complete reaction of the double bonds. Methyl 17-trichlorosilyl heptadecanoate was codeposited with *n*-octadecyl-trichlorosilane (C18) following the same procedure as with bromo/methyl binary SAMs.

SAM Characterization. The binary SAMs were examined by contact angle, ellipsometry, FTIR and XPS. Advancing contact angles were measured by the captive-drop method with a Ramé-Hart Model 100 contact angle goniometer. The contact angles were determined for a 0.05ml drop of fresh ultra-pure (18 MW) water or spectroscopic grade hexadecane at room temperature. The angle was determined by keeping the needle in the drop and adding liquid until the boundary of the drop moved. The highest angle made by the drop before movement is reported as the advancing contact angle. This method yields the maximum advancing contact angle and is known to result in 3-6° higher contact angles than the sessile drop method in which the needle is removed *(43,47)*.

FTIR measurements were made with a BIORAD Digilab FTS-60A single-beam spectrometer equipped with a He-Ne laser. The spectra were recorded in transmission mode at a resolution of 4 cm^{-1} for 256 scans after at least 5 minutes of purging. A reference spectrum was taken for each experiment for background subtraction. FTIR measurements were performed only on fused silica samples due to the absorption interference found with the silicon wafer samples.

Ellipsometry was performed using a Gaertner L116B ellipsometer with a fixed angle of incidence of 70° on silicon wafers. The index of refraction of the film and the silicon oxide layer was fixed at 1.46. Since oxide thickness varies significantly from batch to batch, and sometimes even from sample to sample, it was measured for each individual substrate prior to monolayer formation. Typical measurements consist of five samplings from different parts of the substrate, and an average is reported.

X-ray photoelectron spectroscopy was performed on a Surface Science Instruments S-probe Surface Spectrometer having a monochromated Al K_a source, which is located in the Stanford Center for Materials Research. Measurements were made with spot size of 250 x 1000 mm and a 35° take-off angle on silicon wafer supported samples.

Results and Discussion

Bromo/Methyl-Terminated Binary SAMs. Advancing and receding water contact angles or their cosines for C11Br/C20 binary SAMs are plotted in Figure 1 as a function of the relative C11Br concentration in the deposition solution, where the remaining component is C20. We note that in the range we are examining (60°-120°) the cosine of the contact angle varies almost linearly with respect to changes in the angle. Therefore, we will use the two variables interchangeably even though the cosine is the thermodynamically relevant variable. According to Young's equation, the cosine of the contact angle is a function of interfacial energy between a liquid and a solid (48):

$$\gamma_L \cos\theta = \gamma_s - \gamma_{SL}$$

(1)

The error bars on all contact angle measurements are ±2°. We note that our contact angles are somewhat higher than those reported for C16Br by Balachander and Sukenik (82°) (45), Lander et al. for C11Br (84°) (49), and Fryxell et al. for C17Br (50). This difference is within the 3-6° range attributed to the different measurement method, as previously discussed.

The surface composition may be quantified by using the phenomenological Cassie's equation (51) in an approach similar to Silberzan's discussion for vinyl/methyl surfaces (3,42). Cassie's equation considers a two-component surface to consist of domains of each of the two components. Therefore, the contact angle can be approximated by a linear combination of the two types of surface islands:

$$\cos\theta = \beta\cos\theta_1 + (1-\beta)\cos\theta_2 \qquad (2)$$

where β is the concentration of component 1 on the surface and θ_1 and θ_2 are the contact angles of surfaces that contain only component 1 or 2, respectively. This equation is derived from Young's equation (eqn. 1) combined with the assumption that the total work of adhesion is a linear combination of contributions from the two components

$$W_a = \gamma_L(1+\cos\theta) = \beta W_1 + (1-\beta)W_2 \qquad (3)$$

because the surface is made of domains of each of the two components.

An alternate approach to looking at contact angle measurement of chemically heterogeneous surfaces has been reported by Israelachvili (52), who suggests instead that an arithmetic average of the polarizabilities or dipole moments is more appropriate when the domains are of molecular dimensions. These molecular-level interactions are expressed by a geometric, composition-weighted mean of the contribution from the two components to the work of adhesion:

$$W_a = \gamma_L(1+\cos\theta) = [\beta W_1 + (1-\beta)W_2]^{1/2} \qquad (4)$$

$$W_1 = (w_1 w_L), \ W_2 = (w_2 w_L) \qquad (5)$$

where w_1, w_2 and w_L are the polarizabilities, dipole moments, etc. of the single-component surface 1, 2 and the liquid. Combining these two equations leads to

$$(1 + \cos\theta)^2 = \beta(1 + \cos\theta_1)^2 + (1- \beta)(1 + \cos\theta_2)^2 \tag{6}$$

Comparing both Cassie's and Israelachvili's equations to the advancing contact angle results for the C11Br/C20 binary SAMs, we see from Figure 1 that both fit equally well within the experimental error typical of the contact angle measurement. Cassie's equation does show a slight negative deviation from linear, but, given the error range for contact angles, we interpret the data as indicating no detectible preferential deposition. This interpretation of linear contact angle results as support for no detectible preferential deposition is based on our previous work (1,44) as well as other literature on binary SAMs (3,42,53). A similar conclusion was drawn by Fryxell et al. (50) However, there are specific differences in their observations of contact angle and XPS results compared to ours. These may be related to differences in measurement techniques.

C16Br/C18 binary SAMs were deposited on silicon wafers and glass microscope slides. The contact angle results for C16Br/C18 on both substrates, shown in Figure 2, appeared similar to those for the C11Br/C20, appeared to vary. Again, the linear Cassie's equation and Israelachvili's equation both fit the data equally well. However, there is a positive deviation of the contact angle versus solution concentration plot at the high C16Br concentration end for the advancing contact angle. Similar curvature can be seen in the low bromine concentration end for the receding contact angle. This is different from the negative deviation reported by Fryxell et al. (50) and is more likely indicative of the sensitivity of advancing contact angles to the hydrophobic moieties at the surface. The same curvature is seen in the low C16Br concentration range for receding contact angle of C16Br/C18, showing its sensitivity to the hydrophilic functional groups. These two types of sensitivity have been noted and discussed by other researchers (54,55). Thus, we do not believe the slight curvatures in contact angle are due to preferential deposition; rather, we assign them to these secondary effects.

A key difference between the contact angles for C16Br/C18 binary SAMs on silicon wafers and on glass is the contact angle hysteresis, defined as the difference between the advancing and receding contact angles for the two substrates (Figure 3). Generally, contact angle hysteresis is sensitive to monolayer homogeneity and roughness. It is well established that surface roughness will change effective contact angles due to the change in contact area (56). Bain and Whitesides have reported dramatically higher contact angle hysteresis for binary thiol SAMs of long and short chains in comparison to single component SAMs (31). If the two substrates are compared, it can be seen that highly polished silicon wafers (RMS roughness = 1.5Å) should produce monolayers with much less hysteresis than the glass microscope slides (RMS roughness = 25Å). These type of results have been reported by Lander et al., who found that the contact angle hysteresis for water on methyl-terminated SAMs on glass and silicon wafers to be 19°±3 and 9°±3 respectively (55). The contact angle hysteresis shown here has similar trends but is significantly larger. This may be due to the additional chemical heterogeneity created by the presence of two distinctively different chemical components at the surface.

Ellipsometry measurements of thickness show a linear relationship between thickness and solution composition for the C11Br/C20 binary SAMs, reaffirming the lack of detectable preferential deposition for the C16Br/C18 (Figure 4). The thickness stays constant for C16Br/C18 binary SAMs since both components are similar in length. In addition, the thicknesses for pure C18 and pure C20 monolayers are 24±2Å and 30±2Å respectively, which is in agreement with the 26Å and 29Å thicknesses reported for these two species (43). The error in the ellipsometry data, derived from the standard deviation taken from five or more measurements on each sample, is about ±2Å.

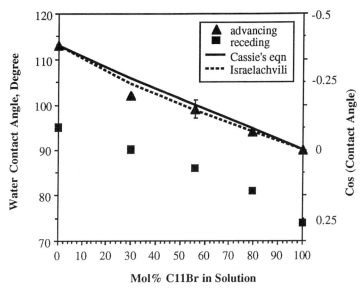

Figure 1. Advancing and receding water contact angles as a function of solution composition for C11Br/C20 binary SAMs on silicon wafers

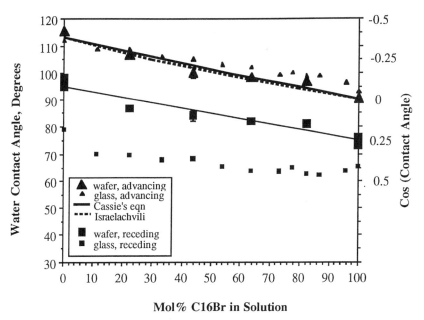

Figure 2. Contact angle results for the C16Br/C18 binary SAMs

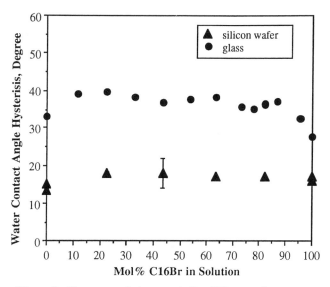

Figure 3. Contact angle hysteresis for different substrates

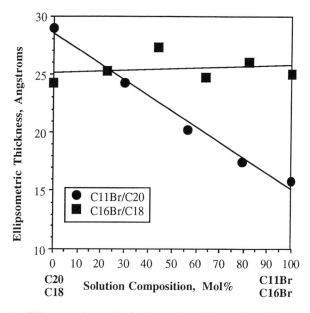

Figure 4. Ellipsometric results for bromo/methyl-terminated binary SAMs

Since the C-Br stretch band in the 600 cm^{-1} range of IR is too weak to measure in the monolayer, XPS was used instead to verify the presence of bromine in these binary SAMs. Binary SAMs formed from as little as 25 mol% C11Br in solution exhibit distinct bromine 3p, 3d, and sometime 3s signals. In Figure 5, the bromine signals from the 3p and 3d levels are averaged and ratioed to the Si signal, averaged for 2p and 2s components. The Si signal is corrected for the difference in overlayer thickness in these samples using a SAM mean free path of 29Å calculated from the data *(57)*. From the calculated film mean free path l_{film} , the SAM thickness dependence of the Si signal is factored out using:

$$I_{measured}/I_{corrected} = \exp{(-d/\sin\alpha)} \qquad (7)$$

where α is the take-off angle (35°) and d is the film thickness.

Figure 5 shows that once the layer thickness difference is taken into account for the Si signal, the Br/Si signal ratio becomes linear with respect to the solution concentration of C11Br. The excellent linear fit of the thickness-corrected intensity ratio confirms contact angle and ellipsometry results which indicate that no significant preferential deposition occurs in the co-deposition of C11Br/C20. This thickness correction described above does not account for the attenuation of the Br signal by the longer C20 chain in binary SAMs. The Br signal is expected to be attenuated by the longer C20 chain in a binary mixture, and this overlayer can be as much as 10Å thick. However, the attenuation of the Br signal by this overlayer is calculated and shown in Figure 5 by the dotted line to be rather small.

Analysis of the XPS results for the C16Br/C18 system does not require thickness correction since the two components are of similar lengths. Therefore, a direct examination of the Br/Si signal ratio reveals the monolayer Br concentration, as shown in Figure 6. Additional corrections for the differences in mean free path for alkyl bromide and alkyl monolayer are not necessary since the mean free paths are similar. The Br/Si signal ratio for the C16Br/C18 system appears to show some negative deviation from linearity in regions of low bromine concentration. We believe this behavior to be due to a documented reductive C-Br cleavage by the secondary electrons, which has been observed to decrease the signal by at least 20% *(58)* over the measurement period. It is reasonable that this reduction would be more apparent in the low concentration region, which pushes the detection limit for the Br. We would not expect this effect to be seen in the low Br concentration range of C11Br/C20 because of the longer alkyl chain coverage, but we can see that at higher Br concentration where the alkyl coverage would be less significant, the Br/Si signal ratio does fall slightly below linear. This observation supports our explanation of the negative curvature in the C16Br/C18 XPS levels. In summary, contact angle, ellipsometry, and XPS data for C11Br/C20 and C16Br/C18 binary SAMs reveal no significant preferential deposition in the co-deposition due to the presence of the mildly polar bromine tail group. This finding is in agreement with the results reported by Fryxell et al. for C17Br/C16 binary trichlorosilane SAMs *(50)* and Bain and Whitesides bromo/methyl-terminated alkanethiol binary SAMs coadsorbed from ethanolic solutions *(53)*.

Methyl Ester/Methyl-Terminated Binary SAMs. The high polarity of the methyl ester group results in coadsorbed SAMs which show substantial preferential deposition, unlike the bromine group (Figure 7). We see that for binary SAMs of octadecyl-trichlorosilane(C18) codeposited with methyl 17-trichlorosilyl heptadecanoate (C16ES), the water contact angles suggest that silanes with the methyl ester end group will deposit in higher concentrations than the solution concentration. This is in agreement with previous studies of hydroxyl/methyl-terminated alkanethiol

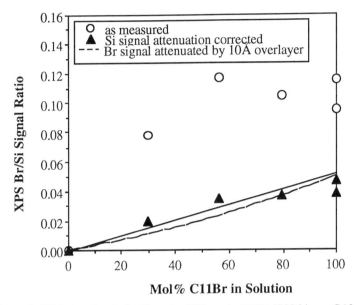

Figure 5. XPS signal ratio for Bromine/Silicon in C11Br/C20 binary SAMs

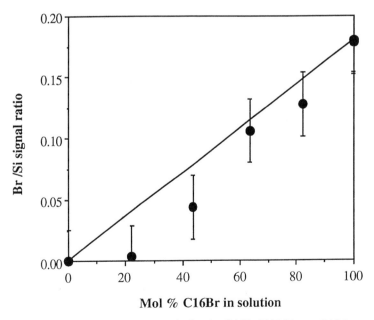

Figure 6. Br/Si XPS signal ratio for the C16Br/C18 Binary SAMs

binary SAMs, where the hydroxyl-terminated material showed strong preferential adsorption *(22,31,53)*. This may be attributed to the additional stabilization from polar bonding between adjacent methyl ester groups, as has been suggested for the hydroxyl group. This polar bonding can be a factor in the domains in methyl ester/methyl alkanethiol SAMS discovered by scanning tunneling microscopy *(30)*. The water contact angle of the pure methyl ester monolayer is slightly higher than the 73° previously reported, most likely due to the different measurement methods.

Ellipsometry measurements show monolayer thickness to be also skewed toward the C16ES monolayer thickness. The average ellipsometric thickness for pure C18 was 23.8Å, for C16 methyl ester, 19.4Å and for a 50/50 mixture in solution, 19.6Å. Thus, the ellipsometric data appear to support the contact angle data and indicate that the C16ES adsorbs preferentially over C18. The pure C16ES monolayer thickness of 19.4Å is much smaller than the 23Å reported for pure alkyl trichlorosilane SAMs with 16 carbons *(43)*. The smaller thickness suggests that there may be significant tilt in the monolayer, perhaps to accommodate greater polar interactions between adjacent methyl ester groups.

FTIR data for the CH_2 asymmetric stretch band position, which may be related to the state of order, and absorbance, which is a measure of methyl content, are consistent with C16ES depositing in larger concentrations than in solution, as shown in Figure 8. The CH_2 absorbance for the pure C16ES monolayer is 0.024, which is slightly higher than the 0.022 measured for pure C15 monolayer and much less than the 0.034 measured for C18 SAM. This is again consistent with the ellipsometry data suggesting that there is more tilt than for pure alkyltrichlorosilane SAMs.

Finally, XPS measurements for the pure C16 methyl ester monolayer on a silicon wafer show a clear shoulder for the C1S peak at binding energies 2 eV higher than for alkyl carbon. We assign this shoulder to the carbonyl carbon. The ratio of the carbonyl carbon signal to total carbon signal is 0.048, which is close to the compositional ratio of 0.059. The carbon to silicon signal ratio is 1.56, which is less than the 1.92 measured for C18, but more than the 1.35 measured for C11.

Thus, the methyl ester/methyl binary SAM system shows significant preferential deposition during coadsorption. The preference toward the methyl ester terminated silanes is consistent with previous work with hydroxyl/methyl binary SAMs, and is explained by greater interaction among the adjacent methyl ester groups. Contact angle, ellipsometry, and FTIR measurements all show this preference. It may be possible to reduce or remove this preference by using a more polar solvent. For example, trifluoroethyl ester terminated silane has been codeposited successfully without preference with methylene-terminated silanes by switching the solvent from cyclohexane to dichloromethane *(50)*. With a less polar end group than methyl ester, pure trifluoroethyl ester terminated silane SAMs have a water contact angle of 83°.

Conclusion

We examined binary SAM formation for two different pairs of terminal groups and found that the methyl ester group, whose monolayers have water contact angles of 73°, showed substantial preferential deposition when co-adsorbed with simple alkyls. On the other hand, three groups whose single-component monolayers have water contact angles of about 90°- bromine from this study, naphthyl *(44)* and vinyl *(3,42)* from other studies- all showed no preferential deposition with simple alkyls. Bromo/methyl-terminated binary SAMs will be easier to use as model surfaces because the surface composition is predetermined by the solution, but the methyl ester/methyl-terminated SAMs offer a larger contact angle range allowing studies with more expanded interfacial energies.

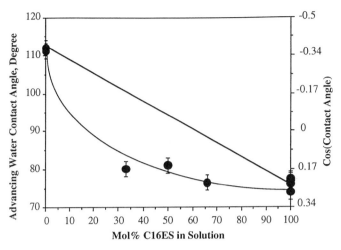

Figure 7. Advancing water contact angles of the C16ES/C18 binary SAMs

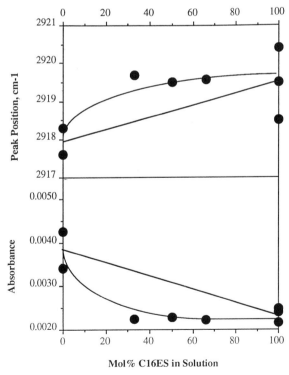

Figure 8. FTIR measurements for the CH_2 asymmetric stretch in C16ES/C18 SAMs

Literature Cited

1) J. Y. Yang, K. Mathauer, and C. Frank, W., in *JRDC-KUL Joint International Symposium on Spectroscopy and Chemistry in Small Domains*, edited by H. Masuhara and e. al. (Elsevier Science Publishers, B.V., Brussels, Belgium, 1993), pp. 441-454.
2) R. A. Drawhorn and N. L. Abbott, *J. Phys. Chem.* **99**, 16511-15 (1995).
3) P. Silberzan and L. Leger, *Physical Review Letters* **66**, 185-188 (1991).
4) A. Ulman, S. D. Evens, Y. Shnidman, R. Sharma, J. E. Eilers, and J. C. Chang, *Journal of American Chemical Society* **113**, 1499-1506 (1991).
5) S. E. Creager and J. Clarke, *Langmuir* **10**, 3675-3683 (1994).
6) K. L. Prime and G. M. Whitesides, *Journal of American Chemical Society* **115**, 10714-10721 (1993).
7) K. L. Prime and G. M. Whitesides, *Materials Research Society Symposium Proceedings* 237, 311-315 (1992).
8) Y. W. Lee, J. Reed-Mundell, C. N. Sukenik, and J. E. Zull, *Langmuir* **9**, 3009-3014 (1993).
9) M. Mrksich, J. R. Grunwell, and G. M. Whitesides, *Journal of American Chemical Society* **117**, 12009-12010 (1995).
10) L. Bertilsson, K. Potje-Kamloth, and H.-D. Liess, *Thin Solid Films* **284-285**, 882-887 (1996).
11) I. Engquist, I. Lundström, and B. Liedberg, *Journal of Physical Chemistry* **99**, 12257-12267 (1995).
12) I. Engquist, I. Lundström, and B. Liedberg, *Journal of Physical Chemistry* **99**, 14198-14200 (1995).
13) P. Rieke, R. Weicek, B. D. Marsh, L. L. Wood, J. Liu, L. Song, G. E. Fryxell, and B. J. Tarasevich, *Langmuir* **12**, 4266-4271 (1996).
14) J. D. Swalen, D. L. Allara, J. D. Andrade, E. A. Chandross, S. Garoff, J. Isrealachvili, T. J. McCarthey, R. Murray, R. F. Pease, J. F. Rabold, K. J. Wynne, and H. Yu, *Langmuir* **3**, 932-950 (1987).
15) A. Ulman, *An Introduction to Ultrathin Organic Films: from Langmuir-Blodgett to Self-Assembly* (Academic Press, Inc., San Diego, 1991).
16) L.-H. Lee, *Journal of Colloid Interface Science* **27**, 751-760 (1968).
17) J. Sagiv, *Journal of the American Chemical Society* **102**, 92-98 (1980).
18) R. Maoz and J. Sagiv, *Langmuir* **3**, 1034-1044 (1987).
19) C. D. Bain, E. B. Troughton, Y. T. Tao, J. Evall, G. M. Whitesides, and R. G. Nuzzo, Journal of American Chemical Society **111**, 321-335 (1989).
20) S. R. Cohen, R. Naaman, and J. Sagiv, *Journal of Physical Chemistry* **90**, 3054-3056 (1986).
21) M. Calistri-Yeh, E. J. Kramer, R. Sharma, W. Zhao, M. H. Ragailovich, J. Sokolov, and J. D. Brock, *Langmuir* **12**, 2747-2755 (1996).
22) C. D. Bain, J. Evall, and G. M. Whitesides, *Journal of American Chemical Society* **111**, 7155-7164 (1989).
23) A. Ulman, *Thin Solid Films* **273**, 48-53 (1996).
24) D. J. Orbis, A. Ulman, and Y. Shnidman, *Journal of Chemical Physics* **102**, 6865-6873 (1995).
25) A. Ulman, S. D. Evans, Y. Shnidman, R. Sharma, and J. E. Eilers, *Advances in Colloid and Interface Science* **8**, 175-224 (1992).
26) L. Bertilsson and B. Leidberg, *Langmuir* **9**, 141-149 (1993).
27) J. P. Folkers, P. E. Laibinis, and G. M. Whitesides, *Langmuir* **8**, 1330-1341 (1992).
28) J. P. Folkers, P. E. Laibinis, G. M. Whitesides, and J. Deutch, *Journal of Physical Chemistry* **98**, 563-571 (1994).
29) S. V. Atre, B. Liedberg, and D. L. Allara, *Langmuir* **11**, 3882-3893 (1995).

30) S. J. Stranick, A. N. Parikh, Y.-T. Tao, D. L. Allara, and P. S. Weiss, *Journal of Physical Chemistry* **98**, 7637-7646 (1994).
31) C. D. Bain and G. M. Whitesides, *Journal of American Chemical Society* **111**, 7164-7175 (1989).
32) D. A. Offord, C. M. John, M. R. Linford, and J. H. Griffin, *Langmuir* **10**, 883-889 (1994).
33) D. A. Offord, C. M. John, and J. H. Griffin, *Langmuir* **10**, 761-766 (1994).
34) P. E. Laibinis, R. G. Nuzzo, and G. M. Whitesides, *Journal of Physical Chemistry* **96**, 5097-5105 (1992).
35) T. Takami, E. Delamarche, B. Michel, and C. Gerber, *Langmuir* **11**, 3876-3904 (1995).
36) L. Zhang, T. Lu, G. W. Gokel, and A. E. Kaifer, *Langmuir* **9**, 786-791 (1993).
37) H. Schönherr, H. Ringsdorf, M. Jaschke, H.-J. Butt, E. Bamberg, H. Allinson, and S. D. Evans, *Langmuir* **12**, 3898-3904 (1996).
38) H. Schönherr and H. Ringsdorf, *Langmuir* **12**, 3891-3897 (1996).
39) J. B. Brzoska, N. Shahidzadeh, and R. Rondelez, *Nature* **360**, 719-721 (1992).
40) J. B. Brzoska, I. B. Azouz, and R. Rondelez, *Langmuir* **10**, 4367-4373 (1994).
41) D. A. Offord and J. H. Griffin, *Langmuir* **9**, 3015-3025 (1993).
42) P. Silberzan, L. Leger, D. Ausserre, and J. J. Benattar, *Langmuir* **7**, 1647-1651 (1991).
43) S. R. Wasserman, Y.-T. Tao, and G. M. Whitesides, *Langmuir* **5**, 1074-1087 (1989).
44) K. A. Mathauer and C. W. Frank, *Langmuir* **9**, 3002-3008 (1993).
45) N. Balanchander and C. N. Sukenik, *Langmuir* **6**, 1621-1627 (1990).
46) S. H. Chen, Doctoral, Stanford University, 1991.
47) L. M. Lander, L. M. Siewierski, M. D. Foster, and E. A. Vogler, *Polymer Preprints* **34**, 606-7 (1993).
48) W. A. Zisman, in *Contact angle wettability and adhesion*, Vol. 43 (American Chemical Society, Washington, D. C., 1964), pp. 1-51.
49) L. M. Lander, William J. Brittain, and V. Tsukruk, *Polymer Preprints* **35**, 488-9 (1994).
50) G. E. Fryxell, P. C. Rieke, L. L. wood, M. H. Engelhard, R. E. Williford, G. L. Fraff, A. A. Capbell, R. J. Wiacek, L. Lee, and A. Haverson, *Langmuir* **12**, 5064-5075 (1996).
51) A. B. D. Cassie, *Discussions of the Faraday Society* **3**, 11 (1948).
52) J. N. Israelachvili and M. L. Gee, *Langmuir* **8**, 288-289 (1989).
53) C. D. Bain and G. M. Whitesides, *Journal of American Chemical Society* **110**, 6560-6561 (1988).
54) R. E. Johnson and R. H. Dettre, in *Contact Angle, Wettability and Adhesion*, Vol. 43, edited by F. Fowkes (American Chemical Society, Washinton D. C., 1964), pp. 113-135.
55) L. M. Lander, William J. Brittain, M. D. Foster, and E. A. Vogler, *Polymer Preprints* **33**, 1154-5 (1992).
56) A. W. Adamson, *Physical Chemistry of Surfaces*, 4th ed. (John Wiley & Sons, New York, 1982).
57) V. I. Nefedov, *X-ray Photoelectron Spectroscopy of Solid Surfaces* (VSP BV, Utrecht, The Netherlands, 1988).
58) P. C. Rieke, D. R. Baer, G. E. Fryxell, M. H. Engelhard, and M. S. Porter, *Journal of Vacuum Science and Technology* **11**, 2292-2297 (1993).

Chapter 7

Orientations of Liquid Crystals on Self-Assembled Monolayers Formed from Alkanethiols on Gold

Nicholas L. Abbott, Vinay K. Gupta, William J. Miller, and
Rahul R. Shah

Department of Chemical Engineering and Materials Science, University
of California, Davis, CA 95616

Recent investigations of the anchoring of liquid crystals on self-assembled monolayers (SAMs) formed from alkanethiols on gold are discussed. The structures of alkanethiols used to form SAMs (e.g., odd and even alkanethiols or long and short alkanethiols), the manner of deposition of films of gold used to support SAMs (uniform or oblique deposition), as well as the procedures used to deposit alkanethiols (spontaneous adsorption or microcontact printing) can all be exploited to control the orientations of liquid crystals on these surfaces.

A liquid crystal (LC), when placed into contact with a surface, will generally assume a restricted set of orientations defined with respect to the surface (1-6). This phenomenon, referred to as the "anchoring" of a LC by a surface, is the result of orientation-dependent interactions between the surface and LC. Because control of the orientations of LCs near surfaces is central to the principles of operation of almost all devices based on these optically anisotropic fluids, a substantial effort in the past has been directed towards both elucidation of fundamental forces that control the orientations of LCs near surfaces as well as the development of procedures for the fabrication of surfaces that permit manipulation of these forces in a predictable and systematic manner (4,7-21).

In this chapter, we describe recent advances in the anchoring of LCs on surfaces which have been enabled by an experimental system that permits the design and synthesis of surfaces with a remarkable level of control over structure and chemical functionality: this system is based on the chemisorption and self-assembly of monolayers of organomercaptans on the surface of evaporated films of gold. Although some questions regarding the details of the structure of these surfaces remain to be answered (see below), in comparison to surfaces used in the past for studies of the anchoring of LCs, self-assembled monolayers (SAMs) formed from organomercaptans and organodisulfides on the surface of gold offer a level of control, stability and reproducibility that is not possible when using procedures such as the rubbing of films of polymers, deposition by the method of Langmuir and Blodgett or the self-assembly of organosilanes on metal oxide surfaces (22,23). Surfaces prepared by the self-assembly of sulfur-containing molecules on thin films of gold are, in our opinion, ideally-suited for use in studies of the anchoring of LCs by surfaces.

Our investigations *(24-30)* of the anchoring of LCs by surfaces have been motivated by two principal questions. The first of these questions deals with the use of liquid crystalline fluids for the study and characterization of surfaces. Whereas recent investigations of the anchoring of LCs by surfaces have focused on the issue of the design and fabrication of surfaces for manipulation of the orientations of LCs, we believe the "inverse" pursuit - that of using LCs to probe the structure of organic surfaces - holds substantial promise as a surface-analytical tool.

Measurement of the contact angle of a droplet of an isotropic fluid placed on a surface (Figure 1a) forms the basis of one of the oldest and most widely used methods for the characterization of surfaces *(31,32)*. This method is an ideal one for the characterization of organic surfaces because it is generally non-destructive, simple to perform and can yield quantitative information about the thermodynamic state of a surface *(32,33)*. We believe, for reasons discussed below, that the incorporation of orientational or positional order within a fluid (Figure 1b) - such as found in LCs - can extend the use of fluids for characterization of surfaces to types of surfaces that cannot be readily distinguished by measurement of contact angles of an isotropic fluid. Three facts lead us to believe that LCs can form the basis of a useful system for the characterization of organic surfaces.

First, the orientations of mesogens within LCs are correlated over distances of micrometers *(4)*. The influence of a surface on the orientation of a LC can, therefore, propagate from the near-surface region (first few nanometers from the surface) into the bulk of the LC (as far as 100 micrometers from the surface). These micrometer-scale orientational correlation lengths of LCs can form the basis of a mechanism by which information about the structure of a surface can be amplified into the properties of a bulk medium. Because LCs possess anisotropic optical properties (birefringence), measurement of the bulk orientation of the LC can be achieved by using any one of a variety of well-established experimental techniques. These techniques are based on the use of polarized light and are typically no more complicated than a standard goniometer used for measurements of contact angles. Unlike contact angles of fluids supported on surfaces, which reflect the relative properties of three interfaces (vapor-solid, vapor-liquid and liquid-solid), LCs can, in principal, be used to yield information about the state of a single interface (LC-substrate).

Second, it is well known that LCs can support elastic deformations (bend, splay and twist) over distances ranging from a few nanometers to micrometers *(3,4)*. Isotropic fluids, in contrast, can not support such distortions. The capability of LCs to store energy in these deformations can provide a sensitivity to the structure of surfaces on the mesoscale (nanometers to micrometers) that is not possible when using isotropic fluids. The effects of roughness of surfaces are typically reflected in the hysteresis of advancing and receding contact angles on surfaces: the contribution of roughness on scales of tens of nanometers to hysteresis of contact angles is, however, vanishingly small.

Third, the molecular-level nature of interactions between a LC and an organic surface can differ substantially from that of an isotropic fluid. The orientational order within a liquid crystal can, for example, impose itself on disordered organic surfaces and thereby permit differentiation of disordered surfaces that are indistinguishable when characterized by using isotropic fluids *(26)*. Alternatively, an ordered surface can have a pre-existing structure that is complementary or not to the structure within a LC: the matching or mismatching of the molecular-level structure of molecules supported on surfaces with LCs can, we believe, lead to a level of discrimination between surfaces that is not possible when using isotropic fluids.

The second question that motivates our studies of the orientations of LCs on SAMs formed from organomercaptans and organodisulfides is related to the broadly explored issue of how to design surfaces to anchor LCs in prespecified orientations. Even though a great deal of effort in the past has been directed towards this issue, a variety of types of anchoring of LCs have eluded easy preparation (e.g., high angles of pretilt from a surface). Furthermore, the development of new classes of optical devices demand types of anchoring that have not been required in the past: for example, in recent years, optical displays with wide viewing angles have created the need for procedures leading to patterned alignment of LCs on surfaces with micrometer-scale resolution *(11-14,34-37)*. Because a variety of methods for patterning alkanethiols on the surface of gold have been reported *(38-46)* over the past five years - including microcontact printing, micromachining and photoassisted oxidation - this system is, we believe, an ideal one for use in studies of patterned alignment of LCs on surfaces. The second objective of the studies reported herein is, therefore, to assess the utility of patterned SAMs as templates for the fabrication of complex optical structures from LCs.

Planar Anchoring on SAMs Formed from $CH_3(CH_2)_nSH$ on Gold

The first experiments described in this chapter revolve around our investigation of the anchoring of LCs on SAMs formed from $CH_3(CH_2)_nSH$ on the surface of gold. These surfaces have been widely studied by a variety of surface analytical techniques and thus they form a logical starting point for an investigation of the anchoring of LCs *(22,47)*. In the following section, we summarize what is known about the structures of SAMs formed from alkanethiols on gold and then we compare the structure of these surfaces to SAMs formed from alkyltrichlorosilanes supported on substrates of metal oxides. This comparison is a useful one because past studies have reported measurements of both contact angles of isotropic fluids and orientations of LCs on SAMs formed from organosilanes on metal oxides. Whereas the contact angles of isotropic fluids on these two types of SAMs are not measurably different, the orientations of LCs are found to be strikingly different. Knowledge of differences in the structures of these two types of SAMs (alkanethiols on gold and organosilanes on oxide surfaces) is used to explore the question of how the structure of a SAM influences the anchoring of a LC.

Long-chain alkanethiols (for example, $CH_3(CH_2)_{17}SH$; Figure 2) and alkylsilanes (for example, octadecyltrichlorosilane or OTS) can react with the surfaces of gold and hydroxylated SiO_2, respectively, and thereby form monolayer-thick hydrocarbon films (SAMs) tethered to the surfaces of their respective substrates *(23)*. The structures of these SAMs, however, differ from one another in three important ways at least. First, surfaces of evaporated films of gold used to support SAMs formed from alkanethiols are predominantly Au(111) *(48)*. In contrast, the surface of SiO_2 is amorphous. Second, the sulfur head groups of alkanethiols chemisorb onto the gold surface to form a densely-packed $(\sqrt{3}x\sqrt{3})R30°$ lattice that is commensurate with the underlying Au(111) *(49)*. Periodicity present in the surface of the gold is thereby communicated to the monolayer of organic molecules. The silane headgroups of SAMs formed from OTS are, in contrast, believed to polymerize in the plane of the SiO_2 surface to form a network of $-Si-O-Si-$ bonds which likely attaches to the surface at randomly spaced surface sites (hydroxyl groups) *(50)*. Whereas the lateral structure of the surface of gold is imposed on SAMs formed from alkanethiols, there appears to be relatively little lateral coupling between metal oxide surfaces and SAMs formed from organosilanes. Third, alkyl chains within SAMs formed from long-chain alkanethiols on gold spontaneously organize into nearly all-trans conformations that tilt by $\approx30°$ away from the normal of the surface *(51-53)*. The twist angles of the chains within these SAMs are not all the same, and variation in the twist from one chain to the next gives rise to a c(4x2) superlattice that is commensurate with the underlying lattice of sulfur atoms *(54,55)*. The alkyl chains within carefully prepared SAMs formed from OTS on SiO_2, in contrast, assume nearly all-trans conformations that typically tilt by

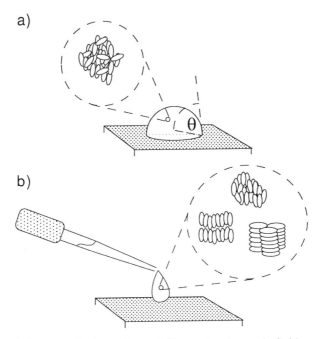

Figure 1. (a) Cartoon of a drop of a partially wetting, isotropic fluid on a surface. The contact angle (θ) is governed by a balance of interfacial tensions at the three-phase contact line. (b) Droplet of a liquid crystalline fluid: nematic, smectic and columnar phases are depicted within the droplet.

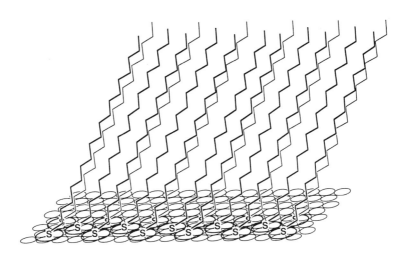

Figure 2. Cartoon of the structure of a SAM formed from $CH_3(CH_2)_{17}SH$ on gold. Details of the bonding between the sulfur and gold (and the superlattice) are not shown.

$\approx 10°$ away from the surface normal *(50)*. No superlattice (or diffraction patterns) have been reported for SAMs formed from OTS.

A number of past reports describe measurements of contact angles of isotropic fluids on SAMs formed from long-chain alkanethiols on gold and OTS on SiO_2 *(56-59)*. Contact angles measured on these two types of SAMs are similar and consistent with the exposure of methyl groups at their surfaces. Contact angles of hexadecane and water, for example, are 45-47° and 110-113°, respectively, on both types of SAMs, and the critical surface energies of both surfaces are approximately 20 ± 1 mJ/m^2 *(56,59)*. Although differences in the structure of SAMs formed from OTS on SiO_2 and long-chain alkanethiols on gold are known to exist (see above), these structural differences do not appear in contact angles of isotropic fluids measured on these surfaces.

The behavior of liquid crystals, in contrast to isotropic liquids, is remarkably different on SAMs formed from OTS on SiO_2 and SAMs formed from long-chain alkanethiols on gold. Whereas past studies *(1,60,61)* have reported the axis of symmetry (the so called "director") of nematic phases of 5CB anchored on SAMs formed from OTS on SiO_2 to be normal to these surface (so called "homeotropic anchoring"), the anchoring of nematic phases of 5CB on SAMs formed from $CH_3(CH_2)_{17}SH$ on gold is planar (parallel to surface) *(28)*. The optical texture of a nematic phase of 5CB confined between two SAMs formed from $CH_3(CH_2)_{17}SH$ on semi-transparent films of gold is grainy when viewed between crossed polarizers (Figure 3a). The grainy appearance of the optical texture reflects a non-uniform azimuthal orientation (orientation of the director in the plane of the surface of the sample) of the LC director. The azimuthal domains within the LC extend over distances of approximately 5 µm or less. Because standard methods for measurement of the tilt angle of the director away from the surface (e.g., crystal rotation method) generally require samples with a uniform azimuthal orientation over a macroscopic area, the tilt angle of the director cannot be determined in samples of the type shown in Figure 3a by using these methods *(2)*. In order to measure the tilt angles of liquid crystals anchored with textures of the type shown in Figure 3a, in collaboration with Prentiss and coworkers, we used a reflection interferometric technique based on circularly polarized light *(29)*. Estimates of the birefringence obtained by using this method allowed us to conclude that the director of a nematic phase of 5CB anchored on a SAM formed from $CH_3(CH_2)_mSH$ ($7<m<17$) on gold lies in the plane of the surface (planar anchoring). In summary, whereas SAMs formed from OTS on metal oxide surfaces cause homeotropic anchoring of 5CB, SAMs formed from $CH_3(CH_2)_mSH$ on gold cause planar anchoring under the same conditions (e.g., temperature).

Thermodynamic theories of the anchoring of liquid crystals at surfaces have attempted to predict the anchoring of LCs on surfaces (planar or homeotropic) by using the critical surface energies of surfaces in combination with knowledge of the orientations of LCs at their free surfaces (LC-vapor interface) *(10,17)*. Because the critical surface energies of SAMs formed from $CH_3(CH_2)_{17}SH$ on gold and SAMs formed from OTS on SiO_2 are not measurably different whereas the anchoring of LCs on these two surfaces is orthogonal, we conclude that some thermodynamic theories, at least, are not yet sufficiently developed to be able to describe the anchoring of liquid crystals on these SAMs. Below we resort, therefore, to a discussion of how known differences in the structures of SAMs (as discussed above) may give rise to planar anchoring of 5CB observed on SAMs formed from long-chain alkanethiols on gold and the homeotropic anchoring on SAMs formed from OTS on SiO_2. First we consider the possible influence of the tilts of the alkyl chains within these SAMs, and second we consider the role of the polarizability of the substrates that support the SAMs (gold is more polarizable than SiO_2).

Because past studies have concluded that the tilts of alkyl chains within Langmuir-Blodgett films can influence the orientations of LCs supported on these surfaces *(62)*, we have considered the possibility that planar anchoring of 5CB on SAMs formed from alkanethiols is caused by the ≈30° tilt (away from normal) of chains within these SAMs. To test the possible role of the tilt of chains within SAMs on the anchoring of LCs, we investigated the anchoring of LCs on SAMs formed from alkanethiols on silver and semifluorinated thiols on gold. Chains within these SAMs are tilted from the normal by 0-15° *(53,63,64)*. In short, we observed LCs to anchor in planar orientations on both types of SAMs (Figure 3b) *(30)*. Because the anchoring of 5CB was found to be planar on SAMs formed from semifluorinated thiols on gold and alkanethiols on silver - SAMs for which the tilt angles of alkyl chains are similar to SAMs formed from OTS on SiO_2 - we conclude that differences in the tilts of chains within SAMs formed on SiO_2 and SAMs formed on gold do not cause the orthogonal anchoring of liquid crystals on observed on these two types of surfaces.

Because planar anchoring of 5CB was found on all densely-packed SAMs supported on metallic substrates (silver and gold), and not on SAMs formed on metal oxides (SiO_2), we have considered the possibility that the planar anchoring of 5CB on SAMs on gold is caused by van der Waals dispersion forces that act between the metallic substrates supporting the SAMs and the LCs *(65,66)*. Differences in the polarizability of metals and metal oxides would then account for the different types of anchoring observed on SAMs supported on these substrates. This proposition - namely that dispersion forces associated with the substrates supporting SAMs influence the orientations of LCs supported on SAMs - is supported both by our observations of planar anchoring of 5CB on "bare" gold as well as calculations of dispersion forces using the theory of Lifshitz, as described below. Below we report first on the use of Lifshitz theory to describe the influence of dispersion forces between metallic substrates and fluids on contact angles of isotropic fluids measured on SAMs. We then extend our discussion to the more complicated case of the anchoring of LCs on SAMs.

A number of past reports have described measurements of contact angles of fluids such as hexadecane on SAMs formed from $CH_3(CH_2)_mSH$ on gold where $2<m<17$ *(56)*. These investigations report contact angles of hexadecane to decrease with decreasing length of the alkanethiol used to form the SAM. Because the decrease in length of the alkanethiol causes the thickness of the SAM to decrease *(67)*, fluids supported on SAMs formed from short-chain alkanethiols will be in close proximity to the gold supporting the SAM. We have used the theory of Lifshitz to evaluate the dispersion forces that act between gold and hexadecane as a function of the decreasing thickness of SAMs. These estimates of dispersion forces were then used to calculate the influence of gold substrates on the energies of the liquid and vapor interfaces of the SAM. Estimates of these energies, when combined with Young's equation *(68)*,

$$\cos\theta = \frac{\gamma_{sv} - \gamma_{sl}}{\gamma_{lv}} \tag{1}$$

(where γ_{sv}, γ_{sl} and γ_{lv} are the interfacial free energies of the solid–vapor, solid–liquid and liquid–vapor interfaces) permit evaluation of the change in contact angle of hexadecane on SAMs formed from $CH_3(CH_2)_mSH$ on gold as a function of m *(69)*. Figure 4 shows that calculated (solid line) and measured (filled circles) contact angles of hexadecane are, indeed, in close agreement. The close agreement of theory and experiment suggests that the decrease in contact angle with m (and thickness of SAM) is due to dispersion interactions associated with the presence of gold beneath the SAM (and probably not due to an m-dependent change in the structure of the SAM). The

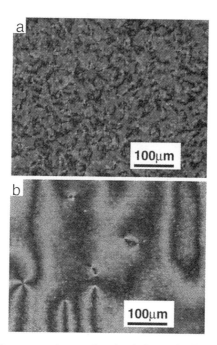

Figure 3. Optical textures (crossed polars) formed when light is transmitted through optical cells containing LC. The optical cells were assembled by pairing two films of gold that supported SAMs. The surfaces of the cell were spaced apart by using 2-25μm Mylar film. The LC was drawn into the cells by using capillarity. The optical images correspond to cells with surfaces supporting SAMs formed from either (a) $CH_3(CH_2)_{17}SH$ or (b) $CF_3(CF_2)_7CONH(CH_2)_2SH$.

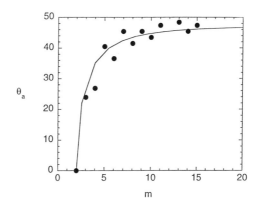

Figure 4. Contact angles of hexadecane supported on SAMs formed from $CH_3(CH_2)_mSH$. The solid line corresponds to contact angles that were calculated using Lifshitz theory and the circles correspond to contact angles measured after advancing drops of hexadecane across SAMs.

decrease in contact angle with decrease in m mainly reflects an increase in γ_{sv}. The increase in the free energy of the solid–vapor interface promotes the spreading of the hexadecane over the surface of the solid and leads, therefore, to low contact angles. Because the theory of Lifshitz is a based on a continuum description of matter, and thus contains a number of well-documented approximations when used to describe a SAM of molecules, we further tested our conclusion by calculation of contact angles of perfluorononane (under an "atmosphere" of liquid hexadecane) on SAMs formed from alkanethiols on gold as well as contact angles of hexadecane on SAMs formed from alkanethiols on Ag and Cu. In these cases too, good agreement was found between calculation and experiment *(69)*.

Our estimates of the order of magnitude of dispersion forces that act between LCs and gold substrates supporting SAMs also demonstrate that these forces should be important in determining the anchoring of liquid crystals on SAMs formed from $CH_3(CH_2)_{17}SH$ (and longer alkanethiols) *(29)*. Whereas past studies demonstrate that interaction energies of 0.01 to 0.001 mJ/m^2 typically control the out-of-plane alignment of liquid crystals on surfaces *(2,6)*, we calculated the dispersion energy associated with the interaction between gold and isotropic 5CB through a SAM formed from $CH_3(CH_2)_{17}SH$ to be 0.2 mJ/m^2. In contrast, the dispersion energy of interaction between SiO_2 and isotropic 5CB through a SAM formed from OTS is calculated to be 0.01 mJ/m^2. The strength of interactions between a metallic substrate and isotropic 5CB through a SAM of given thickness is, therefore, 20 times stronger for gold than for SiO_2. It is plausible, therefore, that dispersion forces acting between LCs and substrates of SAMs do account for the orthogonal alignments of liquid crystals observed on SAMs on gold and SAMs on SiO_2. The results of on-going experiments aimed at measurements of the strength of anchoring of LCs on SAMs formed from alkanethiols on gold, and our use of the theory of Lifshitz to calculate the orientations of liquid crystals on SAMs, will be reported elsewhere in the near future.

In summary, our investigation of the anchoring of LCs on SAMs formed from alkanethiols on gold leads to several conclusions. First, in contrast to past studies where organic surfaces composed of aliphatic chains have been found to orient LCs homeotropically, SAMs formed from alkanethiols on gold are low-energy, structurally well-defined surfaces on which LCs assume planar orientations. This observation is a useful one because most applications of LCs are based on near planar alignment of the LCs. Second, because the anchoring of 5CB on SAMs formed from $CH_3(CH_2)_nSH$ on gold is orthogonal to that on SAMs formed from OTS on SiO_2, and because contact angles of hexadecane (and water) on these two surfaces are indistinguishable, our results clearly demonstrate that liquid crystals can be used to distinguish between surfaces that are reported to be the same by contact angles of isotropic fluids. Third, we conclude that dispersion interactions between fluids in contact with SAMs and metallic substrates supporting SAMs play an important role in determining the behavior of fluids on these surfaces.

Anchoring of Liquid Crystals on Films of Gold Deposited Obliquely

A principal conclusion that emerges from the discussion above is that SAMs formed from densely packed chains on the surface of metallic substrates cause LCs to orient parallel to the surface. This planar anchoring of LCs on SAMs on metals is a general phenomenon and was observed by us for SAMs formed from alkanethiols on gold and silver as well as semifluorinated SAMs on gold. In view of the fact that this variety of SAMs - SAMs which have known differences in molecular-level structure - all cause planar anchoring of LCs, the question that we focus on here is whether or not LCs are indeed useful fluids for probing the *molecular level structure* of SAMs supported on surfaces.

Because LCs orient parallel to the surface of SAMs formed from a variety of organomercaptans on metallic substrates (planar anchoring), we have investigated the effect of the structure of these SAMs on the azimuthal anchoring of LCs. Characterization of the azimuthal anchoring of LCs is, however, made difficult when the optical texture of the LC consists of scattered bright and dark regions with numerous string-like defects, such as shown in Figure 3a. The non-uniform texture in Figure 3a is caused by the absence of a preferred direction of alignment of the LC in the plane of the surface. To facilitate a study of the effect of the structure of SAMs on the in-plane alignment of LCs, it is necessary, therefore, to first introduce a bias into the surface and thereby cause LCs to orient along a preferred direction. The effect of the structure of SAMs on the in-plane alignment of the LCs can then be measured as the orientation of the LC with respect to a preferred direction of alignment in a reference sample.

The azimuthal degeneracy of the in-plane alignment of the LC shown in Figure 3a arises because the SAMs are supported on films of gold that also lack a preferred direction (over macroscopic areas). The glass slides onto which gold was deposited were rotated in an epicyclic manner with respect to the evaporating source of gold (24). The epicyclic rotation of the substrates causes changes in both the angle of incidence (angle between the incoming stream of gold atoms and the normal of the surface of the slide) and the direction of incidence (direction of the incoming stream of gold atoms relative to a reference line in the plane of the surface of the slide) of gold during deposition (Figure 5a). This type of deposition introduces no preferred direction into these films of gold (referred to hereafter as "uniformly" deposited films of gold). In order to break the azimuthal symmetry of the surface of the gold films and thereby orient LCs along a preferred direction, we prepared films of gold by using an "oblique" deposition (24). The glass microscope slides were mounted in a customized holder that was held stationary with respect to the source. The gold was deposited onto the surface of the microscope slides at a constant angle of incidence of 50° and from a single direction (Figure 5b). We refer to films of gold prepared by this scheme as "obliquely" deposited films of gold.

We have characterized gold films prepared by "oblique" and "uniform" deposition by using scanning tunneling microscopy (STM) and ellipsometry. Differences in the structure and properties of these two types of films of gold are subtle. Whereas past studies (70) of uniformly deposited and obliquely deposited films of silicon oxide have shown the latter to contain columnar deposits arranged with periods of hundreds of Ångstroms, STM of our films of obliquely deposited gold on spatial scales from 1 to 500 nm has not yet revealed measurable differences between the surfaces of gold deposited uniformly or obliquely (24). A measurable difference between these two types of gold films was observed, however, when the films were characterized by using ellipsometry. The reflectivity (Δ) and phase of polarized light (ψ), when reflected in orthogonal directions from the surface of the gold, were found to differ more when using gold films deposited obliquely than when using gold films deposited uniformly (Figure 5a).

We characterized SAMs supported on films of gold deposited obliquely and uniformly by using Fourier transformed infrared spectroscopy (FTIR) and contact angles of isotropic fluids. The manner of deposition of the gold was not found to cause measurable differences in the properties of the SAMs. The intensities of the methyl and methylene stretches of the IR absorption spectra for SAMs formed from alkanethiols were found to be the same on gold deposited uniformly and obliquely (Figure 6). Contact angles (Figure 5c and 5d) of fluids that have largely dispersive interactions with SAMs (for example, hexadecane), and contact angles of fluids having chemical functionality similar to those of 5CB (for example, benzonitrile or phenyl hexane), were found to be the same when measured on SAMs supported on gold deposited obliquely or uniformly (26).

	Uniformly deposited films of gold	Obliquely deposited films of gold
Ellipsometric characterization of the thin films of gold	a) 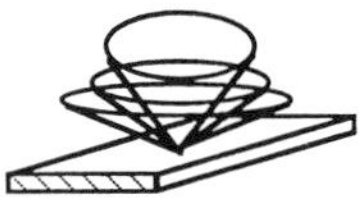	b) 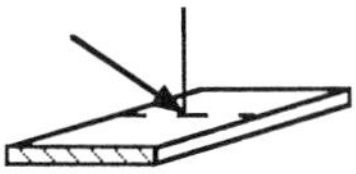
$\Delta_{\parallel} - \Delta_{\perp}$	0.12	0.74
$\Psi_{\parallel} - \Psi_{\perp}$	0.13	-0.59
Contact Angles on SAMs Formed From $CH_3(CH_2)_{15}SH$	c) $\theta_a\ (\theta_r)$	d) $\theta_a\ (\theta_r)$
Hexadecane 	46 (42)	46 (42)
Benzonitrile 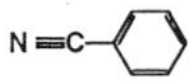	67 (62)	66 (63)
PhenylHexane 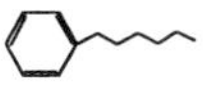	52 (50)	52 (50)
Textures of 5CB oriented on SAMs Formed From $CH_3(CH_2)_{15}SH$	e)  100µm	f) 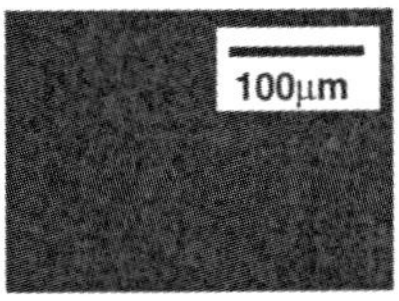 100µm

Figure 5. Comparison of the properties of SAMs supported on "uniformly" and "obliquely" deposited films of gold. (a) Illustration of the procedure used to deposit gold films "uniformly". (b) Illustration of the procedure used to deposit gold films "obliquely". The quantities Δ and ψ (see text) are ellipsometric parameters obtained by reflecting light from the gold films in directions parallel or perpendicular to the direction of deposition of the gold (for obliquely deposited films of gold). (c) and (d) Advancing (θ_a) and receding (θ_r) contact angles of hexadecane, benzonitrile, and phenylhexane on SAMs formed from $CH_3(CH_2)_{15}SH$ on gold. (e) and (f) Typical optical textures observed by placing a LC cell, assembled using surfaces supporting SAMs formed from alkanethiols, between crossed polarizers.

In contrast to STM and contact angles of isotropic fluids, the anchoring of nematic 5CB is distinctly different when 5CB is supported on a SAM formed on uniformly deposited gold as compared to a SAM on obliquely deposited gold (Figures 5e and 5f). Whereas the optical texture of 5CB is non-uniform when anchored on SAMs formed on uniformly deposited gold (Figure 5e), the optical texture of 5CB anchored on obliquely deposited gold has a strongly preferred direction (Figure 5f). Liquid crystals are, therefore, able to detect differences in the structure of these surfaces that are not reflected in measurements of contact angles. Furthermore, films of gold deposited obliquely provide the uniform azimuthal anchoring that is required for a study of the effects of the structure of SAMs on the azimuthal anchoring of LCs.

We have found the preferred azimuthal orientation of a LC on a SAM supported on gold to be an alternating function of the length of the alkanethiol used to form the SAM: an increase or decrease in the length of the alkanethiol by a single methylene group leads to a 90° rotation in the azimuthal orientation of a LC *(25)*. In particular, the azimuthal anchoring of a nematic LC such as 5CB was found to be parallel to the direction of deposition of gold on SAMs formed from even alkanethiols (e.g., $CH_3(CH_2)_{15}SH$, even number of total carbons in alkyl chains) but orthogonal to the direction of deposition of gold on SAMs formed from odd alkanethiols (e.g., $CH_3(CH_2)_{14}SH$).

Past reports have demonstrated that the orientations of the methyl groups within SAMs on gold differ for alkanethiols formed from odd and even chain lengths (Figure 6) *(51)*. The odd-even dependence of the orientation of the methyl group is evident in the IR intensities of the antisymmetric (r_a^-) methyl stretch at 2964 cm^{-1} and the symmetric (r^+) methyl stretch at 2878 cm^{-1} (e.g., $CH_3(CH_2)_{14}SH$ and $CH_3(CH_2)_{15}SH$, Figure 6). Our preliminary results show no odd-even effect in the anchoring of LCs on SAMs formed from alkanethiols on silver. Because the tilt of the alkyl chains in SAMs formed from alkanethiols on silver is 10-15°, the methyl orientation is similar within SAMs formed from odd and even alkanethiols on silver. We conclude, therefore, that the orientation of the methyl groups within SAMs formed on gold plays an important role in determining the azimuthal orientations of nematic LCs on these surfaces.

One possible form of coupling between the methyl orientation of the SAM and the orientation of the LC is through a flexoelectric polarization (deformation-induced polarization) of the LC. Obliquely deposited films of metals are generally believed to possess a statistical roughness that is greatest in a direction parallel to the direction of incidence of the metal during its deposition. A nematic LC, when oriented on a rough surface, undergoes deformation and suffers an elastic penalty. This elastic penalty typically causes the LC to orient in a direction that is parallel to the direction of minimum roughness (that is, perpendicular to the direction of incidence of the metal during deposition) *(15,70-72)*. In a recent theoretical analysis, however, Barbero and Durand predict that substrates with a periodic roughness can induce a flexoelectric polarization within a LC which, in turn, can stabilize the orientations of the LC that are parallel to the direction of maximum roughness (parallel to the direction of deposition of the metal film) *(7)*. This stabilization, which is caused by the electrostatic energy of the system, requires, however, that the dielectric properties of the substrate lie below a critical value. Because we observe the azimuthal orientations of nematic LCs on SAMs to alternate between directions parallel and perpendicular to the direction of incidence of the gold during deposition, we have considered the possibility that flexoelectric effects may control the azimuthal orientations of LCs on SAMs on gold. If a flexoelectric polarization of the LC is the cause of the odd-even effect seen in the orientations of LCs on SAMs, a modulation in the dielectric constant is inferred for odd-even SAMs. Because differences between the chain-end order of odd and even alkyl chains *(73)* are known to cause changes in macroscopic properties such as transition temperatures,

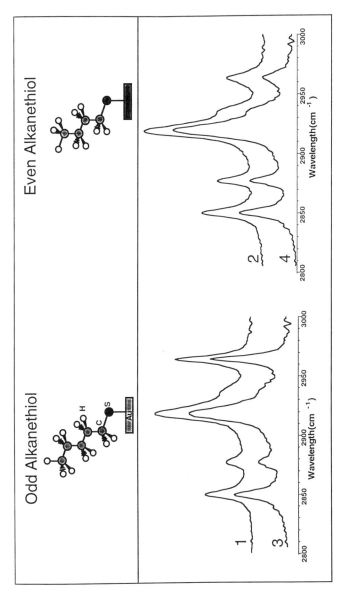

Figure 6. Cartoons of alkanenthiols with odd and even numbers of carbon atoms showing the orientation of the methyl group. The C-H stretching region of FTIR spectra measured using SAMs formed from $CH_3(CH_2)_{14}SH$ on obliquely deposited gold (curve 1) and uniformly deposited gold (curve 3), and of SAMs formed from $CH_3(CH_2)_{15}SH$ on obliquely deposited gold (curve 2) and uniformly deposited gold (curve 4). The spectra were acquired using external reflection of unpolarized IR light at an angle of incidence 84° and by coadding 1000 scans at 2cm^{-1} resolution.

Kerr constants and orientational relaxation times *(74,75)*, the dielectric properties of SAMs in contact with LCs could, plausibly, change with the orientation of methyl groups within SAMs formed from odd and even alkanethiols. To our knowledge, however, past measurements of the capacitance of SAMs formed from alkanethiols in aqueous solutions of electrolyte do not show an odd-even modulation in capacitance *(67,76)*. Studies of capacitance or surface potential of SAMs in contact with LCs have not been performed.

In summary, our measurements of the azimuthal anchoring of nematic LCs on SAMs supported on films of gold deposited obliquely demonstrate that the azimuthal orientation of LCs are sensitive to the molecular-level structure of SAMs. Although we do not yet fully understand how the orientation of methyl groups within SAMs affect the anchoring of LCs, the fact that a change in the orientation of the methyl group at the surface of SAMs is amplified into a change in orientation of bulk LCs demonstrates that LCs are, indeed, highly sensitive to some structural properties of SAMs. Furthermore, comparison of the anchoring of LCs on SAMs on gold to contact angles of isotropic fluids on the same surfaces leads us to conclude that LCs permit discrimination between surfaces with different mesoscale structures (as controlled by the manner of deposition of gold) that are reported to be the same by isotropic fluids. Below we demonstrate that LCs can also be used to distinguish between certain types of molecular-level order within SAMs formed from long and short alkanethiols, SAMs for which the contact angles are measured to be indistinguishable.

Anchoring on SAMs Formed from Long-Short Alkanethiols

Mixed SAMs formed from long and short alkanethiols can be prepared by immersion of films of gold into solutions that contain two alkanethiols *(77-79)*. The composition of mixed SAMs can be manipulated through the ratio of the concentrations of long and short alkanethiols within the solution used to form the SAM. The structure of mixed SAMs, including the lateral distribution of components within the SAMs, depends upon the relative lengths of the long and short chains as well as the procedure used to form the SAMs. When prepared under conditions that give rise to statistically distributed species on the surface, mixed SAMs have a densely packed foundation of thickness comparable to the short alkanethiol and an outer, less densely packed region formed from the long alkanethiols. Figure 7 shows a schematic representation of a SAM formed from equal parts $CH_3(CH_2)_{15}SH$ and $CH_3(CH_2)_9SH$ on gold.

Past studies have reported measurements of contact angles of isotropic fluids on mixed SAMs formed from long and short chains *(77)*. The wetting of isotropic fluids on these mixed SAMs is consistent with the presence of a level of disorder (and degree of exposure of methylenes) within the SAM that exceeds that of SAMs formed from long-chain alkanethiols. Figure 8 shows contact angles of hexadecane on mixed SAMs formed from $CH_3(CH_2)_{15}SH$ and $CH_3(CH_2)_{13}SH$ or $CH_3(CH_2)_9SH$. The contact angles of hexadecane are lower on the mixed SAMs than on single-component SAMs. The decrease in contact angles is likely caused by the dispersion forces that act between hexadecane and the protruding segments of the long alkanethiol.

We have investigated the anchoring of LCs on mixed SAMs formed from $CH_3(CH_2)_{15}SH$ and $CH_3(CH_2)_mSH$ on gold where m=4, 6, 9, 11, and 13 *(26,28,80)*. Whereas the anchoring of LCs on SAMs formed from a single type of alkanethiol is planar, we have observed homeotropic and tilted anchoring of LCs on mixed SAMs of certain compositions. We observe that two criteria must be satisfied in order to observe homeotropic anchoring on a mixed SAM formed from long and short alkanethiols. First, the difference in lengths of the long and short chain must exceed 2 carbons. For example, mixed SAMs formed from $CH_3(CH_2)_{15}SH$ and $CH_3(CH_2)_{13}SH$ on gold do not cause homeotropic anchoring at any composition of the mixed SAM. Second, the mole fraction of long chains within mixed SAMs must range between approximately 0.3 and 0.7. Whereas SAMs formed from $CH_3(CH_2)_{15}SH$ and $CH_3(CH_2)_mSH$ (m=4, 6, 9, 11) with mole fractions of long chains of either 0.1 or 0.9 cause planar anchoring of 5CB, SAMs with mole fractions of long chains between 0.3 and 0.7 cause homeotropic anchoring of 5CB (data not shown).

Inspection of Figure 8 reveals that SAMs (marked as SAMs A and B) formed from solutions of $CH_3(CH_2)_{15}SH$ and $CH_3(CH_2)_9SH$ having different compositions cause the same contact angle of hexadecane (26). The percentage of long alkyl chains on the surface of SAMs A and B were determined by XPS to be 25±5% and 61±8%, respectively. Although the contact angles of hexadecane were identical on these two surfaces, nematic 5CB was found to anchor with a planar orientation on SAM A and homeotropic orientation on SAM B. Clearly, liquid crystals can be used to distinguish between surfaces that can not be distinguished by contact angles of isotropic fluids. Why it is that the LC orients differently on the two SAMs with the same contact angles is not understood, although it is plausible that the LC causes a degree of restructuring of the surface that is absent when using an isotropic fluid.

Stegemeyer and coworkers have reported that some LCs supported on Langmuir-Blodgett (LB) films of zwitterionic phospholipids undergo an anchoring transition from planar to homeotropic as a function of the packing density of the lipids within the film *(81,82)*. These past observations appear, at first sight, to be similar to our measurements of the anchoring of 5CB on mixed SAMs formed from long and short chain alkanethiols on gold. We point out, however, that the anchoring of 5CB on LB films formed from phospholipids was, in fact, reported to be independent of the packing density of the lipids within these films *(82)*.

We infer that the transition from planar anchoring of LCs on single component SAMs to homeotropic anchoring of LCs on mixed SAMs is due to a change in the *density* of alkyl chains – as defined by both the length and number of the protruding segments of the long alkanethiol - in mixed SAMs *(28,30)*. Because long and short alkanethiols that differ in length by 4 carbons can cause homeotropic anchoring of 5CB, interdigitation of at least the alkyl segment of 5CB (5 carbons long) may lead to the homeotropic anchoring. Overall, the change in anchoring of 5CB from planar to homeotropic likely reflects a shift in the balance of two opposing forces: (i) the influence of dispersion forces between the gold substrate and LCs, which favor planar alignment of LCs and (ii) steric forces between mesogens and the protruding segments of the long alkyl chains of the mixed SAMs, which favor homeotropic alignment. We have also observed tilted alignment of LCs when using mixed SAMs formed from $CH_3(CH_2)_{15}SH$ and $CH_3(CH_2)_9SH$ (Figure 8, half filled squares). A variety of tilt angles has been observed, ranging from near planar to near homeotropic. Tilted anchoring of liquid crystals on SAMs formed from alkanethiols on gold most likely represents a near balance of forces between those favoring homeotropic and those favoring planar anchoring.

Our preliminary measurements indicate that mixed SAMs formed from long and short alkanethiols can also be used to manipulate the orientations of smectic LCs. Whereas the anchoring of the smectic A phase of 8CB was found to be planar on single component SAMs formed from alkanethiols, mixed SAMs formed from long and short alkanethiols on gold cause homeotropic anchoring of 8CB over a broad range of monolayer concentrations *(26)*. Figure 8 shows the range of solution concentrations of long and short chains that cause homeotropic alignment of smectic A phases of 8CB. The anchoring of 8CB was found to be homeotropic on SAMs formed from $CH_3(CH_2)_{15}SH$ and $CH_3(CH_2)_9SH$ at solution compositions between approximately 2% and 40% long alkyl chains. It is noteworthy that the anchoring of the smectic A phase of 8CB is especially sensitive to the introduction of small levels of disorder into SAMs (as measured by a slight decrease in contact angles of hexadecane). This result

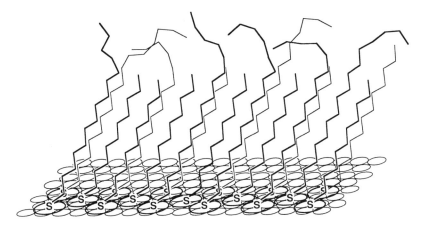

Figure 7. Cartoon of the structure of a mixed SAM formed by coadsorption of $CH_3(CH_2)_{15}SH$ and $CH_3(CH_2)_9SH$ on a film of gold.

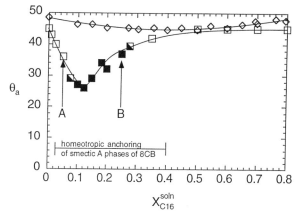

Figure 8. Contact angles of hexadecane measured on mixed SAMs formed by coadsorption of $CH_3(CH_2)_{15}SH$ and $CH_3(CH_2)_{13}SH$ (diamonds) or $CH_3(CH_2)_{15}SH$ and $CH_3(CH_2)_9SH$ (squares) as a function of the mole fraction of $CH_3(CH_2)_{15}SH$ in the solution used to form the SAM. The SAMs were formed for two hours in ethanolic solutions containing a total thiol concentration of 1mM. Filled squares correspond to SAMs that caused homeotropic anchoring of 5CB and half-filled squares correspond to the SAMs that caused tilted anchoring of 5CB. Self-assembled monolayers formed from solutions with compositions given by the horizontal bar caused homeotropic alignment of smectic A phases of 8CB. SAMs marked A and B are discussed in the text.

suggests that smectic LCs might form the basis of useful probes of imperfection within SAMs.

Patterned SAMs as Templates

The results described in the previous two sections of this chapter demonstrate that SAMs formed from alkanethiols on gold can be used to manipulate both the in-plane and out-of-plane orientations of LCs. In the section below, we describe recent results based on the micrometer-scale patterning of surfaces of gold with mixed SAMs formed from long and short alkanethiols, and single-component SAMs, and the use of these surfaces as templates for patterned alignment of LCs. Because a variety of methods now exist to pattern SAMs formed from alkanethiols over spatial scales ranging from a few hundred nanometers to centimeters *(44)*, patterned SAMs do, we believe, provide convenient and general templates for the creation of patterned orientations of LCs.

The procedure we used to pattern the alignment of LCs on surfaces is illustrated by the example shown in Figure 9a. An optical cell containing LC was assembled using SAMs supported on obliquely deposited films of gold. The bottom surface of the cell was patterned by microcontact printing with regions of SAMs formed from an odd $[CH_3(CH_2)_{14}SH]$ alkanethiol and an even $[CH_3(CH_2)_{15}SH]$ alkanethiol. The top surface of the cell was a gold film supporting a SAM formed from $CH_3(CH_2)_{15}SH$. Because the bulk LC was oriented uniformly in regions where the top and bottom surfaces supported SAMs formed from $CH_3(CH_2)_{15}SH$, the polarization of linearly polarized light was not changed by transmission through these regions. In regions where the LC was sandwiched between SAMs formed from $CH_3(CH_2)_{14}SH$ and $CH_3(CH_2)_{15}SH$, the LC was twisted by 90°. The twist in the LC caused the polarization of linearly polarized light to rotate by 90° upon transmission through the LC. When viewed through crossed polars, therefore, the twisted regions of LC appeared bright (light was transmitted by the analyzer) and the uniform regions of LC appeared dark (light was extinguished by the analyzer) (Figure 9b). We have also extended this procedure to the patterning of LCs on surfaces with complex geometric features (Figure 9c) and used this capability to fabricate linear diffraction gratings of the types shown in Figure 10. Each grating differs in the manner of distortion of the LC and thus in its polarization sensitivity.

•Grating A: Grating A (Figure 10a) was formed by planar anchoring along one direction of the top surface of the cell and patterned planar anchoring in two orthogonal directions on the bottom surface. We fabricated the grating of type A by using SAMs formed from $CH_3(CH_2)_{14}SH$ and $CH_3(CH_2)_{15}SH$ on obliquely deposited gold as the bottom surface of the cell. The top surface supported a SAM formed from $CH_3(CH_2)_{15}SH$. Regions of twisted and undistorted LC were thereby replicated periodically across the cell (in direction along Y), thus causing a spatially periodic change in refractive index. Linearly polarized laser light incident on the grating (along Z) was diffracted.

•Grating B: Grating B (Figure 10b) was formed by homeotropic anchoring of the LC on the top surface and patterned planar and homeotropic anchoring of the LC on the bottom surface. Mixed SAMs formed by coadsorption of $CH_3(CH_2)_{15}SH$ and $CH_3(CH_2)_9SH$ on gold were used to homeotropically anchor the LC. Light with polarization along X, when incident on the grating of type B, experienced n_0 (ordinary refractive index) in regions with homeotropic anchoring of the LC. In regions with planar anchoring of the LC on the bottom surface and homeotropic anchoring of the LC on the top surface, the light experienced an effective refractive index that lies between n_0 and n_e (extraordinary refractive index). This periodic change in refractive index caused the light with polarization along X to be diffracted by grating B. Light with

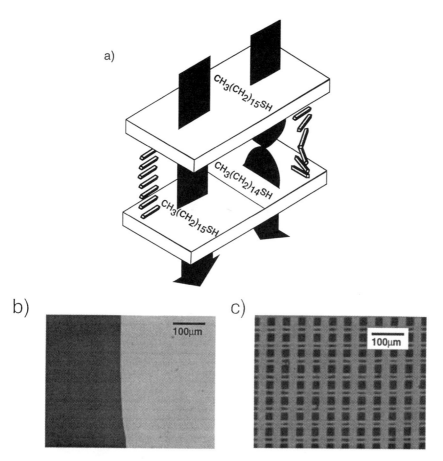

Figure 9. (a) Schematic illustration of a LC cell formed using two films of gold. The top surface supports a SAM formed from $CH_3(CH_2)_{15}SH$ and the bottom surface is patterned with regions of SAMs formed from $CH_3(CH_2)_{15}SH$ and $CH_3(CH_2)_{14}SH$. In the regions where the rod-like LC molecules are oriented uniformly (not twisted), light transmitted through the cell does not change its polarization. In regions where the LC molecules are twisted by 90° (regions where the LC is sandwiched between SAMs formed from $CH_3(CH_2)_{15}SH$ and $CH_3(CH_2)_{14}SH$), the polarization of the light follows the twist of the LC as it is transmitted through the cell. (b) Optical image of the LC cell assembled using the scheme shown in (a) and viewed under crossed polarizers. (c) An optical image of a LC cell fabricated using a bottom surface patterned with SAMs formed from $CH_3(CH_2)_{15}SH$ and $CH_3(CH_2)_{14}SH$.

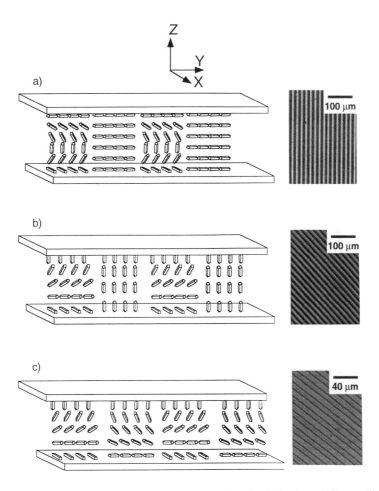

Figure 10. Schematic illustrations of the distortion of LCs in three different linear diffraction gratings, and optical images of these gratings when viewed between crossed polarizers. See text for description.

polarization along Y, however, experienced only n_0 across the grating and was not diffracted.

•Grating C: Grating C was formed by homeotropic anchoring of the LC on the top surface (mixed SAMs formed from long and short alkanethiols), and patterned planar anchoring on the bottom surface (using SAMs formed from odd and even alkanethiols). Because light incident with polarization along either X or Y experienced a similar periodic change in the index of refraction when using grating C, diffraction of incident light from grating C was polarization insensitive.

The method described above for the anchoring of LCs with patterned orientations on surfaces differs from procedures reported in the past. Past studies have fabricated patterned LC structures with spatially non-uniform electric fields produced from patterned electrodes; the resolution of these methods is limited by fringing of electric fields from the electrodes (13,37). Surfaces for patterned anchoring of LCs have also been prepared by using localized, mechanical rubbing of spin-coated polymer films (83). This method is complex and suffers from problems associated with the generation of dust and static charge during rubbing. Because SAMs formed from alkanethiols on gold have low surface energies (~20mJ/m^2), these surface are also less prone to contamination by molecular adsorbates and dust particles than the high energy polymeric surfaces.

In summary, we note that procedures based on SAMs for patterning the alignment of LCs on surfaces are simple and flexible, and permit the fabrication of optical structures with dimensions ranging from micrometers to centimeters. By patterning SAMs on curved surfaces, we have also fabricated patterned LC structures on non-planar surfaces (not shown) (27). Because diffraction of light from the patterned LC structure can be combined with refraction of light at the curved surface, the capability to pattern LCs on curved surfaces may enable easy fabrication of hybrid optical devices that combine diffraction and refraction.

Concluding Comments

The results described in this chapter demonstrate that SAMs formed from alkanethiols on gold do form the basis of a versatile experimental system for studies of the anchoring of LCs by surfaces. This system permits control of the structure of surfaces on spatial scales that range from the molecular (nanometer) to the microscopic (micrometers), all of which influence the anchoring of LCs. Our studies demonstrate that LCs supported on SAMs can be used to distinguish between organic surfaces that differ in subtle ways, including surfaces that are reported to be indistinguishable when characterized by using contact angles of isotropic fluids. These results support our belief that LCs can form the basis of simple and sensitive procedures for the characterization of surfaces. Our results also reveal three challenges that need to be addressed in future work if LCs are to become routinely used for characterization of surfaces. (i) Convenience: our studies have been based on confinement of a film of LC between two semi-transparent surfaces. Methods based on the placement of a droplet of LC on a surface, and characterization of the deformation of the LC within the droplet, will be easier to perform and will permit characterization of optically opaque samples by use of reflecting polarized light. (ii) Quantitativeness: our measurements of the orientations of liquid crystals are qualitative. Interpretation of the deformation of a LC within a droplet might also lead to quantitative ways of characterizing surfaces using LCs. (iii) Averaging of information: because liquid crystals on surfaces respond to the structure of a surface over a wide range of length scales, it may, in some cases, be difficult to identify the cause of a particular orientation adopted by a LC supported on a surface.

100

Literature Cited

1) Cognard, J. *Mol. Cryst. Liq. Cryst. Supp.* **1982**, *78*, 1-77.
2) Blinov, L. M.; Chigrinov, V. G. *Electrooptic Effects in Liquid Crystal Materials*; Springer-Verlag: New York, 1994.
3) Chandrasekhar, S. *Liquid Crystals*; Cambridge University Press: Cambridge, 1992.
4) de Gennes, P. G.; Prost, J. *The Physics of Liquid Cystals*; Oxford University Press: New York, 1993.
5) Jérôme, B. *Rep. Prog. Phys.* **1991**, *54*, 391-452.
6) Sonin, A. A. *The Surface Physics of Liquid Crystals*; Gordon and Breach: U.S.A., 1995.
7) Barbero, G.; Durand, G. *J. Appl. Phys.* **1990**, *68*, 5549-5554.
8) Barbero, G.; Petrov, A. G. *J. Phys. Cond. Matt.* **1994**, *6*, 2291-2306.
9) Bernasconi, J.; Strassler, S.; Zeller, H. R. *Phys. Rev. A* **1980**, *22*, 276-281.
10) Creagh, L. T.; Kmetz, A. R. *Mol. Cryst. Liq. Cryst.* **1973**, *24*, 59-68.
11) Ikeda, T.; Tsutsumi, O. *Science* **1995**, *268*, 1873-1875.
12) Gibbons, W. M.; Shannon, P. J.; Sun, S. T.; Swetlin, B. J. *Nature* **1991**, *351*, 49-50.
13) Gibbons, W. M.; Sun, S. T. *Appl. Phys. Lett.* **1994**, *65*, 2542-2544.
14) Gibbons, W. M.; Kosa, T.; Palffy-Muhoray, P.; Shannon, P. J.; Sun, S. T. *Nature* **1995**, *377*, 43-46.
15) Janning, J. L. *Appl. Phys. Lett.* **1972**, *21*, 173-174.
16) Naemura, S. *J. Appl. Phys.* **1980**, *51*, 6149-6159.
17) Naemura, S. *Mol. Cryst. Liqu. Cryst.* **1981**, *68*, 1131-1146.
18) Ohgawara, M.; Uchida, T.; Wada, M. *Molecular Crystals and Liquid Crystals* **1981**, *74*, 1827-1842.
19) Okano, K.; Murakami, J. *J. Physique* **1979**, *C3*, 525-528.
20) Uchida, T.; Ishikawa, K.; Wada, M. *Molecular Crystals and Liquid Crystals* **1980**, *60*, 37-52.
21) *Liquid Crystals: applications and uses*; Bahadur, B., World Scientific: Teaneck (NJ), 1990; Vol. 1-3.
22) Dubois, L. H.; Nuzzo, R. G. *Annu. Rev. Phys. Chem.* **1992**, *43*, 437-463.
23) Ulman, A. *An Introduction to Ultrathin Organic Films: From Langmuir-Blodgett to Self Assembly*; Academic Press: San Diego, CA, 1991.
24) Gupta, V. K.; Abbott, N. L. *Langmuir* **1996**, *12*, 2587-2593.
25) Gupta, V. K.; Abbott, N. L. *Phys. Rev. E* **1996**, *54*, R4540-R4543.
26) Gupta, V. K.; Miller, W. J.; Pike, C. L.; Abbott, N. L. *Chem. Mats.* **1996**, *8*, 1366-1369.
27) Gupta, V. K.; Abbott, N. L. *Science* **1997**, 276, 1533-1536.
28) Drawhorn, R. A.; Abbott, N. L. *J. Phys. Chem.* **1995**, *99*, 16511-16515.
29) Miller, W. J.; Abbott, N. L.; Paul, J. D.; Prentiss, M. G. *Appl. Phys. Lett.* **1996**, *69*, 1852-1854.
30) Miller, W. J.; Gupta, V. K.; Abbott, N. K.; Tsao, M.-W.; Rabolt, J. F. *Liquid Crystals* **1997**, *23*, 175-184.
31) Adamson, A. W. *Physical Chemistry of Surfaces*; Interscience: New York, 1960.
32) *Wettability*; Berg, J. C., Eds.,;Surfactant Science Series;Marcel Dekker: New York, 1993; Vol. 49.
33) Whitesides, G. M.; Laibinis, P. E. *Langmuir* **1990**, *6*, 87-96.
34) Yang, K. H. *Jpn. J. Appl. Phys. Pt. 2* **1992**, *31*, L1603-L1605.
35) Shannon, P. J.; Sun, S. T.; Gibbons, W. M. *Nature* **1994**, *368*, 532-533.
36) Schadt, M.; Seiberle, H.; Schuster, A. *Nature* **1996**, *381*, 212-215.
37) Patel, J. S.; Rastani, K. *Opt. Lett.* **1991**, *16*, 532-534.
38) Abbott, N. L.; Kumar, A.; Whitesides, G. M. *Chem. Mater.* **1994**, *6*, 596-602.
39) Huang, J.; Hemminger, J. C. *J. Am. Chem. Soc.* **1993**, *115*, 3342-3343.

40) Huang, J.; Dahlgren, D. A.; Hemminger, J. C. *Langmuir* **1994**, *10*, 626-628.
41) Jackman, R. J.; Wilbur, J. L.; Whitesides, G. M. *Science* **1995**, *269*, 664-666.
42) Kumar, A.; Whitesides, G. M. *Appl. Phys. Lett.* **1993**, *63*, 2002-2004.
43) Kumar, A.; Biebuyck, H. A.; Whitesides, G. M. *Langmuir* **1994**, *10*, 1498-1511.
44) Kumar, A.; Abbott, N. L.; Biebuyck, H. A.; Whitesides, G. M. *Accounts Chem. Res.* **1995**, *28*, 219-226.
45) Tarlov, M. J.; Burgess, D. R. F.; Gillen, G. *J. Am. Chem. Soc.* **1993**, *115*, 5305-5306.
46) Xia, Y.; Whitesides, G. M. *J. Am. Chem. Soc.* **1995**, *117*, 3274-3275.
47) Delamarche, E.; Michel, B.; Biebuyck, H. A.; Gerber, C. *Adv. Mats.* **1996**, *8*, 719-729.
48) Nuzzo, R. G.; Fusco, F. A.; Allara, D. L. *J. Am. Chem. Soc.* **1987**, *109*, 2358-2368.
49) Chidsey, C. E. D.; Bertozzi, C. R.; Putvinski, T. M.; Mujsce, A. M. *J. Am. Chem. Soc.* **1990**, *12*, 4301-4306.
50) Allara, D. L.; Parikh, A. N.; Rondelez, F. *Langmuir* **1995**, *11*, 2357-2360.
51) Nuzzo, R. G.; Dubois, L. H.; Allara, D. L. *J. Am. Chem. Soc.* **1990**, *112*, 558-569.
52) Fenter, P.; Eisenberger, P.; Burrows, P.; Forrest, S. R.; Liang, K. S. *Physica B* **1996**, *221*, 145-151.
53) Ehler, T. T.; Malmberg, N.; Noe, L. J. *J. Pjys. Chem. B* **1997**, *101*, 1268-1272.
54) Camillone, N.; Chidsey, C. E. D.; Liu, G.; Scoles, G. *J. Chem. Phys* **1993**, *98*, 4234-4245.
55) Poirier, G. E.; Tarlov, M. J. *Langmuir* **1994**, *10*, 2853.
56) Bain, C. D.; Troughton, B. E.; Tao, Y.-T.; Evall, J.; Whitesides, G. M.; Nuzzo, R. G. *J. Am. Chem. Soc.* **1989**, *111*, 321-335.
57) Maoz, R.; Sagiv, J. *J. Colloid Interface Sci.* **1984**, *100*, 465-496.
58) Wasserman, S. R.; Tao, Y.-T.; Whitesides, G. M. *Langmuir* **1989**, *5*, 1074-1087.
59) Brzoska, J. B.; Shahidzadeh, N.; Rondelez, F. *Nature* **1992**, *360*, 719-721.
60) Yang, J. Y.; Mathauer, K.; Frank, C. W. *Microchemistry* **1994**, 441-454.
61) Peek, B.; Ratna, B.; Pfeiffer, S.; Calvert, J.; Shashidar, R. *Proceedings of the SPIE* **1994**, *2175*, 42-48.
62) Alexe-Ionescu, A. L.; Barberi, R.; Barbero, G.; Bonvent, J. J.; Giocondo, M. *Appl. Phys. A* **1995**, *61*, 425-430.
63) Laibinis, P. E.; Whitesides, G. M.; Allara, D. L.; Tao, Y.-T.; Parikh, A. N.; Nuzzo, R. G. *J. Amer. Chem. Soc.* **1991**, *113*, 7152-7167.
64) Lenk, T. J.; Hallmark, V. M.; Hoffman, C. L.; Rabolt, J. F. *Langmuir* **1994**, *10*, 4610-4617.
65) Israelachvili, J. *Intermolecular and Surface Forces, 2nd ed.*; Academic Press: San Diego, 1992.
66) Mahanty, J.; Ninham, B. W. *Dispersion Forces*; Colloid Science; Academic Press: New York, 1976.
67) Porter, M. D.; Bright, T. B.; Allara, D. L.; Chidsey, C. E. D. *J. Am. Chem. Soc.* **1987**, *109*, 3559-3568.
68) Young, T. *Philos. Trans. R. Soc. London* **1805**, *95*, 65-87.
69) Miller, W. J.; Abbott, N. L. , submitted.
70) Goodman, L. E.; McGinn, J. T.; Anderson, C. H.; Digeronimo, F. *IEEE Trans. Electron. Devices* **1977**, *24*, 795.
71) Berreman, D. *Phys. Rev. Lett.* **1972**, *28*, 1683-1686.
72) Urbach, W.; Boix, M.; Guyon, E. *Appl. Phys. Lett.* **1974**, *25*, 479-481.
73) Marcelja, S. *J. Chem. Phys.* **1974**, *60*, 3599-3604.
74) Hanson, E. G.; Shen, Y. R. *Mol. Cryst. Liq. Cryst.* **1976**, *36*, 193-207.

75) Yamamoto, R.; Isihara, S.; Hayakawa, S.; Morimoto, K. *Phys. Lett. A* **1978**, *69A*, 276-278.

76) Walczak, M. M.; Chung, C.; Stole, S. M.; Wildrig, C.; Porter, M. D. *J. Amer. Chem. Soc.* **1991**, *113*, 2370-78.

77) Bain, C. D.; Whitesides, G. M. *J. Am. Chem. Soc.* **1989**, *111*, 7164-7175.

78)Folkers, J. P.; Laibinis, P. E.; Whitsides, G. M. *J. Phys. Chem.* **1994**, *98*, 563-571.

79) Laibinis, P. E.; Nuzzo, R. G.; Whitsides, G. M. *J. Phys. Chem.* **1992**, *96*, 5097-5105.

80) Drawhorn, R. A.; Miller, W. J.; Pike, C. L.; Abbott, N. L., unpublished results.

81) Hiltrop, K.; Stegemeyer, H. *Ber. Buns. Phys. Chem.* **1984**, *85*, 582-588.

82) Hiltrop, K.; Hasse, J.; Stegemeyer, H. *Ber. Bunsenges. Phys. Chem.* **1994**, *98*, 209-213.

83) Chen, J.; Bos, P. J.; Vithana, H.; Johnson, D. L. *Appl. Phys. Lett.* **1995**, *67*, 2588-2590.

SUPRAMOLECULAR ASSEMBLIES

Chapter 8

Polymer-Supported Biomembrane Models

Claudia Heibel[1], Steffen Maus[1], Wolfgang Knoll[1,2], and Jürgen Rühe[1]

[1]Max-Planck-Institute for Polymer Research, Ackermann Weg 10, D-55128 Mainz, Germany
[2]Frontier Research Program, The Institute of Physical and Chemical Research (RIKEN), Wako, Japan

The preparation of polymer supported lipid layers attached to planar solid substrates is studied. The polymer monolayers are generated at silicon oxide surfaces either through chemisorption of α,ω-functionalized polymers or oligomers or through a stepwise chemisorption procedure. In the latter approach, linear or branched polymers are chemically attached to self-assembled monolayers of bifunctional silanes followed by a deposition of a Langmuir monolayer of a lipid or a lipid mixture. The characterization of the surface-attached supramolecular assemblies by various surface analytical techniques is described.

Polymer supported biomembrane models are a promising approach for the development of biosensor devices (*1-10*). Such model membrane systems try to mimic the plasma membrane of a biological cell in many respects. Proteins, which can act as highly specific receptors for molecules present in the contacting medium, are incorporated into a biomolecular lipid layer. The lipid bilayer provides a hospitable environment for the receptor proteins and holds them at a specific distance to the device surface. The polymer layer, which supports the membrane, acts as a molecularly thin buffer between the membrane and the surface of the substrate. It prevents contact between the incorporated proteins and the solid substrate, because such a contact could cause severe denaturing of the receptor molecules (*3,11*). Additionally, the polymer "cushion" allows for the adjustment of the mechanical properties of the attached layers. For a membrane system coupled to a solid surface a delicate balance between rigidity and mobility of the lipid layer is required. On the one hand, the tethering to the solid substrate anchors the membrane to the support and holds the supramolecular assembly in place. Additionally, it provides mechanical stability to the system. In freestanding membranes, such as the so-called "black lipid membranes" (BLM), the mechanical stability is a factor strongly limiting the size of the membrane assembly (*12*). If the free-spanning area of the BLM becomes too large the membrane becomes unstable as the lipids are held together only by comparably weak intermolecular forces. Therefore, such model systems might collapse even if only small mechanical forces are exerted. This limits practical applications of devices constructed with membranes deposited by this technique.

On the other hand, the fluidity of the membrane must not be reduced too strongly as the mobility of the components (lipids, proteins) within the membrane is an essential requirement for the response of the receptors to molecules in the environment (13). When the membrane is too strongly coupled to the substrate and becomes too rigid, the whole system does not function any longer. How strong the optimal coupling to the substrate should be is a yet totally open question.

In this paper we describe the construction of supramolecular assemblies consisting of "anchor" molecules, which allow a covalent binding to a solid substrate, monomolecular polymer layers and lipid mono- and bilayers, in which some of the molecules are covalently linked to the supporting polymer. The polymers used in these investigations are based on poly(2-ethyl-2-oxazoline)s and poly(ethylene imine)s. The polymers were chosen as they are hydrophilic, protein compatible and allow the reaction with lipids and groups that immobilize the polymer at the surface of the substrate. We will mainly focus on the influence of structure, molecular weight, basicity and other parameters on the properties of the attached layers. The structure of the polymers is varied in small increments and the influence of these changes on the performance of the films is studied. We report on the synthesis and characterization of the molecularly thin films, on the swelling of the attached layers in an aqueous environment and on the covalent attachment of reactive lipids to these monolayers either by self-assembly procedures or by transfer of Langmuir layers.

Synthetic strategies:

The different synthetic concepts for the immobilization of the polymer layers onto the solid substrates, which were followed in this study, are schematically depicted in Figure 1 and 2.

Essentially two different strategies have been employed. Following the first one, the polymers have been immobilized at the surfaces of the substrates by end group attachment of telechelic oligo- and poly(ethyl-oxazoline)s carrying a silane anchor group on one end of the molecule and a reactive group or a lipid on the other one. In this case the whole supramolecular assembly is attached to the substrate requiring only one surface reaction step.

In a second approach, poly(ethyloxazoline)-stat-poly(ethylene imines) (14) and branched or linear poly(ethyleneimines) (15) have been attached to functionalized surfaces in a step-wise fashion. In successive reaction steps, firstly an α,ω-bifunctional "anchor" molecule was attached to the surface of the solid support. In a second step the polymer was covalently coupled to the thus obtained functionalized monolayers. Finally, the lipids were covalently linked to the polymer layer. This procedure allows the attachment of polymers which cannot be obtained by a "living" polymerization process and of polymer molecules, which contain groups that are not compatible with the anchor group chemistry. For example, polymers which carry amino groups cannot be anchored to the surfaces by chlorosilane groups as these two groups mutually exclude each other.

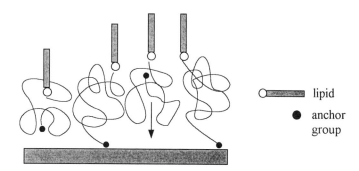

Fig. 1: Polymer supported lipid layers through end group attachment of telechelic oligo- and poly(ethyloxazoline)s

In both synthetic approaches additional lipids were deposited after the assembly of the polymer layers by Langmuir-Blodgett or Langmuir-Schäfer technique or fusion of vesicles onto the chemically attached layers.

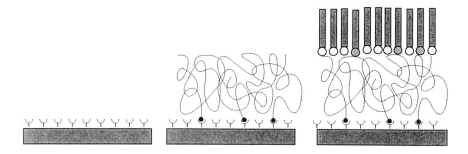

Fig. 2: Sequential build-up of polymer-supported lipid layers: chemisorption of (partially) hydrolyzed linear poly(ethyloxazoline)s and branched poly(ethylene imine)s to functionalized self-assembled monolayers followed by deposition of lipid layers

Selfassembly of Telechelics:

The preparation of surface-attached telechelics is shown in Figure 3. Details of the synthetic procedures have been described elsewhere (*14*). Briefly, allyltosylate is used as an initiator of a living cationic ring opening polymerization of 2-ethyl-2-oxazoline. The living chains are terminated with dialkylamines such as dioctadecyl amine or (for model studies) piperidine. Additionally, the growing chains can be terminated by addition of potassium hydroxide, so that a hydroxyl-terminated polymer is obtained, to which additional functional groups can be attached in subsequent reaction steps. The terminal double bonds in the polymers can be transformed into chlorosilyl groups by a hydrosilation reaction using hexachloro platinic acid as a catalyst. Compounds with only one chlorosilane moiety were used to avoid the problem of multilayer formation. Silanes with two or more reactive head groups can be transformed during the surface reaction into an unspecific, surface attached network of siloxanes. When a monofunctional compound is used such side reactions can only lead to the formation of unreactive dimers, which can be readily extracted from the layer during work-up of the reaction medium.

The chlorosilyl-terminated oligomers or polymers can be attached to SiO$_x$ surfaces in a base catalyzed reaction. A tertiary base such as triethylamine was used as an acid scavenger and catalyst. After the completion of the surface reaction, the substrates with the attached monolayers were extracted in a Soxhlet-extractor with a good solvent for the polymer for at least 14 hours. During simple rinsing procedures some polymer remains in the film, which is not covalently linked to the surface, but is only physically attached.

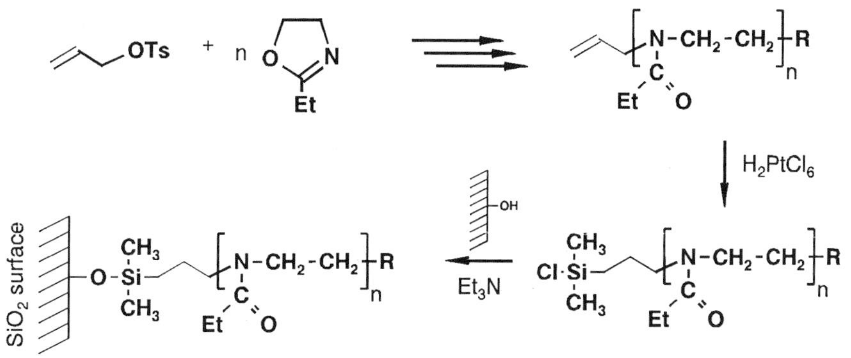

Fig. 3: Synthesis of telechelic oligo- and poly(ethyl oxazoline)s with a monochloro silyl anchor group for immobilization onto SiO$_2$ surfaces; Ts represents a tosyl group; R was a dioctadecyl amine, piperidine or hydroxyl group.

It is observed that the conformation of the polymer molecules within the layer to some extent is a function of the degree of polymerization. When only short chain oligomers are attached to the surfaces (up to degrees of polymerization of about 20) the layer thickness increases with increasing chain length (fig. 4). At higher degrees of polymerization, however, the thickness increase levels off and the thickness of the polymer layer becomes independent of chain length.

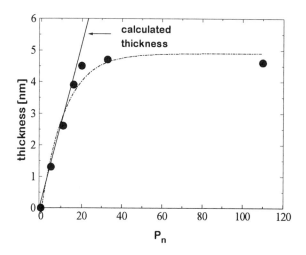

Fig. 4: Thicknesses of layers of poly(ethyloxazoline)s with terminal chlorosilane groups attached to silicon oxide surfaces at various degrees of polymerization as determined by surface plasmon spectroscopy

At low chain lengths of the attached polymer, the graft density (number of attached chains per surface area) is only limited by the size of the silyl head group. It has been shown that the number of silyl molecules attached to SiO_2 surface is not limited by the number of hydroxyl groups present at the surface, but only by the size of the dimethylsilyl head group (*16*). The solid line drawn in Figure 4 is the maximum thickness that can be theoretically reached if all reactive sites on the surface of an amorphous silicon dioxide, which can possibly react with the dimethyl chlorosilyl group of the polymer (under the given geometric constraints), would indeed react. This maximum graft density, the molecular weight of the attached chains and the density of the bulk polymer are then used for the calculation of the maximum layer thickness. It can be seen that the measured layer thickness follows this relation up to degrees of polymerization of about 20.

At higher degrees of polymerization, however, the layer thickness levels off and becomes independent of the length of the polymer molecules. The reason for this behavior is an intrinsic limitation of the thickness of polymer layers obtained by such "grafting to" techniques, where preformed polymers are attached to solid surfaces:

When a certain number of polymer segments is attached to the surface, additional polymer molecules have to diffuse against the concentration gradient built up by the already attached polymer chains. As more and more polymer segments become attached to the surface, the gradient becomes steeper and steeper making the diffusion of additional chains into the layer increasingly difficult. This diffusion barrier leads to a strong kinetic hindrance of further layer growth (*17*) and additional polymer molecules can be added to the layer only on a logarithmic time scale. Thus, with a "grafting-to" technique only layer thicknesses of about 3-5 nm (in the dry, solvent-free state) are experimentally accessible. This diffusion barrier has also a strong influence on the graft density of polymers when polymers with different molecular weight are compared. The situation for the attachment of polymer molecules of different chain lengths to a surface is schematically depicted in Figure 5. If the rate of attachment of polymer molecules to the surface is governed by the diffusion barrier (i.e. a critical segment density at the surface is reached) any increase in the molecular weight of the polymer will lead only to the fact, that this critical segment density of the attached chains at the surface will be reached at shorter reaction times and the graft density of the attached chains deposited within same time period decreases.

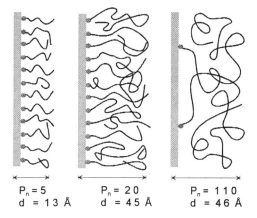

$P_n = 5$ $P_n = 2 0$ $P_n = 1 1 0$
$d = 1 3$ Å $d = 4 5$ Å $d = 4 6$ Å

Fig. 5: Schematic depiction of the intrinsic limitation of the layer thickness in "grafting to" procedures

The Sequential Approach

The reaction sequence is depicted in Figure 6. In the first step a self-assembled monolayer of the "anchor molecules" was deposited onto the surface of the planar silicon oxide substrate (Fig. 6 (A)).

Chlorosilanes with a monofunctional head group were used to establish a covalent bond to the substrate. As in the case described above, monochloro silane compounds were used to avoid the problem of formation of surface-attached networks of the anchor molecules.

110

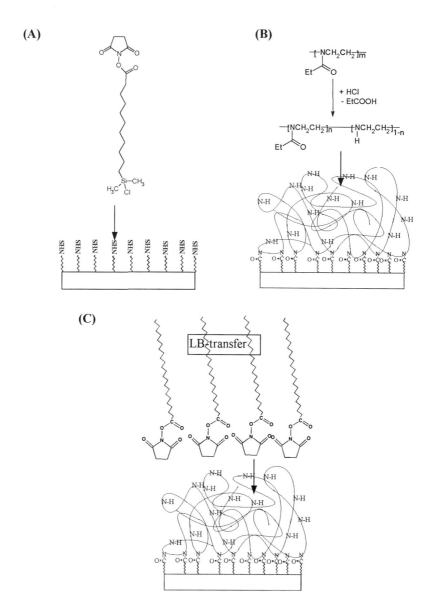

Fig. 6: A sequential approach to polymer supported lipid monolayers; firstly, NHS-ester silanes are attached to the surfaces (A), in subsequent reaction steps poly(ethylene-imine) or poly(ethyloxazoline-stat-ethyleneimine) polymers (B) and LB-films of lipids (C) are bound to the monolayers

Additionally, the use of multifunctional silanes opens the possibility that amino groups in the polymer react with residual silyl groups, which were not successful in binding to the surface. Such a surface attachment would also lead to the establishment of a covalent linkage between the polymer and the surface. However, in contrast to the surface reactions described here, such a linkage is not hydrolytically stable as silicon-nitrogen bonds would be formed. Such bonds slowly hydrolyze in a humid environment or take part in one of the subsequent reaction steps making the characterization of the layers difficult.

The reactive ω-terminus was a N-hydroxysuccinimide (NHS) group or an iodo alkyl group. The coupling of the anchor molecules to the substrate was followed on high surface area model substrates (silica gels) by IR, XP and ^{29}Si-CP-MAS-NMR spectroscopy. Especially the latter technique gives direct evidence for the establishment of a covalent bond between the surface groups of the substrate and the silane compound.

On planar surfaces the layer thickness was determined by surface plasmon spectroscopy. With an assumed value for the refractive index of the layer of $n = 1.5$ a layer thickness d_{anchor} of about 1.1 nm was obtained. The reflectivity curves are shown in Figure 9 together with the results of calculations according to Fresnel equations. Scanning force microscopy (AFM) images show that the substrates are homogeneously covered with the silanes and the surface morphology remains unchanged during the silanization reaction.

In a second reaction step the polymer buffer layer was bound to the monolayer of the anchor molecules (Fig. 6(B)). Linear poly(ethyloxazoline-stat-ethylene imine) (PEOX-LPEI) were attached to the NHS groups. The polymers were prepared by cationic ring opening polymerization followed by partial acidic hydrolysis. Polymers with different molecular weights and degrees of hydrolysis between 15 and 100 % were employed. The reaction for the surface attachment was carried out in a polar solvent (methanol or ethanol) and triethylamine was used as a catalyst. The addition of a base, which catalyses the reaction between the surface attached active ester and the amino groups of the polymer, was found to be an important factor. Model studies in solution have shown that without base the reaction between active ester and amine is very slow and many of the NHS-groups do not react in the intended manner. This agrees well with results obtained for the reaction of low molecular weight amines with active esters with and without addition of a base (*18,19*).

The surface attachment of the polymer can be followed by surface plasmon spectroscopy in real time (*20*). A typical reflectivity curve is shown in Figure 7. Due to the increase in the layer thickness the resonance signal of the surface plasmons shifts to higher angles. If the measurements are carried out at a fixed angle slightly off from the resonance minimum (depicted by the arrow in Fig.7 (A)) the reflectivity increases with increasing layer thickness (Fig. 7 (B)).

As substrates glass slides coated with 50 nm gold and ca. 30 nm SiO_x to which a monolayer of the NHS-silane had been attached were employed. From the changes in the reflectivity it can be concluded that the chemisorption process is relatively rapid.

(A)

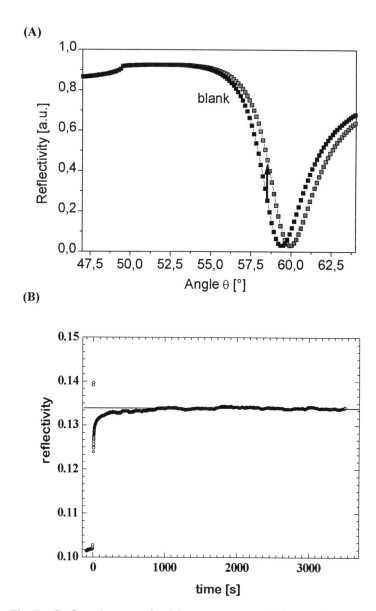

(B)

Fig. 7: Surface plasmon reflectivity measurement of the attachment of PEOX-LPEI monolayers to NHS-ester functionalized surfaces as a function of reaction time; (A) angular scan (B) reflectivity as a function of adsorption time, the angle at which the measurements have been carried is similar to the one marked with an arrow in (A)

It can be seen form Figure 5 (b), that already after a few minutes reaction time the layer thickness increases only very little and after about 30 minutes the final layer thickness of d_{Peox} is obtained. Such kinetic measurements are very useful in order to check whether the reaction time has been long enough to reach a stable situation and avoid a situation where only small differences in reaction time cause strong differences in layer thicknesses.

The absolute layer thickness of the chemisorbed polymer films, however, can only be obtained from such kinetic measurements if the assumption is made that the layer is very homogenous and no broadening of the SPS signal, e.g. by scattering or changes in the homogeneity, occur. As this might not always be the case, the final layer thicknesses were determined by angular reflectivity scans in air. It should be noted that all samples were extracted with a good solvent for the polymer (CH_2Cl_2) in a Soxhlet setup for several hours and dried carefully.

It was found that one important factor for the final thickness of the polymer layer was the concentration of the polymer solution during the chemisorption process (Fig. 8). This concentration dependence can be explained if a situation is envisaged, where one molecule is attached to a certain reactive site at the substrate. Now other reactive groups of the same polymer molecule could get into contact with the surface and attach to similar reactive surface sites. If these reactions are irreversible, the polymer is "frozen" into a non-equilibrium, pancake-like structure at the surface. In a very simplistic picture the situation of the polymer can be compared to that of a fly on a fly-paper. Once the fly sticks to the paper with each additional movement more parts of the body come into contact with the sticky surface, which restricts the number of further movements even stronger. In a similar fashion the polymer essentially "rolls out" on the surface and many different segments become attached to the surface. This, however, hinders the attachment of further molecules from solution as more and more surface sites become blocked. Consequently the layer thickness remains low even if high reaction times are employed. However, if a higher concentration of the chemisorption solution is employed, other polymer molecules efficiently compete for the reactive sites on the surface. Thus, the number of surface-bound segments per molecule decreases and the conformation of the polymer molecules becomes less flat. As the number of attached molecules increases the film thickness increases accordingly. This increase, however, is expected to saturate at a certain value as at that point the thickness of the layer is no longer limited by the number of available reactive sites at the surface, but diffusion of the polymer molecules through the attached chains becomes the limiting factor, as already discussed above. That the proposed model is plausible can be seen if polymer molecules with only one reactive site are attached to surfaces. In this case no dependence of the film thickness on the concentration of the chemisorption solution can be observed (*21*)

In a final reaction step (step (C) in fig. 6), the phospholipids (dimethylphosphatidyl-choline, DMPC) were attached to the polymer layer (Fig. 9). The lipids were preoriented as a Langmuir monolayer at the air/water interface and were subsequently transferred onto the solid substrate with the attached monolayer. To achieve some

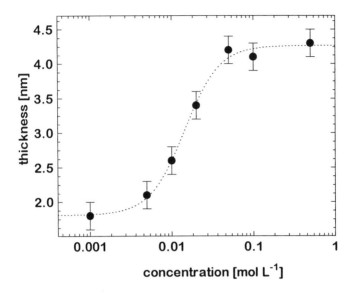

Fig. 8: Thicknesses of grafted layers of an PEOX-LPEI polymer attached to an NHS-ester functionalized monolayer as a function of the polymer concentration during deposition; degree of hydrolysis = 40 %

covalent bonding between the lipids and the polymer layer, about 20 mol% of an active ester of myristic acid (C14-NHS) were added to the lipid before spreading. The monolayers were immediately transferred after spreading to avoid excessive hydrolysis of the active ester. After transfer of the monolayer, some of the active ester groups of the lipids can react to the amine group-containing polymer. The thickness increase obtained was d_{Lipid} = 1.6 nm, which corresponds to a monomolecular layer.

In a second series of experiments, the amine group-containing polymers were reacted to silane monolayers containing iodoalkyl groups. As the alkylation of the amines during the surface attachment of the polymer is relatively slow (determined by NMR spectroscopy of model reactions in solution), the chemisorption reactions were carried out for at least 10 hours. Through this method, linear (obtained by complete hydrolysis of poly(2-ethyloxazoline)) and commercially available branched poly(ethylene imines) were attached to the surfaces.

It was found that very similar film thicknesses could be obtained compared to those of the NHS ester-attached polymers. However, for the performance of further reactions it has to be considered that the basicity of the polymers is higher and the polymer is almost completely protonated in aqueous solution, which will certainly have an influence on the behavior of the films in an aqueous environment. When the preparation of polymer layers from the two polymers (branched vs. linear polyethylenimine) is compared, it should be noted that the two polymers attach differently at low concentrations of the chemisorption solution and that different film thicknesses are obtained when the same reaction conditions are employed (Fig. 10).

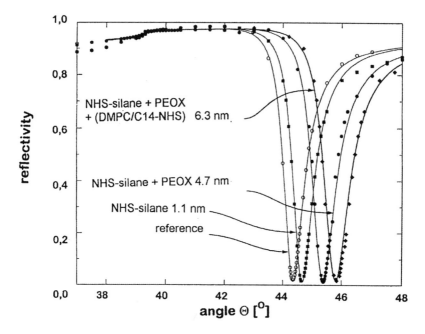

Fig. 9: Surface plasmon resonance measurements of surface attached monolayers with increasing complexity; symbols represent data points, full curves are Fresnel calculations

Swelling behavior of the attached monolayers

A critical factor for the performance of a polymer "cushion" is its swelling behavior in an aqueous/humid environment. The swellability determines to a large extend the mobility within the layer and how easily large molecules such as proteins can be incorporated into the layer. The swelling of the polymer layers in air having different relative humidities was also determined by surface plasmon optical measurements using a controlled environment setup (Fig. 11). All swelling experiments were carried out at a constant temperature (25 $^{\circ}$C) to avoid the condensation of moisture on the sample. The relative humidities within the measurement chamber were adjusted by adding a small reservoir with different concentrated salt solutions and allowing the equilibrium water vapor pressure above these solutions to develope.

As a starting point the samples were first dried in an atmosphere of 7% relative humidity. Compared to the same sample in ambient humidity, the reflectivity measured at a fixed angle of observation decreased slightly because the resonance angle for the plasmon excitation shifted to lower angles upon the loss of some water

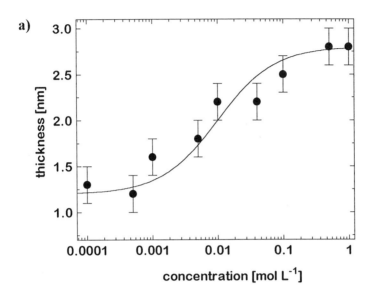

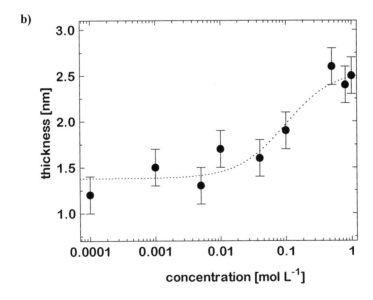

Fig. 10: Film thicknesses of monolayers of a) LPEI and b) BPEI as a function of the polymer concentration in the chemisorption process

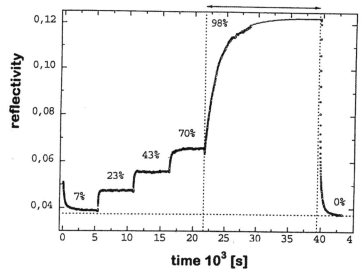

Fig. 11: Swelling behavior of a monolayer of partially hydrolyzed PEOX (degree of hydrolysis 40%) attached to a SiO_2-NHS monolayer as a function of the relative humidity at 25 °C. Onto this layer a monolayer of DMPC/C14-NHS had been transferred by the LB-technique.

from the multilayer system and an according thickness decrease. A stepwise increase of the relative humidity, on the other hand, resulted in a stepwise increase of the reflectivity caused by the incorporation of water into the polymer layer. At 98% relative humidity the observed thickness increase (measured by an angular scan) is equivalent to an effective water layer of d = 2.2 nm. The swelling of the polymer layer is fully reversible, as can be seen in Figure 11, where the water is removed again from the film through exposure to a dry atmosphere.

That the water is indeed incorporated into the polymer film and that not just droplets of water are condensed at the surface of the polymer layer, which would give a similar angular shift in the optical measurements, could be shown with the help of neutron reflectivity measurements using deuterated water. Preliminary calculations from the neutron measurements show that only profiles where the water is homogenously distributed within the film are in agreement with the measured reflectivity data.

Conclusions:

The data presented describe some examples for structures prepared with the aim of forming a polymer-tethered lipid monolayer. Through adjustment of the chemical composition of the polymer the properties of the layers can be tailored over a wide range. The results of our studies show that the conditions of the deposition of the polymer "cushion" play an important role on the structure of the polymer film and

118

only small differences in the conditions during the deposition of the polymer layers can cause significant changes in the film properties. The characteristics of the layers could be proven by surface analytical techniques and also some preliminary functional investigations have been made. In the following communications we will describe the swelling behavior of the attached layers and structural investigations of the attached tethered lipid layers in more detail. Although the chemical approach described here is restricted to silicon oxide surfaces, the same principles can be transferred to thiol based compounds, which can be used for the attachment of model membranes to gold surfaces.

Acknowledgments:

The authors would like to thank S. Lingler and A. Offenhäuser for many helpful discussions. Financial support by the Bundesministerium für Bildung, Wissenschaft Forschung und Technologie (BMBF) (Project 0310852) is gratefully acknowledged.

Literature Cited:

(1) E. Sackmann, *Science* **1996**, *271*, 43
(2) H. Bader, K. Dorn, B. Hupfer, H. Ringsdorf, *Adv. Mater.* **1985**, *64*, 1
(3) H. Ringsdorf, B. Schlarp, J. Venzmer, *Angew. Chem.* **1988**, *100*, 117
(4) A. Arya, U.J. Krull, M. Thompson, H.E. Wong, *Anal. Chim. Acta* **1985**, *173*, 331
(5) C. Dietrich, R. Tampe, *Biochim. Biophys. Acta* **1995**, *1238*, 183
(6) M. Uto, M. Araki, T. Taniguchi, S. Hoshi, S. Inoue, *Analyt. Sci.* **1994**, *10*, 943
(7) R. Naumann, A. Jonczyk, R. Kopp, J. Esch, H. Ringsdorf, W. Knoll, P. Gräbner, *Angew. Chem.* **1995**, *107*, 2168
(8) S. Heyse, H. Vogel, M. Sänger, H. Sigrist, *Protein Sci.* **1995**, *4*, 2532
(9) J.T. Groves, N. Ulman, S. Boxer, *Science* **1997**, *275*, 651
(10) B.A. Cornell et al., *Nature*, **1997**, *387*, 580
(11) H.M. McConnell, T.H. Watts, R.M. Weis, A.A. Brian, *Biochim. Biophys. Acta* **1986**, *864*, 95
(12) P. Müller, D.O. Rudin, H.T. Tien, W.C. Wescott, *Nature*, **1962**, *196*, 979
(13) M. Seul, H.M. McConnell, *J. Physique* **1986**, *47*, 1587
(14) C. Heibel, Ph.D. thesis, University of Mainz, 1996; C. Heibel, W. Knoll, J. Rühe (in preparation)
(15) S. Maus, Ph.D. thesis, University of Mainz, 1997
(16) D. W. Sindorf, G.E. Maciel, *J. Phys. Chem.* **1982**, *86*, 5208
(17) R. Zajac and A. Chakrabarti, *Phys. Rev. E*, **1995**, *52*, 6536
(18) G. W. Cline, S.B. Hanna, *J. Am. Chem. Soc.* **1987**, *109*, 308
(19) G. W. Cline, S.B. Hanna, *J. Org. Chem.* **1987**, *53*, 3583
(20) W. Knoll, *MRS Bull.*, **1991**, *16*, 29
(21) T. Lehmann, J. Rühe (submitted for publication)

Chapter 9

Methodologies and Models of Cross-Linking Polymerization in Supramolecular Assemblies

Thomas M. Sisson, Henry G. Lamparski, Silvia Kölchens,
Tina Peterson, Anissa Elayadi, and David F. O'Brien[1]

Carl S. Marvel Laboratories, Department of Chemistry, University of Arizona, Tucson, AZ 85721

Polymerization of monomeric lipids in an assembly proceeds in a linear·or crosslinking manner depending on the number of polymerizable groups per monomer. The gel point for polymerizations in two-dimensional lipid assemblies was determined by correlation of the onset of significant changes in the bilayer properties with bilayer composition. The experimental methods determined changes either in lipid lateral diffusion, polymer solubility, or assembly stability to a nonionic surfactant. Each method used indicated a substantial mole fraction (0.30 ± 0.05) of the bis-substituted lipid was necessary for crosslinking. Two models rationalizing the inefficient crosslinking observed in organized media were proposed and further tested. The experimental data are consistent with an intramolecular macrocyclization of the bis-lipid as a principal pathway responsible for the crosslinking inefficiency. The solubility methodology was employed to study the nature of polymerizations in inverted hexagonal and bicontinuous inverted cubic phases.

Polymerization of supramolecular assemblies is an effective means to modify their chemical and physical properties (1). The polymerization of monomeric lipids in an assembly proceeds in a linear or crosslinking manner depending on the number of polymerizable groups per monomeric lipid. Lipids that contain a single reactive moiety in either of the hydrophobic tails or associated with the hydrophilic head group yield linear polymers. Polymerization of lipids with reactive groups in each hydrophobic tail generally yields crosslinked polymers. Linear and crosslinked polymeric assemblies exhibit significant differences in physical properties, e.g. permeability (2-4), chemical stability (5,6), solubility (5,7), lateral diffusion of components (8), among others. The formation of linear polymer chains in bilayer vesicles causes a moderate reduction in permeability of water soluble solutes, e.g. glucose or sucrose, relative to that of unpolymerized bilayers. In contrast the formation of crosslinked poly(lipid) vesicles is reported to decrease the permeability by two orders of magnitude (2,3). The stability of vesicles to surfactant solubilization increases from unpolymerized vesicles, to linear poly(lipid) vesicles, and finally to crosslinked ilayer vesicles (5). The increased chemical stability can

[1]Corresponding author.

be attributed to the crosslinking of the lipids into a covalently linked monodomain polymeric vesicle. In this report we review three approaches to the characterization of the gel point for polymerizations constrained by the two-dimensional nature of lipid bilayers, and examine the relative effectiveness of unsymmetrical and symmetrical crosslinking agents on the efficiency of crosslinking in organized assemblies. The observations obtained with lamellar assemblies may also be applied to characterize crosslinking in nonlamellar phases, i.e. bicontinuous cubic (Q_{II}) and inverted hexagonal (H_{II}) phases.

Results and Discussion

Fluorescence photobleaching recovery (FPR) has been utilized to determine the lateral diffusion coefficient (D) of a fluorescent probe within the bilayer following the procedures previously developed for unpolymerized lipid assemblies (8). Lateral diffusion coefficients for NBD-PE in linear or crosslinked polymeric hydrated lipid bilayers were measured, where the mole fraction of NBD-PE was 10^{-3} (9,10). The D for the probe decreased moderately as the degree of polymerization of linear poly(lipid) increased. The effect of the bilayer molar ratio of mono-AcrylPC and bis-AcrylPC (Scheme 1) was determined. The relative reactivities of acryloyl group mono-AcrylPC and both acryloyl groups in bis-AcrylPC are expected to be similar since they are chemically equivalent. The positional differences between the sn-1 and sn-2 chains of the bis-AcrylPC should not significantly effect the relative reactivity of the two groups. At low mole fractions of the crosslinker lipid, bis-AcrylPC, in the bilayer the D was relatively unaffected. However at mole fractions of bis-AcrylPC greater than 0.3 the diffusion coefficient for NBD-PE was significantly reduced. The FPR measurements show a significant decrease in the dynamic motion of the NBD-PE probe at a bilayer composition of ca. 2:1 mono- and bis-AcrylPC. This inhibition of small molecule mobility within the bilayer signals a change in the polymerized bilayer structure, which was ascribed to gelation (crosslinking) (8,11).

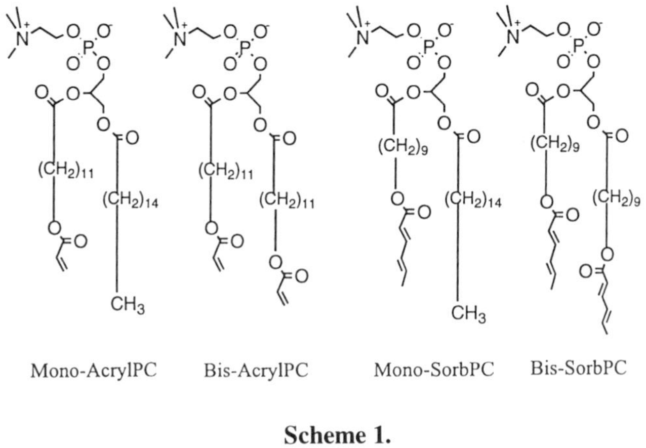

Mono-AcrylPC   Bis-AcrylPC   Mono-SorbPC   Bis-SorbPC

Scheme 1.

Poly(lipid) solubility was measured by attempting to dissolve the dehydrated lipid polymer, removing any insoluble material and weighing the recovered soluble polymer after removal of the solvent. The drastic change in solubility between sample compositions that are below and above the gel point has been usefully employed in isotropic polymerizations as an indication of crosslinking (*12*). The bilayer composition was controlled by varying the molar ratio of mono-substituted and bis-substituted lipids and examining the change in solubility associated with the onset of crosslinking. Because gelation is dependent on the percent conversion to polymer, these experiments were performed at high conversion (>85%) (*13*). A preliminary examination of the effectiveness of various solvents for polymerized zwitterionic PC lipids indicated that 1,1,1,3,3,3-hexafluoro-2-propanol (HFIP) was the most effective solvent for these polymers. HFIP is an excellent hydrogen-bond donor, therefore it can interact strongly with the carbonyl functionalities of the polymeric lipids (*14-16*). Furthermore its low self-association makes it a good solvent for zwitterionic polymers (*17*). The solubility of linearly polymerized phospholipids in HFIP is great enough to permit the effective use of poly(lipid) solubility to estimate the mole fraction of the bis-substituted lipid required for effective crosslinking of the bilayer. The weight percent of the lipid sample that was soluble in HFIP is shown in Figure 1 as a function of mole fraction of bis-AcrylPC in the bilayer at the time of the polymerization.

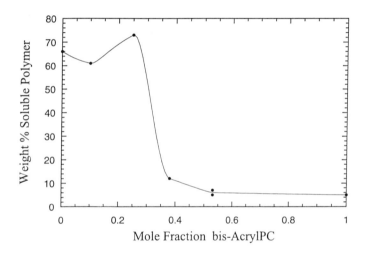

Figure 1. Weight percent solubility of polymerized mono-AcrylPC/bis-AcrylPC bilayers as a function of the mole percent of bis-AcrylPC. Initiation by decomposition of AIBN at 60±2 °C; [M]/[I] = 5.

The polymerized lipid solubility is drastically different if the mole fraction of bis-AcrylPC is 0.25 (soluble) as opposed to 0.38 (insoluble). This physical change suggests the formation of crosslinked polymers when the mono-AcrylPC to bis-AcrylPC ratio was ca. 2:1, whereas at a molar ratio of ca. 3:1 or higher the polymerized lipids do not appear to be crosslinked. The data in Figure 1 also show that even when the bilayer composition has insufficient bis-AcrylPC to effectively crosslink the bilayer, the resulting polymer chains were not completely solubilized by the HFIP. The poly(AcrylPC) chains are relatively large and the reported polydispersity indicates that some of the polymer chains have molecular weights approaching a million (*18,19*). High molecular weight chains, such as the larger poly(AcrylPC) chains, can become entangled thereby preventing complete solubilization even in HFIP. Data for poly(SorbPC) bilayer compositions also indicate a change in solubility at ca. 2:1 ratio of mono-SorbPC to bis-SorbPC (*11*). Radical polymerization of mono-SorbPC yields poly(lipid) with molecular weights one quarter that of mono-AcrylPC. Consequently the solubility of noncrosslinked poly(SorbPC) in HFIP was greater than that of poly(AcrylPC) (*18-20*).

Ease of solubilization can be usefully employed to characterize the chemical stability of lipid bilayer vesicles in aqueous suspension. Although organic solvents are not especially effective for such studies, surfactants are ideal as long as the polymer chains are short enough to be incorporated into a mixed micelle. It has long been known that surfactant micelles are effective at solubilizing lipids and other biomembrane components. The kinetics of solubilization and interaction of nonionic surfactants with phospholipid vesicles has been reviewed (*21-23*). The effective solubilization of lipids into micelles requires that the surfactant concentration be greater than its critical micelle concentration (CMC) and that the surfactant to lipid molar ratio be ca. 3 to 5.

Regen and coworkers reported that stability of vesicles to surfactant treatment increases in the following order: unpolymerized < linearly polymerized < crosslinked (*5*). Surfactant solubilization (lysis) of bilayers initially involves the incorporation of surfactant into the bilayer. When the surfactant concentration predominates, the lipids no longer exist in a vesicle but are incorporated into a mixed micelle of lipid and surfactant. If the lipids are crosslinked in a polymerized vesicle or are incorporated into quite large linear polymers they cannot be readily solubilized by the surfactant.

The chemical stabilities of polymerized vesicles composed of both mono- and bis-SorbPC were examined by surfactant lysis monitored by QELS (Figure 2).

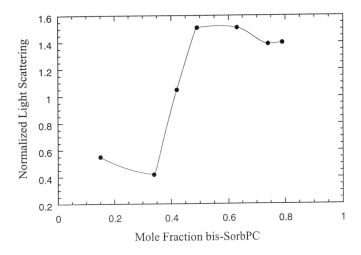

Figure 2. Average mean diameter of polymerized mono-Sorb/bis-SorbPC LUV after addition of 15 equivalents of Triton X-100 as a function of mole fraction bis-SorbPC. Polymerization by direct photoirradiation.

The average particle size after the addition of 15 equivalents of Triton X-100 (TX-100) to the polymerized bilayer samplevaries as a function of the mole fraction of bis-SorbPC in the bilayer prior to polymerization. A change in vesicle stability to the addition of TX-100 occurs for vesicle compositions with greater than a mole fraction of 0.34 bis-SorbPC. The resistance to solubilization of the polymerized vesicles composed of high mole fractions of bis-SorbPC is attributed to effective 2D-gelation of the vesicles during the polymerization, because crosslinking of the lipids prevents the extraction of lipid molecules from the vesicle and into mixed micelles.

Investigation of Crosslinking Inefficiency. These studies indicate that effective bilayer crosslinking requires a high (0.3 ± 0.05) mole fraction of the appropriate bis-PC. This contrasts with the quite low mole fractions of bis-monomers that are usually necessary for crosslinking polymerizations in isotropic media (*13,24*). Inefficiencies in crosslinking reactions could result from low conversions to polymer, significant differences in chemical reactivity, and/or small degrees of polymerization. The first two possibilities appear unlikely because the experimental data reported here were all obtained at high conversions of monomers to polymer, and as noted earlier the relative reactivities of the groups in the mono- and bis-lipids should be quite similar. Moreover differences in the degree of polymerization of AcrylPC monomers ($X_n=233\pm29$) and SorbPC monomers ($X_n=7\pm3$) only resulted in modest differences in the observed crosslinking efficiency (*19,20*). Thus other potential explanations for the crosslinking inefficiency were examined. A reasonable possibility is intramolecular macrocyclization of the *sn*-1 and *sn*-2 groups in the same lipid (Scheme 2). Such a macrocyclization would yield a ring within a linear polymeric backbone. Although the formation of such a large macrocycle is statistically unfavored in an isotropic media, the two-dimensional constrained environment of the bilayer reduces the number of conformations available to the lipid tails and backbone , thereby enhancing the possibility of intramolecular

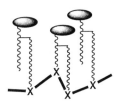 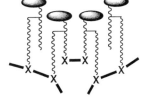

Intramolecular Macrocyclization   AB type crosslinking

Scheme 2.

macrocyclization. The drawing in Scheme 2 implies that the two reactive groups in a bis-PC are positionally inequivalent in a relatively static bilayer (*25*). This suggests an AB type crosslinking with *s n*-1 tail of bis-PC and mono-PC preferentially reacting with each other in a topological fashion rather than with the *sn*-2 tail of bis-PC. In this case a crosslink between polymer chains occurs when a reactive group on the *sn*-2 tail of a bis-PC lipid in one polymer chain reacts with a similar group in a neighboring polymer chain. This mode of crosslinking appears to require reactive *sn*-2 tails to form dimeric links between poly(lipid) chains. A competitive process is the reaction of the groups in *sn*-2 tails with others in the same polymer chain. Such an intrachain reaction modulates the polymer motion without linking polymer chains together and therefore requires a larger mole fraction of the bis-PC to cause gelation. A dynamic bilayer such as that found at temperatures above the main phase transition temperature, T_m, provides a more favorable environment for the reaction of a reactive group in *sn*-1 tails with each other as well as with reactive groups in *sn*-2 tails. Such a process leads to an irregular polymer backbone with an occasional bis-PC bridge between two polymer chains via a *sn*-1 group in one polymer chain and a *sn*-2 group in the second chain. Recent molecular simulations of bilayers in the L_α phase ($>T_m$) reveal a dynamic mixture of lipid tail conformations for the portion of the lipid tail distal to the water interface (26,27).

Further examination of the crosslinking efficiency utilized a symmetrical crosslinking agent, SorbNXL8 (4). The synthesis of SorbNXL8 was accomplished using the efficient reactions described by Figure 3 (28). The synthesis of the starting acid 9-(sorbyloxy)nonanoic acid (1) was accomplished by the methods previously reported (29,30). The stability of poly(lipid) vesicles obtained from polymerization of various bilayer compositions of mono-SorbPC and SorbNXL8 were examined by surfactant lysis. The relative reactivity of mono-SorbPC and SorbNXL8 should be similar since the reactive group is identical. The mono-SorbPC appears to have an equal opportunity to react with either chain of SorbNXL8, since the symmetrical bifunctional monomer places both reactive sorbyl

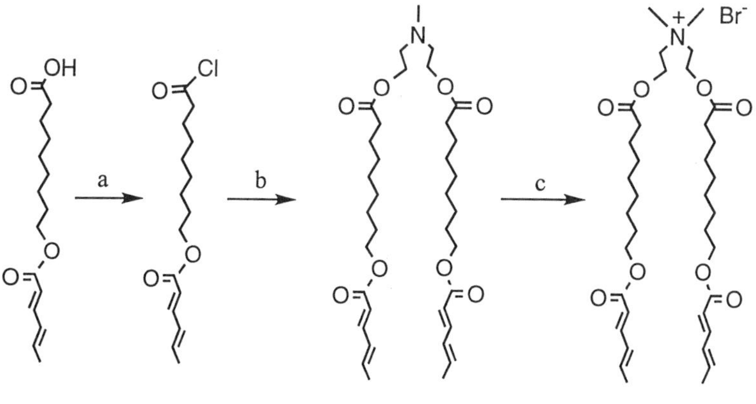

Figure 3. Synthesis of symmetrical crosslinking agent SorbNXL8: a) oxalyl chloride, benzene; b) methyl diethanolamine, DMAP, THF; c) CH₃Br, acetone.

moieties at the same depth within a bilayer (31). Reaction of one SorbNXL8 acyl chain may reduce the reactivity of the second group (24). Note that the high overall conversion insures the reaction of both sorbyl groups on the symmetrical crosslinker. The change in average mean diameter of the vesicle particles after the addition of 8 equivalents of TX-100 to the polymerized bilayer sample is shown in Figure 4 as a function of the mole fraction of SorbNXL8 in the bilayer prior to polymerization. A change in vesicle stability to TX-100 occurs for vesicle samples having between 0.35 and 0.53 mole fraction SorbNXL8. This behavior is similar to that observed for polymerizations of bilayer vesicles composed of mono-SorbPC and bis-SorbPC (Figure 2). If bis-SorbPC was predominantly an AB type of crosslinker, then the crosslinking efficiency of SorbNXL8 and bis-SorbPC should be dissimilar. Since that was not observed these data suggest that intramolecular macrocyclization is an important reaction path in bilayers, that may contribute to the observed inefficiency of crosslinking in bilayer polymerization with bis-substituted lipids.

Crosslinking Polymerizations in Nonlamellar Phases. A polymerizable bicontinuous cubic phase with *Pn3m* symmetry is formed by a 3:1 mixture of the PE:PC lipids shown in Scheme 3 (32). The structure before and after polymerization was confirmed by ^{31}P NMR, X-ray diffraction and transmission electron microscopy (TEM). The bis-PC shown not only acts to facilitate formation of the cubic phase but serves as a crosslinking agent in the polymerization because it

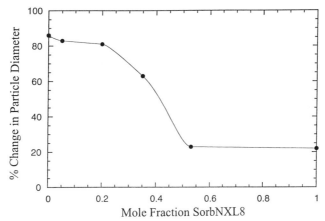

Figure 4. Average mean diameter of polymerized mono-Sorb/SorbNXL8 LUV after addition of 8 equivalents of Triton X-100 as a function of mole fraction SorbNXL8. Polymerization by direct photoirradiation.

is a bifunctional monomer. Since the mole fraction of the bis-PC (0.25) may be sufficient to crosslink this assembly, we examined the solubility of the polymerized Q_{II} phase. Note that previous studies in the lamellar phase for SorbPC lipids revealed the onset of gelation at a lower mole fraction of bis-SorbPC with radical polymerizations compared to UV irradiation (*11*). After polymerization the sample was dehydrated and found to be insoluble in HFIP, thereby indicating the assembly was crosslinked. The assembly retained its architecture after dehydration and in the presence of organic solvent as observed by TEM. Whether the effective crosslinking observed at a 0.25 mole fraction of bis-PC was favored by the placement of the reactive groups near the lipid glycerol backbone and/or due to the nature of the nonlamellar phase still remains to be examined.

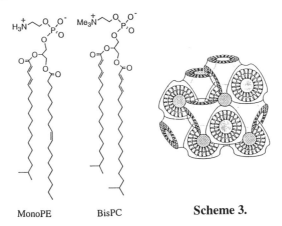

MonoPE   BisPC   **Scheme 3.**

The solubility methodology was also employed to study the nature of polymerizations in the H_{II} phase for the lipid shown in Scheme 4. The bis-PE lipid formed a H_{II} phase before and after polymerization at temperatures from 5 to 80 °C as determined by variable temperature ^{31}P NMR and X-ray diffraction (*33*). This polymerization in the H_{II} phase yielded an assembly which was completely insoluble in all solvents including HFIP. The mole fraction of bis-lipid was unity and in this case a crosslinked polymeric assembly was easily achieved. The insolubility and retention of architecture in the presence of organic media greatly expands the possible utility of these novel nanoporous materials.

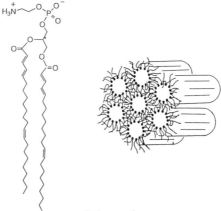

Scheme 4.

General Discussion

The solubility and surfactant lysis data for the AcrylPC bilayers agrees well with the lateral diffusion measurements obtained by FRP, thereby providing a high degree of confidence in both procedures since the errors in each are dissimilar. Both methodologies are insensitive to the molecular weights of the polymeric lipids. The HFIP solubility protocol is most effective with poly(lipid) chains that have a degree of polymerization less than ca. 10^3. As long as this possible limitation is given proper consideration, the HFIP procedure appears to have broad applicability for the determination of gelation of polymerized assemblies, i.e. bilayers, LB multilayers, etc. The solubility of cubic and hexagonal phases has been used to provide evidence for polymerization yielding crosslinked nonlamellar assemblies. The solubility procedure does not require the sophisticated instrumentation necessary for FRP studies, and the amount of lipid required is only dependent on the sensitivity of an available balance. The effect of polymer size limits the applicability of surfactant lysis, but not necessarily the type of assembly since the addition of TX-100 could be expanded into nonlamellar phases with an appropriate change in monitoring of dissolution. For example the dissolution of H_{II} or lamellar phases formed by phospholipids could be monitored by ^{31}P NMR (34).

Experimental

Solvents and Reagents. The lipids were purified using silicic acid specifically treated for lipid chromatography (Biosil-A 200-400 mesh; Biorad). Lipid purity was evaluated by thin-layer chromatography (TLC) with chloroform/methanol/water (65:25:4 by volume) and visualized by a UV lamp. Ion exchange resin AG 501-X8 was obtained from Bio-Rad (Richmond, CA). Flash chromatography silica gel (200-425 mesh) was obtained from Fischer Scientific. Azobis(isobutyronitrile), AIBN, was purified by recrystallization three times from methanol. 4-(N,N-Dimethylamino)pyridine (DMAP) was obtained from Aldrich Chemical Co. (Milwaukee, WI) and was purified by recrystallization from chloroform/diethyl ether (1:1, v/v) three times. Potassium persulfate, sodium bisulfite, Triton X-100 (TX-

100), and 1,1,1,3,3,3-hexafluoro-2-propanol (HFIP) were purchased from Aldrich Chemicals and used as received. The lipids were hydrated in Milli-Q water, Millipore Inc.. Benzene and tetrahydrofuran (THF) was distilled from sodium benzophenone ketyl. Chloroform was distilled from calcium hydride. N,N-Dimethylformamide (DMF) was vacuum distilled and stored over molecular sieves (5A). All other reagents were used as received or purified by standard methods.

Methods. Compounds containing UV-sensitive groups were handled under yellow lights. ^{1}H NMR spectra were acquired on a Varian-Unity 300 spectrometer in chloroform-d with tetramethylsilane as an internal reference. UV-Vis absorption spectra were recorded on a Varian DMS 200 spectrophotometer. Quasi-elastic light scattering (QELS) was performed with a BI8000-autocorrelator from Brookhaven Instruments Corp. and particle size was calculated with the accompanying software. The polymerizable lipids, i.e. mono-SorbPC, bis-SorbPC, mono-AcrylPC, and bis-AcrylPC, were synthesized as described previously (*19,20*).

9-(Sorbyloxy)nonanoic Acid (1).

9-(Sorbyloxy)nonan-1-ol (29,30) (4.2 g, 16.4 mmol) in DMF (20 mL) was added dropwise to pyridinium dichromate (PDC) (21.5 g, 57.3 mmol) in DMF (30 mL) at 0 °C. The reaction was allowed to warm slowly to room temperature and stirred under a positive argon atmosphere for 1 day. The mixture was filtered through silica gel and this process repeated until no trace of PDC was observed. The solution was then concentrated and the crude aldehyde purified by column chromatography using Hex/EtOAc/HCOOH (60/40/1), affording 8-(sorbyloxy)octanoic acid. Yield 2.8 g (64%); TLC: R_f = 0.51 (Hex/EtOAc/HCOOH, 60/40/1); (1)H-NMR (CDCl₃) d 1.20-1.40 ppm (br s, 8H, CH₂), 1.52-1.70 ppm (m, 4H, CH₂CH₂R), 1.78-1.86 ppm (d, 3H, C=CHCH₃), 2.28-2.37 ppm (t, 2H, CH₂CO₂H), 4.05-4.15 ppm (t, 2H, CO₂CH₂), 5.70-5.79 ppm (d, 1H, C=CHCO₂), 6.05-6.22 ppm (m, 2H, =CH-CH=), 7.15-7.27 ppm (m, 1H, CH=CHCH₃).

9-(Sorbyloxy)nonanoyl Chloride (2).

9-(Sorbyloxy)nonanoic acid (0.66 g, 2.6 mmol) in dry benzene (10 ml) was reacted with oxalyl chloride (0.47 g, 2.9 mmol) by dropwise addition. The solution was stirred at room temperature for 30 min beyond gas evolution and then excess oxalyl chloride and solvent were removed by rotary evaporation. Recrystallization from benzene (2x) afforded the acid chloride as colorless needles that were used directly in the esterification. Yield 0.70 g (100%).

Bis-9-(sorbyloxy)nonanoyl Methyl Diethanolamine (3).

9-(Sorbyloxy) nonanoyl chloride (0.70, 2.6 mmol) in THF (10 ml) was added dropwise to methyl diethanolamine (0.15g, 1.2mmol) and 4-dimethylaminopyridine (0.46 g, 3.7 mmol) in THF (20 ml). The reaction was stirred overnight under argon. 4-Dimethylaminopyridinium hydrochloride was filtered off and the filtrate concentrated by rotary evaporation resulting in a viscous oil which was further purified by flash silica gel chromatography using Hex/EtOAc/TEA, 100/100/1. Yield 0.44 g, (59%): TLC R_f = 0.25 (Hex/EtOAc/TEA, 100/100/1); ^{1}H-NMR (CDCl₃) d 1.20-1.40 ppm (br s, 16H, CH₂), 1.52-1.70 ppm (m, 8H, CH₂CH₂R), 1.82-1.88 ppm (d, 6H, C=CHCH₃), 2.12-2.21 ppm (t, 4H, CH₂CO₂R), 2.22 ppm (s, 3H, NCH₃), 4.08-4.18 ppm (m, 8H, NCH₂CH₂, CO₂CH₂), 4.59-4.63 ppm (m, 4H, NCH₂), 5.70-5.79 ppm (d, 2H, C=CHCO₂), 6.05-6.22 ppm (m, 4H, =CHCH=), 7.15-7.27 ppm (m, 2H, CH=CHCH₃).

Bis-9-(sorbyloxy)nonanoyl Dimethyl Diethanolamine (4). Bromomethane in diethylether (0.75 ml, 1.5 mmol) was added to **3** (0.44 g, 0.70 mmol) in dry acetone (10 ml) at 0 °C. The reaction was stirred overnight and a precipitate formed. The solvent and excess bromomethane were removed by rotary evaporation resulting in a white solid. Yield 0.43 g (96%); ^{1}H-NMR (CDCl$_3$) d 1.20-1.40 ppm (br s, 16H, C$\underline{H}_2$), 1.52-1.70 ppm (m, 8H, C$\underline{H}_2$CH$_2$R), 1.82-1.88 ppm (d, 6H, C=CHC$\underline{H}_3$), 2.12-2.21 ppm (t, 4H, C$\underline{H}_2$CO$_2$R), 3.58 ppm (s, 6H, NC$\underline{H}_3$), 4.08-4.15 ppm (t, 4H, NCH$_2$C$\underline{H}_2$), 4.14-4.18 ppm (m, 4H, CO$_2$C$\underline{H}_2$), 4.59-4.63 ppm (m, 4H, NC$\underline{H}_2$), 5.70-5.79 ppm (d, 2H, C=C$\underline{H}$CO$_2$), 6.05-6.22 ppm (m, 4H, =C$\underline{H}$CH=), 7.15-7.27 ppm (m, 2H, CH=C$\underline{H}$CH$_3$). UV (λ_{max} = 262 nm CH$_3$OH, ε = 44,715 L/cm mol)

Thermal AIBN Polymerizations. Large multilamellar vesicles of polymerizable lipid were prepared as follows: approximately 2.5 to 15 mg of lipid was evaporated from stock solutions (10 mg/ml in benzene) by passing a stream of argon over the sample and drying under high vacuum (0.4 mm Hg) for a minimum of 4 h. Each lipid was dried and weighed separately for mixtures of mono- and bis-PC. The lipid weight was determined and each lipid film redissolved in 2 ml of benzene, then the lipids were combined. The appropriate amount of initiator from a freshly prepared AIBN stock solution (1-1.5 mg/ml in benzene) was added to yield a monomer to initiator ratio of 5, where [M] represents the concentration of polymerizable groups. The solvent was evaporated as above and dried under high vacuum for 2 h. All steps involving lipid and AIBN were performed in subdued light. The dried lipid/AIBN film was hydrated with deoxygenated MilliQ water to a final concentration of 7 mM. Samples were vortexed to uniformity and subjected to ten freeze-thaw-vortex cycles (-77°C→T$_m$+5 °C) cycles. Samples were placed in an ampoule sealed with a septum and flushed with argon for 0.5 h. Polymerizations were performed at 60±2 °C in a water circulating bath under a positive argon pressure. All polymerizations were carried out in the absence of light. Polymerizations were monitored by UV absorption spectroscopy of aliquots diluted with Milli-Q water to ca. 80 μM.

Redox Polymerizations. Unilamellar vesicles (LUV) of SorbPC were prepared as above in the absence of AIBN. The lipid suspension was extruded ten times through two stacked Nuclepore polycarbonate filters (0.1 mm diameter) at 37 °C using a stainless steel extruder from Lipex Biomembranes. The redox initiator was prepared from K$_2$S$_2$O$_8$ (300 mg, 1.1 mmol) and NaHSO$_3$ (115 mg, 1.1 mmol) weighed into a 10 ml volumetric flask and diluted. An aliquot (200-300 ml) was pipetted into the vesicle suspension. The sample was placed into an ampoule and sealed with a septum and flushed with argon for 0.5 h. Polymerizations were performed at 60±2 °C in a water circulating bath under a positive argon pressure for 18 h. All polymerizations were carried out in the absence of light. Polymerizations were monitored by UV absorption spectroscopy of aliquots diluted with Milli-Q water to ca. 80 μM.

Direct Photoirradiation. LUV were placed into a 3 ml quartz cuvette equipped with a magnetic stirbar and placed 1 cm from a low pressure Hg vapor Pen Lamp. Samples with a lipid concentration of 100-300 μM were irradiated for 45 min at a temperature of 40 °C. Polymerizations were monitored by UV absorption spectroscopy as above.

Weight Percent Solubility. Only samples with greater than 85% monomer conversion as determined by UV-Vis spectroscopy were used in the solubilty studies. Samples were lyophilized after polymerization. The lipid was weighed and HFIP added until the concentration was 5 mg/ml. The samples were shaken for 2 min and allowed to stand for 5 min. The solution was filtered into a dry test tube through a pipet with glass wool to remove any solid particles. The pipet was then rinsed with 0.5 ml of HFIP. The solvent was removed by passing a stream of argon over the sample then drying under high vacuum for a minimum of 4 h leaving the soluble polymer, which was weighed and used to calculate the percent solubilty.

Surfactant Dissolution of Vesicles. LUV were prepared as described above. After polymerization the LUV were characterized by QELS for 2 ml sample with a lipid concentration of 100-300 µM. An aliquot of 45 mM TX-100 solution was added. Each aliquot was equal to 2 equivalents of lipid. The light scattering intensities were determined again by QELS. TX-100 was added in 2 equivalent increments up to 15 total equivalents. Measurements at each concentration of TX-100 were performed a minimum of three times. The average mean diameter of vesicles/particles was calculated by multiple mathematical procedures, i.e. CUMULANT, exponential sampling, non-negatively constrained least square, and CONTIN. With the exception of exponential sampling, consistent results were obtained.

Acknowledgment. This research was supported by the National Science Foundation.

Literature Cited

(1)   O'Brien, D. F., *Trends Polym. Sci.* **1994**, *2*, 183-188.
(2)   Dorn, K.; Klingbiel, R. T.; Specht, D. P.; Tyminski, P. N.; Ringsdorf, H.; O'Brien, D. F., *J. Am. Chem. Soc.* **1984**, *106*, 1627-1633.
(3)   Stefely, J.; Markowitz, M. A.; Regen, S. L., *J. Am. Chem. Soc.* **1988**, *110*, 7463-7469.
(4)   Ohno, H.; Takeoka, S.; Hayashi, N.; Tsuchida, E., *Makromol. Chem., Rapid Comm.* **1987**, *8*, 215-218.
(5)   Regen, S. L.; Singh, A.; Oehme, G.; Singh, M., *J. Amer. Chem. Soc.* **1982**, *104*, 791-795.
(6)   Tsuchida, E.; Hasegawa, E.; Kimura, N.; Hatashita, M.; Makino, C., *Macromolecules* **1992**, *25*, 207-212.
(7)   Ohno, H.; Takeoka, S.; Iwai, H.; Tsuchida, E., *Macromolecules* **1988**, *27*, 319-322.
(8)   Kölchens, S.; Lamparski, H.; O'Brien, D. F., *Macromolecules* **1993**, *26*, 398-400.
(9)   Axelrod, D.; Koppel, D. E.; Schlessinger, J.; Elson, E. L.; Webb, W. W., *Biophys. J.* **1976**, *16*, 1055-1069.
(10)   Koppel, D. E.; Axelrod, D.; Schlessinger, J.; Elson, E. L.; Webb, W. W., *Biophys. J.* **1976**, *16*, 1315-1329.
(11)   Sisson, T. M.; Lamparski, H. G.; Kölchens, S.; Elayadi, A.; O'Brien, D. F., *Macromolecules* **1996**, *29*, 8321-8329.
(12)   Ulbrich, K.; Ilavsky, M.; Dusek, K.; Kopecek, J., *Europ. Polym. J.* **1977**, *13*, 579-585.
(13)   Storey, B., *J. Polym. Sci.* **1965**, *A3*, 265-282.
(14)   Hamori, E.; Prusinowski, L. R.; Sparks, P. G.; Hugues, R. E., *J. Phys. Chem.* **1965**, *69*, 1101-1105.
(15)   Montroy Soto, V. M.; Galin, J. C., *Polymer* **1984**, *25*, 254-262.
(16)   Pujol-Fortin, M.; Galin, J.-C., *Polymer* **1994**, *34*, 1462-1472.

(17) Ueda, T.; Oshida, H.; Kurita, K.; Ishihara, K.; Nakabayashi, N., *Polymer J.* **1992**, *24*, 1259-1269.

(18) Sells, T. D.; O'Brien, D. F., *Macromolecules* **1991**, *24*, 336-337.

(19) Sells, T. D.; O'Brien, D. F., *Macromolecules* **1994**, *27*, 226-233.

(20) Lamparski, H.; O'Brien, D. F., *Macromolecules* **1995**, *28*, 1786-1794.

(21) Lichtenberg, D.; Rodson, R. J.; Dennis, E. A., *Biochim. Biophys. Acta* **1983**, *737*, 285-304.

(22) Phillippot, J.; Mutaftschiev, S.; Liautard, J. P., *Biochim. Biophys. Acta* **1983**, *734*, 137-143.

(23) Mimms, L. T.; Zampighi, G.; Nozaki, Y.; Tanford, C.; Reynolds, J. A., *Biochemistry* **1981**, *20*, 833-840.

(24) Landin, D. T.; Macosko, C. W., *Macromolecules* **1988**, *21*, 846-852.

(25) Pearson, R. H.; Pascher, I., *Nature (London)* **1979**, *281*, 499-501.

(26) Heller, H.; Schaefer, M.; Schulten, K., *J. Phys. Chem.* **1993**, 8343-8360.

(27) Berendsen, H. J. C., *personal communication*.

(28) Ringsdorf, H.; Schupp, H., *J. Macromol. Sci. Chem.* **1981**, *A15*, 1015-1026.

(29) Lamparski, H.; Liman, U.; Frankel, D. A.; Barry, J. A.; Ramaswami, V.; Brown, M. F.; O'Brien, D. F., *Biochemistry* **1992**, *31*, 685-694.

(30) Lamparski, H.; Lee, Y.-S.; Sells, T. D.; O'Brien, D. F., *J. Am. Chem. Soc.* **1993**, *115*, 8096-8102.

(31) Tyminski, P. N.; Ponticello, I. S.; O'Brien, D. F., *J. Am. Chem. Soc.* **1987**, *109*, 6541-6542.

(32) Lee, Y.-L.; Yang, J.; Sisson, T. M.; Frankel, D. A.; Gleeson, J. T.; Aksay, E.; Keller, S. L.; Gruner, S. M.; O'Brien, D. F., *J. Am. Chem. Soc.* **1995**, *117*, 5573-5579.

(33) Srisiri, W.; Sisson, T. M.; O'Brien, D. F., *J. Am. Chem Soc.* **1997**, *119*, 4866-4873.

(34) Anikin, A.; Chupin, V.; Anikin, M.; Serebrennikova, G.; Tarahovsky, J., *Makromol. Chem.* **1993**, *194*, 2663-2673.

Chapter 10

Factors Influencing the Layer Thickness of Poly-L-glutamates Grafted from Self-Assembled Monolayers

H. Menzel[1], A. Heise[1], Hyun Yim[2], M. D. Foster[2], R. H. Wieringa[3], and A. J. Schouten[3]

[1]Institut für Makromolekulare Chemie, Universität Hannover, Am Kleinen Felde 30, 30167 Hannover, Germany
[2]Institute of Polymer Science, University of Akron, Akron, Ohio 44325-3909
[3]Department of Polymer Chemistry, University of Groningen, Nijenborgh 4, 9747 AG Groningen, Netherlands

Factors influencing the thickness of polypeptide layers grafted from self-assembled monolayers were investigated by varying the initiator site density and the reactivity of the N-carboxyanhydride monomer. To vary the density of initiating sites and to match the steric requirements of the growing polymer chain, mixed self-assembled monolayers were prepared with terminal amino groups. These SAMs were prepared by coadsorption of bromine- and methyl-terminated silanes and a subsequent *in-situ* modification. Polymerization experiments with these SAMs reveal an influence of initiator site density on the polymer layer thickness. SAMs prepared with 60% functionalized silane showed an optimum thickness. The variation in layer thickness due to the variation of initiator site density is smaller than the influence of the monomer reactivity.

α-helical polypeptides like poly-L-glutamates are rigid rod-like polymers and have several interesting properties. Most notably, they have a large dipole moment along the helix axis (*1*). The antiparallel orientation of the rods is the energetically favored arrangement and the dipole moments of the individual polymer rods are compensated. Approaches to an unidirectional orientation, which would result in a net dipole moment of the polymer film, are casting the polymer in a strong magnetic field (*2*) or tethering the polymer backbone to a surface and achieving a high density of polymer chains (*3 - 9*). The tethering of the polymers to a surface has been performed by grafting preformed polymers onto the surface (*3 - 5, 10*) as well as by grafting the polymer from the surface (*6 - 8, 10*). The latter method has the advantage that a higher grafting density can be achieved because only small monomeric molecules have to diffuse to the surface and there is no blocking of binding sites by already tethered material. On the other hand, the surface the polymer is grafted from has to have appropriate initiating sites, which have to be introduced first. Furthermore, the polymer grafted from a surface is hard to characterize, and there is no chance to fractionate the polymer or to separate oligomers with a different secondary structure as they may occur in the polypeptide synthesis (*11*). Therefore, special attention has to be paid to the polymerization conditions in order to yield homogeneous films.

Mechanism of the NCA-Polymerization

In order to understand the requirements and limitations of a grafting from polymerization it is worthwhile to have a closer look at the mechanism. Poly-L-glutamates can be polymerized from the corresponding N-carboxyanhydride **1**. If the polymerization is initiated by primary amines or secondary amines without any steric hindrance, the polymerization proceeds via the "amine mechanism" (Figure 1) (*11 - 13*).

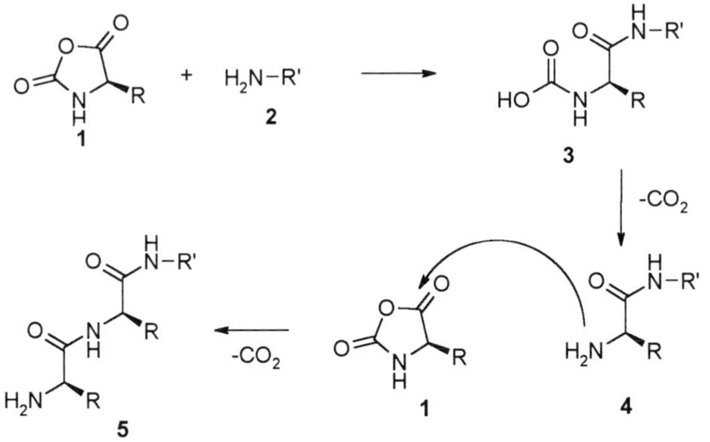

Figure 1: Initiation and propagation in the NCA-polymerization via the amine-mechanism (According to ref. *12*)

The amine attacks the NCA-ring and ring opening occurs. The intermediate **3** eliminates carbon dioxide and a dimer with an amino endgroup is generated, which can attack the next monomer. The new amino group is less reactive than a primary amine. Therefore, the initiation is much faster than the propagation and the degree of polymerization can be adjusted to some extent by varying the monomer-to-initiator ratio (*11 - 13*). In this respect, the NCA polymerization has a kind of "living character". (In a living polymerization all chains are started simultaneously because the initiation is much faster than the propagation and there are no terminating side reactions. The latter condition is not fulfilled for the NCA-polymerization.) The amine polymers can be synthesized (*14, 15*) or tethered to a surface if the amino group is attached.

A further mechanism which is discussed in literature, is the "carbamate mechanism" (Figure 2). The intermediate carbamic acid **3** is deprotonated by the initiating amine or the terminal amino group of the growing chain and further reaction takes place via the deprotonated carbamic acid **3'**. The nucleophilic attack results in the formation of an intermediate anhydride **6**, which decarboxylates. A new peptide bond is formed and the polymer chain has been prolonged by one monomeric unit.

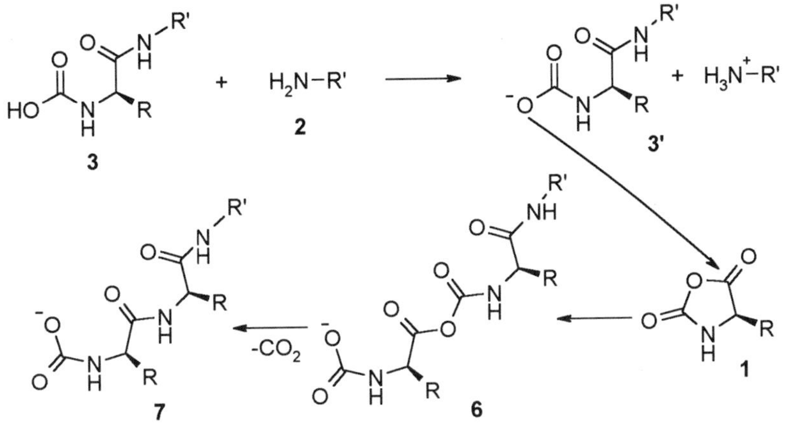

Figure 2: Initiation and propagation in the NCA-polymerization via the carbamate mechanism (According to ref. *12*)

Strong bases can deprotonate the N-carboxyanhydride **1**. The deprotonated anionic monomer **1'** is a strong nucleophilic agent which can attack a monomeric NCA and, therefore, is an "activated monomer". The attack of a NCA ring results in the intermediate **8** (Figure 3). The intermediate **8** has two reactive sites: i) the electrophilic N-acyl group and ii) the nucleophilic carbamate group. The latter one can react according to the mechanism shown in Figure 2 (pathway A in Figure 3). Furthermore, this group can decarboxylate and the generated amine end group can react according to the amine mechanism shown in Figure 1 (pathway B in Figure 3). Since the decarboxylation intermediate **9'** is an amide anion, the decarboxylation also results in deprotonation (activation) of a monomer molecule. This activated monomer can now attack a NCA ring either of a monomer or at the end of a growing chain (pathway C in Figure 3).

The polymerization is much faster if strong bases or sterically hindered amines are used as initiators. Therefore, it was concluded that the attack of the activated monomer (pathway C in Figure 3) is much faster than the other possible polymerization reactions (*11*). The diverse propagation steps encountered with this mechanism have consequences. First, the molecular weight is not determined by the ratio of monomer to initiator. Secondly, the molecular weight distribution is wide. Since the polymer chains are bifunctional, intermolecular coupling can occur which increases the molecular weight (*11 - 13*). If the polymerization of the NCA proceeds via the activated monomer mechanism, the initiator is not incorporated in the polymer. Therefore, this mechanism has to be avoided if the polymer shall be attached to a surface in a grafting from experiment. The activated monomer mechanism is preferred in the case of strong bases as initiators. Unfortunately, amines are strong bases too, and can initiate the NCA polymerization via the activated monomer mechanism, especially when sterically hindered. Therefore, for an efficient grafting from polymerization of NCA the surface should have primary amino groups with low steric hindrance.

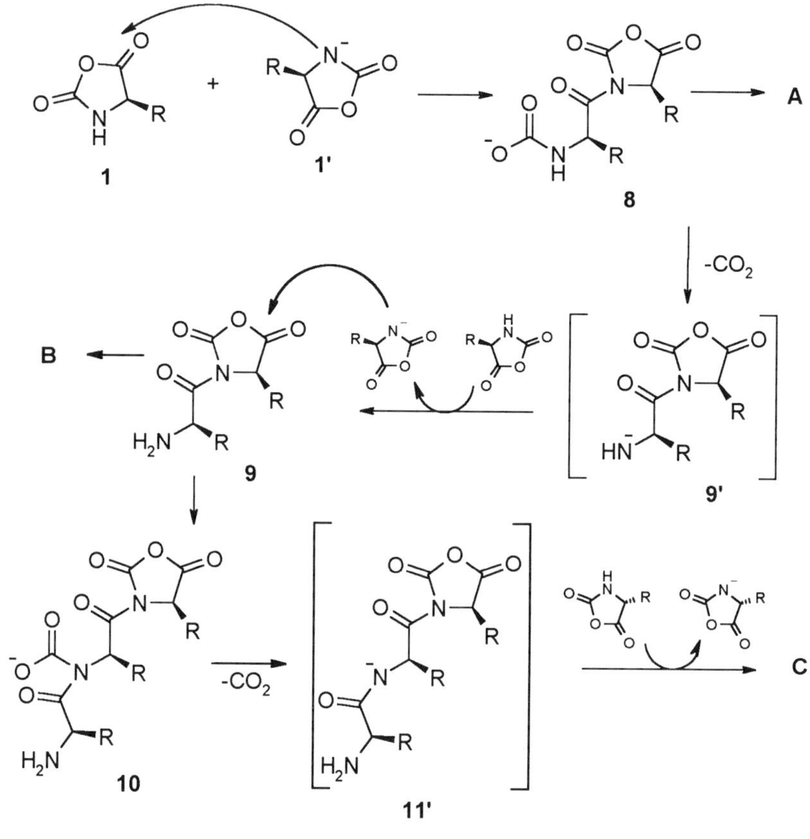

Figure 3: Initiation and the various propagation reactions for the NCA-polymerization via the activated monomer mechanism

The Initiating Layer

Design. WHITESELL and coworkers suggested the use self-assembled monolayers of thiols on gold to prepare an initiating layer in which the steric hindrances are minimized. They prepared specially designed thiols, whose footprint on the surface exactly match the diameter of the polypeptide helix (*6*). We have chosen a method to prepare initiating layers with adjustable density of active sites which is not as synthetically demanding. We prepared mixed self assembled monolayers of functionalized (Br end group) and non functionalized (methyl end group) undecyl-trichlorosilanes in different mixing ratios and subsequently transformed the bromine groups into amino groups. Assuming that the functional groups are moleculary dispersed at the surface of the monolayer, these mixed monolayers should avoid any steric hindrances at the initiating sites, too. Silanes were chosen for self-assembling, because they form very stable monolayers which are well defined (*16*): The headgroups constitute a two-dimensional polysiloxane network with some attachments to the surface (*17, 18*). The alkyl chains form a well-ordered layer and the functional groups, if carefully selected not to compete with the headgroups in interaction with the surface, are located atop this layer (*19*).

Preparation and characterization of the mixed self assembled monolayers. The monolayer preparation is described in detail in a previous paper (*20*). The thickness of the layers was measured by ellipsometry and x-ray reflectivity. The data are in accordance with a monolayer with the above-mentioned structure (*20*). The characterization, in terms of the monolayer composition, has been done by means of contact angle measurements and X-ray photoelectron sprectoscopy (XPS). The results are depicted in Figure 4.

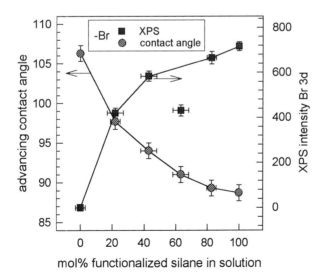

Figure 4: Advancing contact angles and Br 3d XPS intensity for the mixed self-assembled silane monolayers

It is clearly evident that there is a nonlinear dependence of the contact angle and the bromine XPS signal intensity on the composition of the solution. This can be explained by a preferred adsorption of the bromine terminated silane. Such preferred adsorption of slightly different molecules in the formation of self assembled monolayers has been described for thiols on gold (*21*). FRYXELL and coworkers recently reported no preferred adsorption for mixed silane monolayers (C16), although the contact angles in this work show a strong non-linear dependence on solution composition, too (*22*).

The bromine groups are subsequently converted into amino groups (see Figure 5) (*20, 23 - 25*). This *in-situ* modification was followed by means of contact angle and XPS measurements. The contact angle measurements clearly indicate a strong increase in hydrophilicity with the *in-situ* modification, whose extent depends on the density of functional groups at the surface (*20*). Some characteristic XPS spectra are shown in Figure 5.

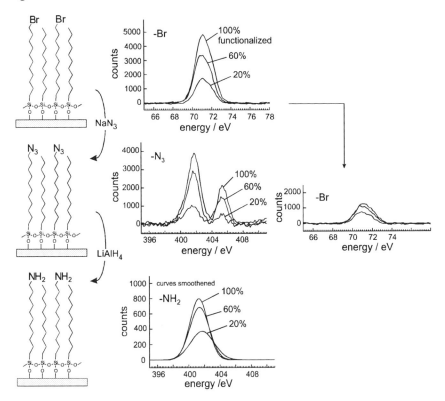

Figure 5: Schematic of the in-situ modification of mixed SAMs and the corresponding XPS-spectra

After treating the monolayers with sodium azide, there is a new signal in the XPS spectrum from the nitrogen. The signal is comprised of two peaks, consistent with the structure of the azide moiety. The substitution is not complete, as indicated by the residual bromine signal. From the decrease in the bromine signal an 80% - 90% conversion can be estimated. On the other hand, the reduction of the azide moieties with lithiumaluminiumhydride seems to be complete, since there is only one nitrogen peak in the XPS spectrum of the amino-terminated SAMs.

Summarizing the characterization of the monolayers and the *in-situ* modification, it can be concluded that we have mixed self-assembled monolayers, which are partially terminated with primary amino groups, and the concentration of these initiating sites can be adjusted via the composition of the adsorption solution. Due to a preferred adsorption and a non-complete substitution the solution composition does not match exactly with the monolayer composition.

Polymerizations

The self-assembled monolayers with terminal amino groups can be used to initiate a polymerization of glutamate-NCA. The grafted polymer layer can be proven by means of FTIR spectroscopy, which gives a spectrum of pure α-helical e.g. poly-γ-benzyl-L-glutamate (*20*). No β-sheet material can be detected, as is the case for the polymerization of methyl-L-glutamate-NCA using for example aminopropyltriethoxy-silane or (4-aminobutyl)dimethylmethoxysilane layers (*9*).

Thicknesses of the polymer layers were determined by ellipsometry and X-ray reflection. The latter method also gives the roughness of the polymer layers (*20*). The data obtained for a polymerization using benzyl-L-glutamate-NCA are compiled in Figure 6. The thickness of the polymer layers grafted from the self-assembled monolayer depends on the concentration of amine moieties at the surface. The thickness has a maximum for monolayers prepared from solutions with approximately 40% - 60% of functionalized silane. In addition, the roughness of the air/polymer interface (depicted as symbol size) has a minimum for this monolayer composition. Therefore, we assume that for this initiator concentration at the surface the polymerization is less interfered. This result is in accordance with the considerations made in the paragraph "Design of the Initiating Layer".

The monomer for the polymerizations discussed in the previous section was prepared by the classical method (*11*, *20*, *26*) and purified by recrystallization. The polymerizations were carried out by dissolving the monomer and placing the substrates into the solution carefully excluding moisture. According to DORMAN, NCAs have a higher reactivity and yield higher molecular weights if the monomer is purified by a special procedure involving a second treatment with triphosgene (*27*, *28*). A methyl-L-glutamate-NCA prepared according to this method and recrystallized several times in a closed apparatus in fact yielded much thicker polymer layers in a grafting from experiment (see Figure 6). The polymer film is thicker for all compositions of the monolayer. The increase in thickness due to the use of the more reactive monomer is more pronounced than the thickness variation due to the differences in initiator site concentration. Therefore, the influence of the monomer purity seems to be more important than adjustment of the steric requirements.

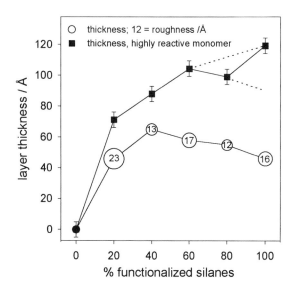

Figure 6: Thickness of the polymer layer grafted from self-assembled monolayers with various density of initiating amine moieties as determined by X-ray reflectivity or ellipsometry (highly reactive monomer) respectively.

The strong influence of monomer purity on the polymer layer thickness can be understood if the polymerization mechanism is taken into account. The NCA-polymerization in solution has some "living character" when initiated with primary amines, as has been discussed in the introduction. Despite this, high molecular weights ($DP_n > 150$) are, in general, not accessible using these initiators (but only by those which initiate via the activated monomer mechanism (*11 - 13*)), because of the nature and frequency of termination reactions. In the case of polyglutamates, cyclization of the terminal monomeric unit (see Figure 7) is assumed to be a termination reaction (*13*). Furthermore, in several polypeptides polymerized from NCAs, hydantoic acid units **13** have been found as the end group. The mechanism of the termination leading to this group is still under discussion (*13*). Impurities like free amino acid, due to an incomplete reaction or hydrolysis of NCA, may result in further side reactions which terminate the polymerization reaction. The activated monomer mechanism is less sensitive to termination reactions since the active site is located on a monomeric unit and every growing chain has two sites which allow growth. Furthermore, the polymerization is much faster for this mechanism than for the amine mechanism. Therefore, the propagation is preferred over side reactions and terminations yielding higher degrees of polymerization.

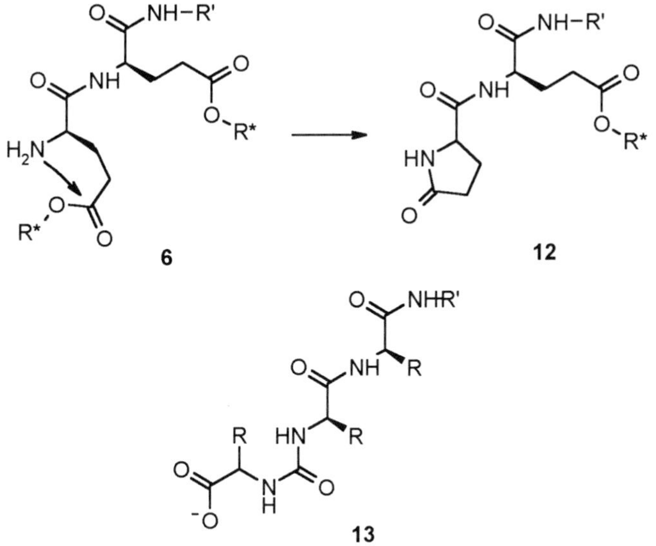

Figure 7: **Cyclization as a termination reaction and hydantoic acid end groups observed in the NCA-polymerization via the amine mechanism**

The finding that more pure and with that more reactive monomers result in a higher layer thickness is in accordance with the finding that special non solution grafting from polymerizations result in thicker polymer films. CHANG and FRANK reported polymer layers up to 400 Å thick obtained by a polymerization from the gas phase (or after sublimation of the monomer onto the substrate, respectively) (29). WIERINGA and SCHOUTEN obtained layer thickness in the range of 250 Å by polymerizing highly reactive benzyl- and methyl-L-glutamate-NCA by heating a spin coated monomer film to temperatures above the melting point (9, 30). In this approach the propagation is accelerated due to the high concentration of monomers and the elevated temperature.

Conclusions

Self-assembled monolayers were prepared with terminal amino groups which can serve as initiators for a grafting from polymerization of N-carboxyanhydrides. The initiator site density was varied by preparing mixed self-assembled monolayers in order to match the steric requirements of the growing chain and minimize the pertubations of the polymerization process. This was done by coadsorption of bromine- and methyl-terminated silanes and a subsequent modification including substitution of the bromine by azide and a reduction to amino groups. The bromine-terminated silanes show a preferred adsorption over the methyl-terminated ones. The substitution of the bromine by azide proceeds with approximately 80%-90% yield under the conditions applied. The partially amino terminated self assembled monolayers can be used to initiate the NCA-polymerization. The initiator site density has an influence on the polymer layer thickness. As expected, there is an optimum density of initiating sites for which the thickness has a maximum and the roughness of the air/polymer interface has a minimum. This optimum corresponds to approximately 40% -60% functionalized silanes in solution. On the other hand, the reactivity of the monomer, which can be controlled to some extent by the preparation and purification procedure, has an even bigger influence on the polymer layer thickness. The thickness variation within a series of experiments due to the different initiator site density is approximately 40%, but it is almost 100% by the change in the monomer purity. For fabrication of polypeptide layers with a thickness that exceeds the limits that have been reported in literature, it is necessary to further reduce side and termination reactions.

Acknowledgment

We thank *Wacker Chemitronic GmbH* for the donation of silicon wafers and the *Volkswagen Stiftung* for financial support.

Literature Cited

(1) Levine, B.F.; Bethea, L.G. *J. Chem. Phys.* **1976**, *65*, 1989
(2) Murthy, N.S.; Samulski, E.T.; Knox, J.R. *Macromolecules* **1986**, *19*, 943
(3) Enriques, E. P.; Gray, K. H.; Guarisco, V. F.; Linton, R. W.; Mar, K. D.; Samulski, E. T. *J. Vac. Sci. Technol. A* **1992**, *10*, 2775
(4) Worley, C. G.; Linton, R. W.; Samulski, E. T. *Langmuir* **1995**, *11*, 3805
(5) Sano, K.; Machida, S.; Sasaki, H.; Yoshiki, M.; Mori, Y. *Chem. Lett.* **1992**, 1477
(6) Whitesell, J. K.; Chang, H. K. *Science* **1993**, *621*, 73
(7) Whitesell, J. K.; Chang, H.K.; Whitesell, C.S. *Angew. Chem. Int. Ed. Engl.* **1994**, *871*, 33
(8) Ying-Chih Chang, van Esbroeck, H.; Frank, C.W. *Polym. Prepr.* **1995**, *36*, 121
(9) Wieringa, R. H.; Schouten, A. J. *Macromolecules* **1996**, *29*, 3032
(10) Ying-Chih Chang, Frank, C.W. *Langmuir* **1996**, *12*, 5824
(11) Kricheldorf, H.R. *α-Aminoacid-N-carboxyanhydrides and related heterocycles*, Springer-Verlag, Berlin, Germany 1987
(12) Block, H. *Poly-γ-benzyl-L-glutamate and other Glutamic Acid Containing Polymers*, Gordon and Breach, New York, NY, 1983

(*13*) H.R. Kricheldorf , in: *Models of Biopolymers by Ring Opening Polymerization*, Ed. S. Penczek, CRC-Press, Boca Raton, Fl, 1990

(*14*) Daly, W.H.; Poche, D. S.; Negulescu, I.I. *Prog. Polym. Sci.* **1994**, *19*, 79

(*15*) Daly, W.H.; Negulescu, I.I.; Russo, P.S.; Poche, D.S. *ACS Symp. Ser.* **1992**, *493*, 292

(*16*) Ulman, A. *An Introduction to Ultrathin Organic Films - From Langmuir-Blodgett to Self-Assembly*, Academic Press, San Diego 1991 and references cited therein

(*17*) Tidswell, I.M.; Ocko, B.M.; Pershan, P.S.; Wasserman, S.R.; Whitesides, G.M.; Axe, J.D. *Phys. Rev. B: Condens. Matter.* **1990**, *41*, 1111

(*18*) Tidswell, I.M.; Rabedeau, T.A.; Pershan, P.S.; Folkers, J.P.; Baker, M.U.; Whitesides, G.M. *Phys. Rev. B: Condens. Matter.* **1991**, *44*, 10869

(*19*) Bierbaum, K.; Hähner, G.; Heid, S.; Kinzler, M.; Wöll, Ch.; Effenberger, F.; Grunze, M. *Langmuir* **1995**, *11*, 512

(*20*) Heise, A.; Menzel, H.; Hyun Kim, Foster, M.D.; Wieringa, R.H.; Schouten, A.J.; Erb, V.; Stamm, M. *Langmuir* **1997**, *13*, 723

(*21*) Bain, C.D. ; Evall, J.; Whitesides, G.M., *J. Am. Chem. Soc.* **1989** *111*, 7155

(*22*) Fryxell, G.E.; Rieke, P.C.; Wood, L.; Engelhard, M.H.; Williford, R.E.; Graff, G.L.; Campbell, A.A.; Wiacek, R.J.; Lee, L.; Halverson, A. *Langmuir* **12**, *5064* (1996)

(*23*) Balachander, N.; Sukenik, C. N. *Langmuir* **1990**, *6*, 1621

(*24*) L.M. Lander, Ph.-D. Thesis, The University of Akron, Akron, OH (1994)

(*25*) Liebmann, A.; Lander, L.M.; Foster, M.D.; Brittain, W.J.; Vogler, E.A.; Satija, S.; Majkrzak, C.F. *Langmuir* **1996**, *12*, 2256

(*26*) Fuller, W.D.; Verlander, M.S.; Goodmann, M. *Biopolymers* **1976**, *15*, 1869

(*27*) Dorman, L.C.; Shiang, W.R.; Meyers, P.A. *Synth. Commun.* **1996**, *22*, 3257

(*28*) Eckert, H.; Förster, B. *Angew. Chem. Int. Ed. Engl.* **1987**, *26*, 894

(*29*) Ying-Chih Chang, C.W. Frank, following chapter this book

(*30*) Wieringa, R.H.; Schouten, A.J.; Erb, V.; Stamm, M. presented at the 7th International Conference on Organized Molecular Films - LB7, 10.-15.9. 1995, Numana (Ancona), Italy

Chapter 11

Chemical Grafting of Poly(L-glutamate) γ-Esters on Silicon (100) Surfaces by Vapor Polymerization of N-Carboxy Anhydride Monomers

Ying-Chih Chang and Curtis W. Frank

Center on Polymer Interfaces and Macromolecular Assemblies (CPIMA) and Department of Chemical Engineering, Stanford University, Stanford, CA 94305–5025

We show that vapor polymerization of the N-carboxy anhydride of an α-amino acid in vacuum is a powerful technique for the fabrication of thin, chemically-grafted polypeptide films. In this study, we demonstrate the application of a vapor polymerization method to poly(γ-benzyl-L-glutamate) (PBLG), poly(γ-methyl-L-glutamate) (PMLG) and their copolymer systems. We investigate the influences of the reaction vessel geometry, vapor concentration, reaction temperature, pressure and the initiator site density on the resulting film properties by ellipsometry, Fourier transform infrared spectroscopy (FTIR) and contact angle goniometry. Under optimal reaction conditions, e.g., 95 ~ 125°C and 0.03 torr, a 42 nm α-helical PBLG film can be synthesized. Furthermore, we also discuss the potential application of vapor polymerization for fabrication of grafted random and block copolymers.

Polypeptides having a regular secondary structure, such as α-helix or β-sheet, are considered to be interesting materials for fabrication of ordered, ultra-thin organic films on solid substrates for potential applications in optics, liquid crystal displays, and biosensors.(*1-4*) In particular, various poly-L-glutamate, γ-esters (PG) have been widely studied as thin films or membranes in which their liquid crystalline behavior, selective permeability, high dielectric constant and birefringence properties are of interest.(*5-15*)

Traditionally, ultra-thin films have been prepared by spin-casting from solution or Langmuir-Blodgett-Kuhn (LBK) deposition.(*16-18*) Such films prepared by both methods are physically adsorbed on the surface. Through the LBK technique, one can obtain in-plane ordered PG multi-layers with anti-parallel pairs of PG molecular backbones oriented along the transfer direction, thus canceling out the individual dipole moment and minimizing the intermolecular energy.

More recently, several efforts have been directed toward covalent attachment of polypeptide chains to solid substrates.(*19-28*) This work has been motivated by the desire to obtain a PG film wherein there is reduced, or minimal cancellation of the individual dipoles. Considerable effort has been invested in the development of chemical grafting techniques in the belief that, under appropriate conditions, a chemical bond formed between the solid substrate and the polypeptides can overcome the intermolecular repulsive forces and stabilize the molecular alignment with a net dipole moment.

Materials Under Study

A major advantage of PG as a model polypeptide is that its molecular characteristics can be varied simply by modifying its ester side chain group. We are particularly interested in two extreme cases of the PG family: poly(γ-benzyl-L-glutamate) (PBLG), **1,** and poly(γ-methyl-L-glutamate) (PMLG), **2**, grafted on silicon (100) native oxide surfaces. The high molecular weight PBLG only adopts an α-helical conformation because its bulky phenyl side group shields the amide backbone from the surroundings, leading to the highest helical stability among all PGs.(*7*) By contrast, PMLG contains the least rigid side chain among the PGs; therefore, its high molecular weight form may also adopt the β-sheet structure after being subjected to mechanical stress or certain solvents. (*29,30*)

Description of Solution and Melt Grafting Techniques

To design a synthetic route to graft PG on solid substrates, one can either adopt the "grafting *to*" technique where the pre-formed PG directly reacts with the surface, (*19,22,25*)or the " grafting *from*" (*20-22,24,26-28,31-33*) protocol where monomers of the amino acid polymerize from a surface initiator group. In our studies, both grafting *to* and grafting *from* methods have been shown to be feasible and complementary.(*22*)

Grafting *to* Technique. We have applied two "grafting *to*" approaches for the PBLG system. (*22*) These involve either reaction between the N-terminus of PBLG with a surface-bound chloroformate group, which is introduced by the reaction of trichlorosilyl propyl-3-chloroformate and the silicon oxide surface (Scheme 2a), or the reaction between a modified PBLG having a triethoxysilane end group and the silanol group on the native oxide surface (Scheme 2b). The advantage of the grafting *to* approach is that the molecular weight and the polydispersity may be measured prior to deposition and thus, may be controlled in advance; therefore, the properties of the subsequently deposited thin films can be readily analyzed. The limitation of the grafting *to* technique is the low probability of the terminal functional group reaching the surface reactive site; the size of the pre-formed polypeptide is considerably larger than the surface linkage group, so the previously tethered polypeptides would block the surrounding linkage groups, thus hindering further reaction. This proposal has been proven by both thickness measurements from ellipsometry and an orientation study using infrared spectroscopy.(*22*) Both sets of data consistently show that the PG

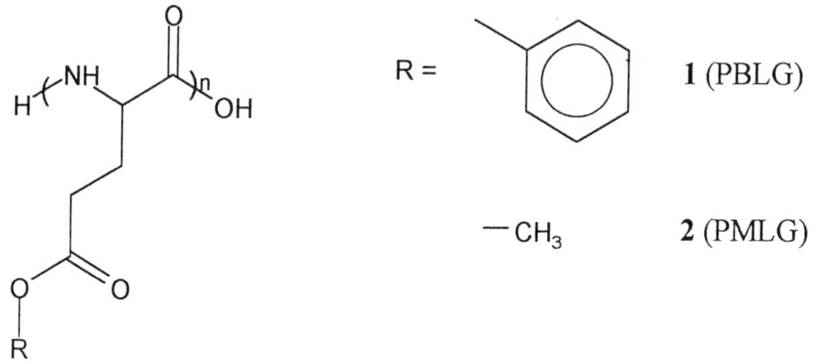

Scheme 1. The molecular formulae of PBLG and PMLG with n repeating units.

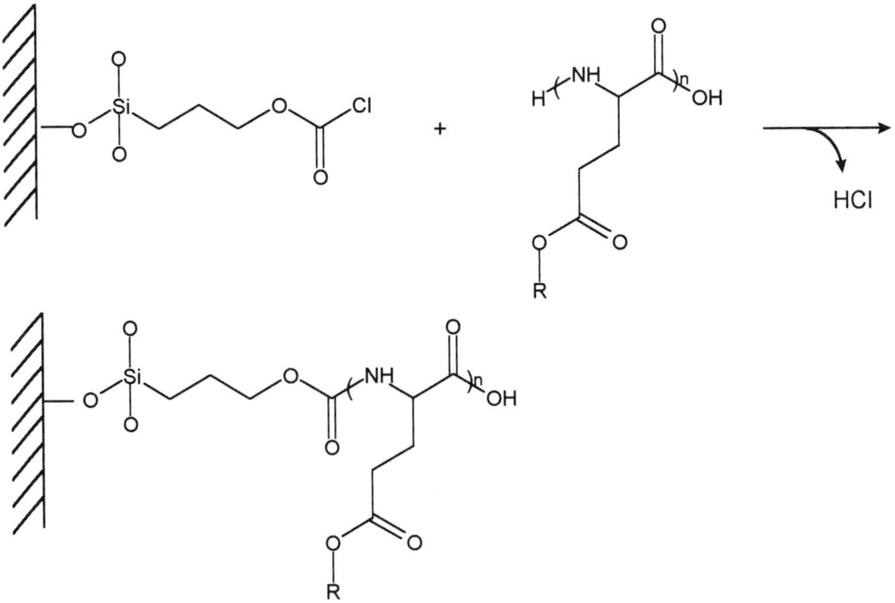

Scheme 2a . Grafting *to-* PBLG grafted on a chloroformate-modified silicon (100) native oxide surface. Note that the surface attachment of the Si of the chloroformate surface modifier is left somewhat ambiguous, due to the complexity of the silane interaction with the silicon oxide surface. Nevertheless, all materials studied are robustly attached to the substrate.

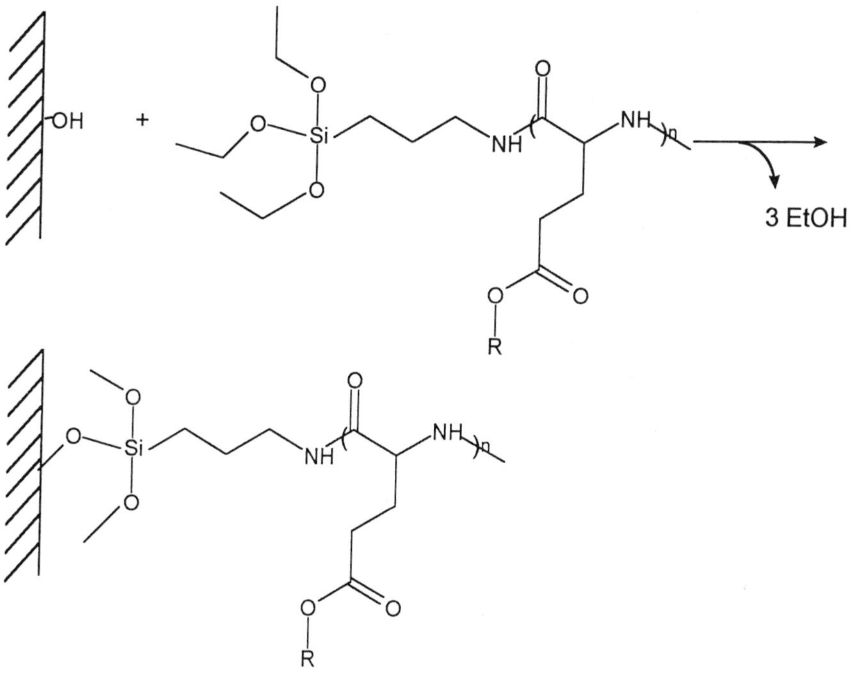

Scheme 2b. Grafting *to-* PBLG with terminal triethoxysilane group grafted onto silicon(100) native oxide surface. The material was prepared by 1-(amino propyl)triethoxysilane initiated bulk polymerization of the N-carboxy anhydride of benzyl L-glutamate. The average degree of polymerization of the material is ca 40 from GPC measurement, based on polystyrene standard. The detailed synthesis procedure was described elsewhere.(*22*)

molecules in the resulting film formed by the procedure of Scheme 2a, after reaction proceeding for various times from 1~2 days at 65°C in THF/dioxane mixture, are preferentially parallel to the surface. The average film thickness is 2.5 nm, for samples with degrees of polymerization of 100 and 130.(22) A similar observation of the orientation of the grafted PBLG on Au surface was also reported by Enriquez et al.(19)

The low reactivity can be improved by the second grafting *to* approach, given in Scheme 2b, in which the PG molecules are designed to have three reactive functional groups which allow the reaction to occur through both surface condensation and intermolecular crosslinking. This strategy does help to improve the resulting film thickness to ca. 5 nm. However, most polypeptides do not dissolve in the commonly used organic solvents- only their monomers do. For example, PMLG can only dissolve in a limited number of solvents, such as trifluoroacetic acid and pyridine; therefore, the grafting *to* method cannot be utilized for the PMLG system. To overcome this technical difficulty, one needs to develop a grafting *from* approach to achieve a more general application.

Grafting *from* Technique. The chemical synthetic route applied in our grafting *from* approach was to conduct a surface polymerization of the monomer N-carboxy anhydride (NCA) of the desired α-amino acid from a surface-immobilized primary amine group, e.g., a condensed 1-(aminopropyl) triethoxysilane (APS) on a silicon(100) native oxide surface (Scheme 3). (21,26) The surface polymerization was first conducted in a solution of NCA. However, we found that the surface reaction is extremely sensitive to moisture and impurities from the NCA material, surface initiator layer and the solvent, thus leading to an early termination of the reaction. (24) More recently, Menzel et al. (28) proposed using a carefully-prepared self-assembled monolayer with amino end groups as the initiator layer and ultra-pure NCA of benzyl-L-glutamate (B-NCA) based on the synthesis method from Dorman et al.(34) With this approach, they were able to obtain the helical conformation PBLG film with maximum thickness ca. 12 nm. Furthermore, Wieringa and Schouten (24) conducted the surface polymerization of the ultra-pure NCA of methyl-L-glutamate (M-NCA) under non-solvent conditions by first spin casting M-NCA solution on a primary amine-modified silicon wafer and then heating up above the melt temperature of the M-NCA to promote the thermal polymerization. By thermal polymerization in the melt state, the resulting PMLG film with maximum thickness ca. 20 nm, as estimated from the FTIR spectra, was obtained.

Experimental

To avoid the impurities and moisture often introduced during the reaction process, we have adopted a completely solvent-free approach to conduct the surface polymerization of both B-NCA and M-NCA from an APS modified silicon wafer at an elevated temperature *in vacuo*.(27,32) The procedure for preparing fully-covered APS modified substrates was reported earlier, and the resulting layer thickness was estimated to be ca. 12(1) Å, based on the ellipsometric measurement (the refractive indices of both APS and silicon native oxide layers are assumed to be 1.46).(22)

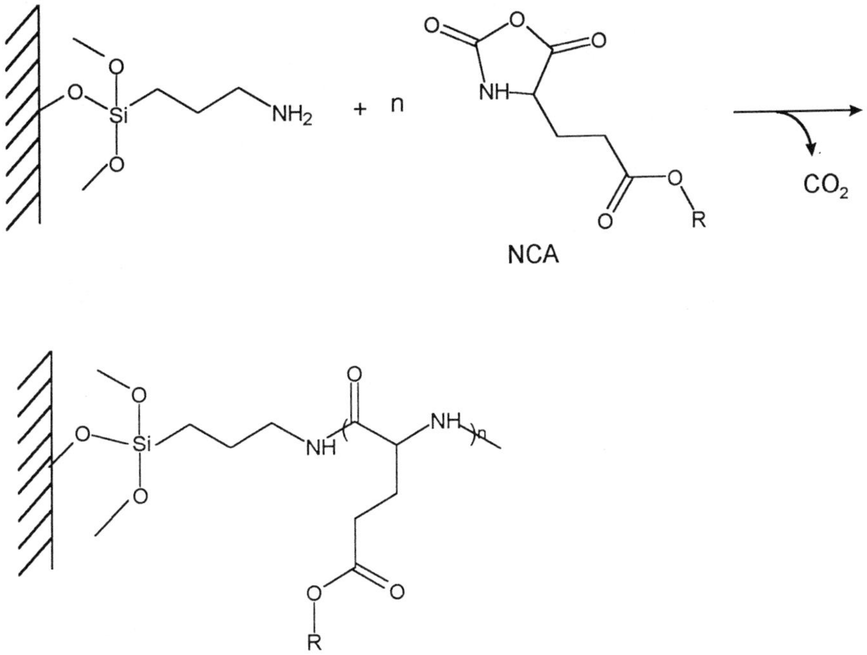

Scheme 3. Grafting *from-* PG via surface polymerization of NCA of L-glutamate from a surface-immobilized primary amino group. The APS molecules form a complex, bonded network on the silicon oxide surface. The ellipsometric thickness of 1 ± 0.2 nm suggests that the APS layer probably exists as a multilayer, disordered structure.

Scheme 4a shows the setup of the vapor polymerization method used in this study. The NCA monomer was placed on the bottom of a Schlenk tube, and a 2 × 5 cm APS modified silicon substrate then was carefully placed in the tube to avoid any direct contact with the NCA powder; in this setup, the minimal displacement from the edge of the substrate to the NCA powder is ca 0.5 cm. The Schlenk tube was evacuated to 0.1~ 0.03 torr, then was tightly sealed and placed into an oven with controlled temperature from 85°C ~ 135°C, and the reaction was allowed to proceed for various time periods. It is well known that PGs readily form side-by-side aggregates, leading to the possible insertion of physisorbed PG in the chemisorbed PG film. To extract such physisorbed material from the surface, we selected either a dichloroacetic acid (DCA)/chloroform (1/9) mixture or formic acid, both of which are known to be effective de-aggregants, as cleaning reagents for PBLG and PMLG surfaces, respectively.(6) The films were immersed in the cleaning reagents for at least 15h, then rinsed with chloroform and ethanol.

Results

In this report, we will first review the results to demonstrate the versatility of the vapor polymerization, mainly based on the studies from the PBLG system. We will also report the results from the PMLG and the random copolymers of the methyl- and benzyl-L-glutamates (PMBLG) and explore the feasibility of fabricating block copolymers by vapor polymerization.

Pressure. One can anticipate that the vacuum condition would play an important role in this dry process because it can reduce the required temperature for producing the vapor reactant and minimize the moisture content in the atmosphere. To monitor the film growth, we measured the resulting PBLG film thickness by ellipsometry (Gaertner variable angle ellipsometer, Model L116A). For the range of pressure examined (0.03 torr ~ 0.15 torr), we found that for two samples with identical experimental conditions, i.e. at 124°C with 10 mg B-NCA, the resulting film thickness varied from 13 nm at 0.15 torr after 5h reaction time to 28 nm at 0.03 torr after only 30 min.

Initiator Site Density. The site density of the initiator has been proposed to be an important factor for successful fabrication of grafted α-helical polypeptides.(20,28,33) One suggestion is that an initiator site density comparable to the cross-section of the helical polypeptide is required to minimize the steric hindrance. We controlled the site density of the APS layer by conducting the APS condensation reaction for various periods of time. As Table I illustrates, the APS surfaces then have different advancing water contact angles, which suggests different APS surface coverage. For the APS surface prepared by dipping in the APS 0.1% solution for 5 s, the resulting contact angle of 5°, which is the same as that of a bare substrate, indicates that only a few APS molecules are grafted on the silicon substrate. Using this sample as the control, we obtained no observable PBLG film after the reaction and the cleaning treatment; thus, there was no chemisorbed PG and all physisorbed PG was removed by the cleaning treatment. For the other two samples with average contact angles 42° and 53°, on the

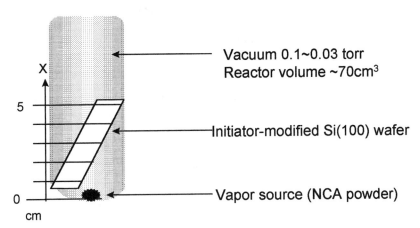

Vacuum 0.1~0.03 torr
Reactor volume ~70cm³

Initiator-modified Si(100) wafer

Vapor source (NCA powder)

Scheme 4a. Diagram of the reaction vessel setup

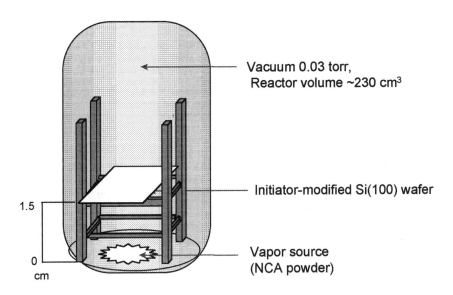

Vacuum 0.03 torr,
Reactor volume ~230 cm³

Initiator-modified Si(100) wafer

Vapor source
(NCA powder)

Scheme 4b. Diagram of the reaction vessel setup for the temperature dependence experiments.

other hand, the film thicknesses are not significantly different. Compared to the pressure difference from 0.15 torr to 0.03 torr, which leads to a doubling of the thickness, we conclude that there is little influence of the initiator density on the subsequent surface polymerization. A similar observation has been made by Menzel et al;(28) they also suggested that in the solution polymerization the purity of the monomer seems to be a more important factor than the site density in leading to a thicker film.

Table I. Vapor polymerization of B-NCA on the APS surfaces with various deposition times, as characterized by the water contact angles.

Sample No.	APS deposition time (s)	CA(°) of APS surface	Thickness of the PBLG layer (nm)*
65-6	3	$5° \pm 1°$	0.3 ± 0.2
65-3	60	$42° \pm 2°$	12.8 ± 2.5
65-1	2400	$53° \pm 1°$	13 ± 1.9

* The thickness of subsequent PBLG deposited films was measured by ellipsometry, with refractive indices assumed to be 1.46 for the SiOx, APS, and PBLG layers.

Monomer Concentration. The effect of the monomer vapor concentration was first studied by using three samples with initial monomer B-NCA quantities of 1 mg, 10 mg and 50 mg. Under the identical reaction conditions, i.e. 124°C, 0.15 torr and 5h reaction time, the resulting PBLG film thicknesses are 15(4), 14(1) and 16(1) nm (standard deviation of the samples is in parentheses). Although there is no difference in the thickness for these three samples, the standard deviations of the thicknesses indicate that for smaller quantities (i.e. 1mg NCA), it is more difficult to produce a uniform film over the whole surface (2x5 cm) , and this is probably due to the existence of a monomer concentration gradient. On the other hand, as the NCA quantity is increased (10 mg and 50 mg), there is more uniform monomer concentration; thus the film is more uniform.

To further examine the existence of the concentration gradient and its effect on the resulting film thickness, we monitored the thickness as a function of the displacement from the NCA vapor source. One can anticipate that the vapor concentration decreases along the radial distance from the vapor source. As Scheme 4a illustrates, we examine the thickness changes along the x-axis by ellipsometry. Figure 1 shows the ellipsometric thickness data for various reaction times before and after the cleaning treatment with DCA/chloroform mixture: The black diamond (solid line) and the gray triangle (dashed line) symbols illustrate the film thicknesses with reaction times 24h and 5h, respectively, before any cleaning treatment. It is straightforward to observe the existence of a vapor concentration gradient along the axis as the film thickness decreases about 50% for 5 cm displacement from the vapor source. Furthermore, the thickness of the 24h reaction is somewhat higher than that of the 5h reaction, indicating that there is residual vapor NCA even after the 5h reaction. However, after an intensive cleaning treatment, the film thicknesses of both samples drop dramatically, as the open diamond (24h) and the open triangle (5h) symbols indicate in Figure 1. The observation of the two samples (24h and 5h reactions) having similar resulting film thicknesses after the cleaning treatment might suggest that most of the surface PBLG chains propagated from the surface initiator layer are terminated within 5h; therefore, there is no significant thickness improvement observed even when the reaction time is prolonged to 24h. Although there is more NCA monomer vapor in the 24 h reaction than in the 5 h reaction, most of the NCA perhaps self-polymerizes on the surface, thus forming physisorbed material that is subsequently washed away by the DCA/chloroform cleaning mixture. We will discuss the chain termination further in a later section.

Temperature. From the preceding discussion, we found that the monomer concentration changes along the radial distance from the NCA powder. To avoid the potential problem associated with a vapor concentration gradient leading to inhomogeneous film thicknesses, we then designed a spacer that would allow us to place the substrate horizontally over the NCA vapor source with a separation distance of 1.5 cm, as Scheme 4b illustrates. We used this setup for the subsequent temperature controlled experiments. Figure 2 shows the film thickness for different temperatures, with vessel pressure 0.03 torr and 1.5 cm displacement from the 10 mg B-NCA to the APS-Si for each experiment. Between 95°C and 125°C, the film thickness can reach ca. 28 nm in 30 min. This implies that the vapor polymerization process is very robust, as the resulting film thickness is quite insensitive to temperature over a wide temperature range but is easy to manipulate by varying the reaction time.

The experimental results presented here strongly support the feasibility of the vapor polymerization of NCA on APS surfaces; through this powerful deposition method, as we have shown, one can easily obtain the desired film thickness by adjusting the reaction parameters, such as the reaction time, pressure, temperature or the vapor concentration (displacement from the vapor source). Figure 3 shows an example prepared under the optimal reaction condition; by conducting the polymerization of 10 mg B-NCA at 105°C and 0.03 torr with a 15 mm horizontal distance from the APS-Si, the reaction proceeds rapidly in the first 1h. Film thicknesses of 22 nm in 15 min and 28 nm in 30 min were obtained. The maximum film thickness of 42 nm was reached within 4h.

Copolymerization. It is also possible to synthesize a random copolymer on the solid substrate by the vapor polymerization process. We fabricated thin films of the random copolymer of γ-benzyl-L-glutamate and γ-methyl-L-glutamate (PBMLG) by simultaneously introducing both B-NCA and M-NCA monomers as the vapor source. Four samples were prepared (Table II) with M-NCA molar percentages of 100%, 67%, 33% and 0%. Figure 4 shows the transmission infrared spectra (BioRad Digilab FTS 60A single beam spectrometer equipped with a TGS detector, resolution 4 cm^{-1}, 2048 scans for each measurement) of these four samples before the cleaning treatment. The spectra show the typical PG signature with the N-H stretching band at ca. 3290 cm^{-1} and carbonyl stretching band mainly from the side chain ester at ca. 1735 cm^{-1}. The amide I and the amide II bands at ca. 1654 cm^{-1} and 1550 cm^{-1}, respectively indicate an α-helical conformation, while those at ca. 1628 cm^{-1} and 1520

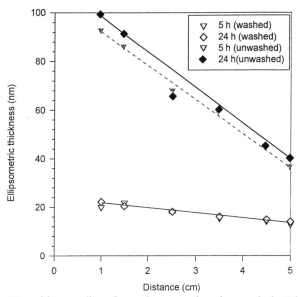

Figure 1. Deposition gradient for various reaction time periods. The reaction was conducted at 120°C, 0.03 torr; the distance was taken from the NCA source, according to the diagram shown in Scheme 4a.

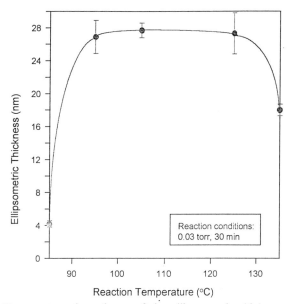

Figure 2. Temperature dependence of the ellipsometric thickness of grafted PBLG thin films for the reaction geometry of Scheme 4b. The reaction conditions are 0.03 torr and 1.5 cm vertical distance between the substrate and the monomer powder. The reaction time is 30 min.

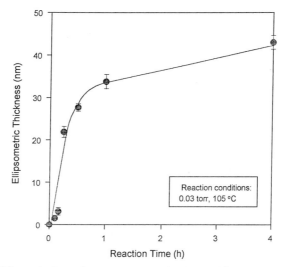

Figure 3. Reaction progress monitored by ellipsometry.

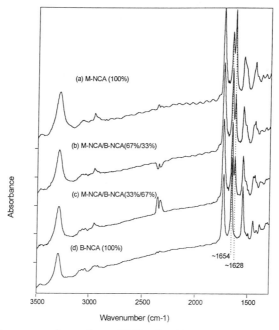

Figure 4. IR spectra of copolymer PMBLG with the initial monomer mixture ratio M-NCA/B-NCA from 100%, 67%, 33% to 0%. Spectra are taken before the cleaning treatment. Peaks at ca 3290 cm^{-1} are due to N-H stretching, 1732 cm^{-1} to side chain carbonyl stretching, 1654 cm^{-1} and 1628 cm^{-1} to amide I in helical and sheet conformation, respectively. Note that the peak ratio of 1654 cm^{-1} to 1628 cm^{-1} increases as the M-NCA molar ratio decreases.

cm^{-1}, respectively, indicate a β-sheet conformation. Based on this information, the amount of the α-helix increases as the composition of B-NCA increases. For the pure M-NCA polymerization, the resulting film before the cleaning treatment has the highest β-sheet content, while the film made only of B-NCA adopts a pure α-helical structure.

Table II. Contact angles (CA) of the copolymer PMBLG with different initial NCA composition after the cleaning treatment.

M-NCA composition(%)	100	67	33	0
CA (°)	53	56	56	66

Figure 5 shows the IR spectra of the resulting films polymerized from the pure M-NCA or B-NCA after immersing in the DCA/chloroform mixture for at least 15h. Interestingly, most of the β-sheet material is washed away, and the α-helical structure is predominant in both PMLG and PBLG samples. This observation is similar to the result reported by Wieringa and Schouten (24) for a PMLG film synthesized by melt polymerization. It is difficult to follow the evolution of the four samples from pure PMLG to pure PBLG, due to the low resolution in IR region for the C-C stretch of the phenyl group of PBLG molecules (ca 1498 cm^{-1}) and the C-H stretch of the methyl group of PMLG molecules (ca 2950 cm^{-1}). Therefore, we measured the advancing water contact angles of the four samples to monitor the surface energy of the systems. Table II shows the corresponding advancing water contact angles for different monomer compositions. The higher value for the pure PBLG film compared to the pure PMLG film is due to the more hydrophobic phenyl side chain group.

The melting temperatures of B-NCA and M-NCA are 97°C and 100°C, respectively,(35) so it is reasonable to assume that there might be a slight preferential polymerization of B-NCA in relation to the M-NCA due to the different sublimation rates between these two monomers. This would cause more benzyl-L-glutamate repeating units than the expected statistical value as found in the bulk polymerization.(36,37) If this is true, when most of the B-NCA is consumed more M-NCA will be found in the residues of the top layer. This may explain the observation that the contact angles for both copolymer samples are closer to the value for the pure PMLG film, regardless of their monomer ratios. Overall, it is possible to fabricate random copolymers on surfaces by conducting the vapor polymerization with the desired monomer mixture; however, it is important to calibrate the monomer ratio against the corresponding composition of the resulting polymer.

Two-Step Polymerization. To test the feasibility of fabricating a block copolymer by sequential vapor polymerization, our first attempt has been to conduct a two-step vapor polymerization of PBLG and monitor the thickness changes. The first vapor polymerization was done at 118°C for 1h, with 30 mm distance from the NCA source, followed by thorough cleaning with DCA/chloroform (1/9), and rinsing with chloroform. We then made the second vapor polymerization with half the amount of

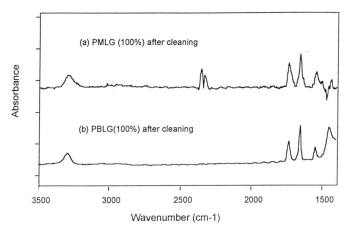

Figure 5. The IR spectra of the resulting PMLG and PBLG grafted films, the spectra 5a and 5b correspond to Figures 4a and 4d, after cleaning with DCA/chloroform (1/9) mixture 15h. Both samples are dominated by the helical conformation.

the B-NCA monomers (5mg) but otherwise the same reaction conditions. The resulting film was cleaned by the same procedure as the first step polymerization in order to remove the physisorbed material. During the first step polymerization, we obtained a PBLG film with thickness 14 nm and the subsequent polymerization increases the film thickness to 21 nm. This result shows the feasibility of making a block copolymer by the vapor polymerization, and proves that the terminal amino group of the PBLG remains accessible and functional for further reaction.

Conclusion

The vapor polymerization of NCA to synthesize polypeptides on a solid substrate has been shown to be a powerful technique to obtain a thin polypeptide film with a controlled thickness from 1 nm to 42 nm, which is mainly determined by the reaction temperature, pressure, vapor concentration and the reaction time. Both homopolymer PBLG and PMLG films obtained here show α-helical character after being washed with DCA/chloroform mixture. Our attempts to synthesize random and block copolymers by this dry process are preliminary but promising. Future studies will emphasize the examination of the reaction mechanism and further improvement of the reaction process, such as adopting highly purified NCA monomers for the experiment or using a high vacuum system.

Acknowledgment

This work was supported by the NSF Materials Research Science and Engineering Center Program through the Center on Polymer Interfaces and Macromolecular Assemblies (CPIMA).

References

1.   Debe, M. K. *Progress in Surface Sci.* **1987**, *24*, 1.
2.   Hakemi, H. *J. Polym. Sci. B: Polym. Phys.* **1986**, *24*, 2445.
3.   Knichel, M.; Heiduschka, P.; Beck, W.; Jung, G.; Gopel, W. *Sensors and Actuators B* **1995**, *28*, 85.
4.   Ballauff, M. *Angewandte Chemie* **1989**, *28*, 253.
5.   Daly, W. H.; Poche, D.; Negulescu, I. I. *Prog. Polym. Sci.* **1994**, *19*, 79.
6.   Block, H. *Poly(γ-benzyl-L-glutamate) and Other Glutamic Acid Containing Polymers*; 1st ed.; Gordon and Breach Sci Publishers, Inc.: New York, 1983.
7.   Samulski, E. T. *Liquid Crystalline Order in Polypeptides*; Samulski, E. T., Ed.; Academic: NY, 1978, pp 167.
8.   Inoue, S.; Maeda, M.; Chung, D.-W.; Aoyama, M. *J. Macromol. Sci.-Chem.* **1991**, *A28*, 1137.
9.   Wada, A. *J. Chem. Phys.* **1959**, *31*, 495.
10.   Wada, A. *J. Phys. Chem.* **1959**, *30*, 328.
11.   Levine, B. F.; Bethea, C. G. *J. Chem. Phys.* **1976**, *65*, 1989.

12. Wada, A. *J. Phys. Chem.* **1958**, *29*, 674.
13. O'Konski, C. T.; Yoshioka, K.; Orttung, W. H. *J. Phys. Chem.* **1959**, *63*, 1558.
14. Tinoco, I. J. *J. Am. Chem. Soc.* **1957**, *79*, 4336.
15. Tsvetkov, V. N. *Rigid-chain Polymers*; 1st ed.; Consultants Bureau: NY, 1989.
16. Blaudez, D.; Buffeteau, T.; Desbat, B.; Escafre, N.; Turlet, J. M. *Thin Solid Films* **1994**, *243*, 559.
17. Takenaka, T.; Harada, K.; Matsumoto, M. *J. Colloid Interface Sci.* **1980**, *73*, 569.
18. Zasadzinski, J. A.; Viswanathan, R.; Madsen, L.; Garnaes, J.; Schwartz, D. K. *Science* **1994**, *263*, 1726.
19. Enriquez, E. P.; Gray, K. H.; Guarisco, V. F.; Linton, R. W.; Mar, K. D.; Samulski, E. T. *J. Vac. Sci. Technol., A* **1992**, *10*, 2775.
20. Whitesell, J. K.; Chang, H. K. *Science* **1993**, *261*, 73.
21. Chang, Y.-C.; Esbroeck, H. v.; Frank, C. W. *Polym. Preprints* **1995**, *36*, 121.
22. Chang, Y.-C.; Frank, C. W. *Langmuir* **1996**, *12*, 5824.
23. Worley, C. G.; Enriquez, E. P.; Samulski, E. T.; Linton, R. W. *Surface Interface Anal.* **1996**, *24*, 59.
24. Wieringa, R. H.; Schouten, A. J. *Macromolecules* **1996**, *29*, 3032.
25. Machida, S.; Urano, T. I.; Sano, K.; Kawata, Y.; Sunohara, K.; Sasaki, H.; Yoshiki, M.; Mori, Y. *Langmuir* **1995**, *11*, 4838.
26. Chang, Y.-C.; Frank, C. W. *Macromol. Symp.* **1997**, *118*, 641.
27. Chang, Y.-C.; Frank, C. W. *Polym. Preprints* **1997**, *38*, 945.
28. Menzel, H.; Heise, A.; Yim, H.; Foster, M. D.; Wieringa, R. H.; Schouten, A. J. *ACS Symposium.*
29. Rinaudo, M.; Domard, A. *Biopolym.* **1975**, *14*, 2035.
30 Bamford, C. H.; Brown, L.; Elliott, A.; Hanby, W. E.; Trotter, I. F. *Nature* **1954**, *173*, 27.
31. Nakajima, A.; Fujiwara, T.; Hayashi, T.; Kaji, K. *Biopolym.* **1973**, *12*, 2681.
32. Whitesell, J. K.; Chang, H. K. *Mol. Cryst. Liq. Cryst.* **1994**, *240*, 251.
33. Chang, Y.-C.; Frank, C. W. *Langmuir (submitted)* **1997**.
34. Heise, A.; Menzel, H.; Yim, H.; Foster, M. D.; Wieringa, R. H.; Schouten, A.J.; Erb, V.; Stamm, M. *Langmuir* **1997**, *13*, 723.
35. Dorman, L. C.; Shiang, W. R.; Meyers, P. A. *Synth. Commun.* **1992**, *22*, 3257.
36. Kricheldorf, H. R. *α-Aminoacid-N-Carboxy-Anhydrides and Related Heterocycles*; 1st ed.; Springer-Verlag: Berlin, 1987.
37. Kanamori, T.; Itoh, K.; Nakajima, A. *Polym. J.* **1970**, *1*, 524.
38. Mohadger, Y.; Holaday, W.; Wilkes, G. L. *J. Polym. Sci.* **1977**, *15*, 2935.

NANOSTRUCTURES

Chapter 12

Quantum Confinement Effects in Thin Films of Block Conjugated Copolymer Heterostructures

X. Linda Chen and Samson A. Jenekhe[1]

Department of Chemical Engineering and Center for Photoinduced Charge Transfer, University of Rochester, Rochester, NY 14627–0166

A new ABA block conjugated copolymer, poly(2,5-benzoxazole)-*block* -poly(benzobisthiazole-2-hydroxy-1,4-phenylene)-*block* -poly(2,5-benzoxazole), has been synthesized, characterized, and used to investigate quantum confinement effects in semiconducting polymer heterostructures. Thin films of the self-organized block copolymer / homopolymer (ABA/A) blends are shown to exhibit quantum confinement effects at room temperature. New optical transition, enhanced luminescence, and enhanced stability of luminescence at high electric fields were observed to be characteristic of low dimensional excitons in the semiconducting polymer quantum-well heterostructures. Efficient excitation energy transfer from the higher energy gap block or homopolymer A to the lower energy gap block B in these self-organized block copolymer/homopolymer heterostructures was shown to lead to enhanced photobleaching, population inversion of B chromophores, and large stimulated emission from block B. Observation of efficient stimulated emission in these block copolymer /homopolymer blend thin films suggests the potential application of organic quantum-well heterostructures in solid state diode lasers.

Low dimensional inorganic semiconductor systems have been of much scientific and technological interest in the past 25 years (*1-3*). Compared to bulk semiconductors, low dimensional semiconductors exhibit novel or modified electronic and optical properties due to spatial confinement of charge carriers or excitons (*1-3*). Increased exciton binding energy, enhanced oscillator strengths, large Stark shifts, and enhanced optical nonlinearities in low dimensional

[1]Corresponding author.

semiconductors have led to device applications such as diode lasers, Stark effect optical modulators, self-electrooptic effect devices, optical waveguide switches, and cavity-less nonlinear optical bistability (1-3). Organic semiconductors with reduced dimensionality are also of growing interest owing to their expected unique features compared to the corresponding inorganic semiconductors (4-22). Defect-free heterostructures can be more readily fabricated due to the weak Van der Waals interactions between organic molecules. The smaller dielectric constant of organic molecules and polymers [~3-4] compared to inorganic semiconductors [>10] could result in strong Coulomb interactions between electrons and holes, leading to strong excitonic effects and large exciton binding energies [~0.5-1.0 eV]. In spite of the many theoretical studies which have predicted quantum confinement effects in heterostructured organic semiconductors (9-13), clear experimental observation of such effects was not reported until recently (5,6, 19-22). One major experimental difficulty is the rather small exciton Bohr radii (a_B) in bulk organic semiconductors (a_B ~ 1.0-2.0 nm) which places severe limitations on suitable techniques for preparing the nanoscale structures(5,6,18,23).

Layered organic quantum-well heterostructures, which are stacks of alternating thin films of different organic dye molecules, have been prepared by ultrahigh vacuum organic molecular beam deposition (OMBD) technique (4-7). Layered nanostructures of naphthalene and perylene derivatives grown by OMBD have been shown to exhibit the characteristics of multiple quantum wells (MQWs) with one-dimensional exciton confinement (4-6). The resulting two-dimensional excitons in such OMBD grown crystalline MQWs have been shown to be spatially extended (Wannier-like), almost spherical with Bohr radii of 1.3 nm, and to have binding energies as large as 150 meV which are larger by an order of magnitude than usually found in inorganic MQWs (5,6). Other approaches to layered organic semiconductor nanostructures, such as sequential electrochemical polymerization, have also been explored (8). Semiconducting polymer heterostructures consisting of alternating layers of polythiophene and polypyrrole have been successfully prepared (8). However, the characteristic new discrete excitonic features in absorption or excitation spectra commonly observed in inorganic semiconductor MQWs and superlattices (2,3) were not observed in the layered organic nanostructures (5,6,8).

It is expected that organic semiconductor systems in which the motion of charges or excitons is confined to one or zero dimension (quantum wires or quantum boxes) will provide materials with even more remarkable new properties compared to layered quantum-well heterostructures. Many theoretical calculations have thus investigated quantum confinement effects in block conjugated copolymer chains (9-13). Theoretical studies of polydiacetylene-polyacetylene-polydiacetylene triblock heterostructures led to the prediction of results similar to those of multiple quantum wells in inorganic low dimensional semiconductors. Discrete split-off exciton states as well as localization of electronic states were predicted(9, 10).

Theoretical studies of various aspects of the electronic and excitonic properties of quasi-one-dimensional semiconducting polymer superlattices which are periodic copolymers $(-A_mB_n-)_x$ have been reported *(11-13)*. Electronic localization phenomena, including split-off subbands and the number of subbands in a potential well were investigated as a function of both the well width and the barrier width *(11)*. The electronic structure was systematically studied as a function of copolymer composition and block sizes A_m and B_n *(12)*. Recent calculations of multiblock copolymer chains showed that alternation of sufficiently long sequences of a lower energy gap polymer (B_n) with those of a higher gap one (A_m) can result in electron confinement in the segments of the chain with lower gap *(13)*. However, the few prior experimental attempts to synthesize such block conjugated copolymers did not produce materials with evidence of quantum size effects *(14,15)*. The experimental studies on conjugated copolymers $(-A_mB_n-)_x$, where the segment lengths were relatively short (m=2-3, n=2-6), could not clearly test the predicted spatial confinement effects *(14,15)*.

Thin films of conjugated Schiff base polymers of chart 1 have been prepared by chemical vapor polymerization of terephthaldehyde and various aromatic diamines *(16)*. Insertion of —O—, —S—, and —CH$_2$— linkages into the parent conjugated polymer poly(1,4-phenylenemethylidynetrilo-1,4-phenylenenitrilomethylidyne) (PPI) was claimed to result in the formation of quantum dot sequences in the chain *(16)*. These authors regarded conjugated polymers such as PPI as "natural quantum wires" and the ether–, thio–, and methylene–linked polymers in chart 1 as quantum-well structures. However, no concrete evidence of quantum confinement effects in the vapor-deposited polymer thin films was presented *(16)*. In fact, similar conjugated Schiff base polymers have been chemically synthesized and their electronic structures and nonlinear optical properties extensively investigated in our laboratory *(17)*. There is no conceptual or physical basis for regarding thin films of these polymers (chart 1) as quantum-well structures.

Recently, we reported a self-assembly approach to preparing semiconducting polymer quantum boxes and quantum wires and the first experimental observation of the discrete exciton states in a low dimensional organic semiconductor *(19-22)*. Here, we briefly discuss aspects of quantum confinement of excitons in these self-assembled semiconducting polymer heterostructures *(19-22)*. The quantum boxes and wires investigated were prepared from the conjugated homopolymers poly(benzobisthiazole-1,4-phenylene)(PBZT) and poly(2,5-benzoxazole)(2,5-PBO) and triblock copolymer poly(2,5-benzoxazole)-*block*-poly(benzobisthiazole-1,4-phenylene)-*block*-poly(2,5-benzoxazole) which was denoted TBA *(18)*. The symmetric $A_nB_mA_n$ triblock copolymer TBA has a middle PBZT block and two outer 2,5-PBO blocks. The homopolymers PBZT and 2,5-PBO have energy gaps, based on absorption band edges, of 2.48 eV and 3.24 eV, respectively *(18)*. Thus, the

Chart 1

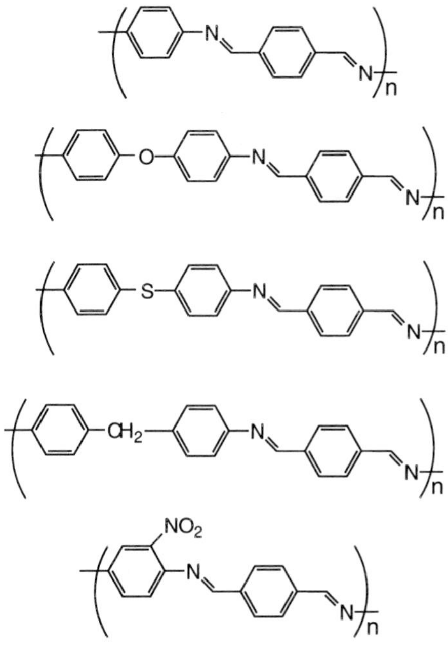

PBZT block of TBA represents a quantum well and the 2,5-PBO blocks constitute electron potential barriers with a height of $\Delta E_g = E_g^A - E_g^B = 0.76$ eV. However, the as-synthesized triblock copolymer was shown to be microphase separated with consequent aggregation of PBZT blocks (18). By blending the triblock copolymer TBA with the higher energy-gap homopolymer (2,5-PBO), self-organized semiconducting polymer heterostructures were obtained in which isolated single chains of the triblock can be obtained. From the known 12.5 Å repeat unit length of PBZT (18), quantum boxes could be obtained when compared to the exciton Bohr radius (a_B) of PBZT ($a_B = 13$ Å) (23) if the block length of the PBZT in TBA is small (2-9 repeat units). On the other hand, PBZT repeat units of 15 or more in TBA ensures that a quantum wire could be obtained. Our initial studies of these low-dimensional organic semiconductors prepared from block conjugated copolymers have revealed remarkable new phenomena due to quantum confinement effects. Of fundamental interest was the detection of new discrete exciton states which confirmed the low dimensional nature of the excitons in these semiconducting polymer heterostructures. Additional evidence confirming the quantum confinement of excitons in these semiconducting polymer heterostructures included large enhancements of photoluminescence quantum efficiency, enhanced exciton lifetime, and the exceptional exciton stability with temperature and large electric field (19-22). These results represent the first experimental evidence for the quantum confinement effects theoretically predicted (9-13) in block copolymer heterostructures.

In this paper, we report quantum confinement effects in thin films of semiconducting polymer heterostructures prepared by self-organization in blends of a new ABA triblock copolymer with its parent homopolymer A. The new triblock copolymer, poly(2,5-benzoxazole)-*block*-poly(benzobisthiazole-2-hydroxy-1,4-phenylene)-*block*-poly(2,5-benzoxazole) which is denoted TBC (Figure 1), was synthesized, characterized, and used to investigate quantum confinement effects. Copolymer TBC-1 consists of a poly(benzobisthiazole-2-hydroxy-1,4-phenylene) (HPBT) middle block and a poly(2,5-benzoxazole) (2,5-PBO) outer blocks with the composition $A_{20}B_9A_{20}$. The electroactive and photoactive properties of HPBT and 2,5-PBO homopolymers have previously been investigated in our laboratory (18, 24). The optical absorption maxima in the homopolymers HPBT and 2,5-PBO are 450 and 479 nm, and 340 and 355 nm, respectively. The corresponding HOMO-LUMO energy gaps based on absorption band maxima are: $E_g^B = 2.59$ eV (HPBT) and $E_g^A = 3.49$ eV (2,5-PBO). Thus, the HPBT segment of triblock TBC-1 chain is expected to form a quantum well while the 2,5-PBO blocks form electron potential barriers with $\Delta E_g = 0.90$ eV. Thin films of pure triblock copolymer TBC-1 and its binary blends with 2,5-PBO were investigated by optical absorption, photoluminescence (PL), and photoluminescence excitation (PLE) spectroscopies, time-resolved PL decay dynamics as well as picosecond transient absorption spectroscopy. The results show clear evidence of quantum confinement effects in TBC-1/2,5-PBO blend system.

Experimental Section

Materials. 2,5-Diamino-1,4-benzenedithiol (DABDT) (Daychem or TCI) was purified by recrystallization from aqueous HCl as previously reported(*24*). The 2-hydroxyterephthalic acid was prepared from bromoterephthalic acid by copper-catalyzed displacement of bromine reported by Miura *et al.*(*25*). 3-Amino-4-hydroxybenzoic acid (TCI, 97 %) was used as received. Poly(phosphoric acid)(PPA) and 85 % phosphoric acid (ACS reagent grade, Aldrich) were used to prepare 77 % PPA, which was used as the polymerization medium. Phosphorus pentoxide (Fluka) was used as received.

Synthesis. The triblock copolymer TBC-1 was synthesized by condensation copolymerization in polyphosphoric acid, similar to previously reported procedures(*18*). Carboxylic acid-terminated HPBT block (HOOC-B_m-COOH) was synthesized by reacting 2,5-diamino-1,4-benzenedithiol dihydrochloride (DABDT) with excess 2-hydroxy-terephthalic acid (HTA) in polyphosphoric acid (PPA). The $A_nB_mA_n$ triblock was obtained by copolymerizing HOOC-B_m-COOH with 3-amino-4-hydroxybenzoic acid (AHBA) in PPA. The block lengths m and n were controlled through the stoichiometric ratios of HTA to DABDT and AHBA to HOOC-B_m-COOH. The condensations were designed for a product content of 10 % w/w in 77 % PPA at the completion of reaction. The polymer product was first washed by a large amount of water and then refluxed with water for 24 hours. The typical yield was ~100%. For TBC-1, the average composition is $A_{20}B_9A_{20}$, where A and B are the 2,5-PBO and HPBT repeat units respectively.

Characterization. The ^1H NMR spectra were taken at 300 MHz using a General Electric Model QE 300 instrument. Solutions for ^1H NMR spectra were prepared in a dry box, using deuterated nitromethane (CD_3NO_2) containing aluminum (III) chloride. FTIR spectra were taken at room temperature using a Nicolet Model 20SXC Fourier transform infrared (FTIR) spectrometer under nitrogen purge. The FTIR samples were in the form of free standing films. Thermogravimetric analysis (TGA) and differential scanning calorimetry (DSC) were done using a Du Pont Model 2100 Thermal Analyst based on an IBM PS/2 Model 60 computer and equipped with a Model 951 TGA and a Model 910 DSC units. The TGA data were obtained in flowing nitrogen at a heating rate of 10 °C/min whereas the DSC thermograms were obtained in nitrogen at a heating rate of 20 °C/min. Intrinsic viscosity [η] of the triblock copolymer was measured in dilute solutions in the range of 0.1-0.2 g/dL in methanesulfonic acid (MSA) at 30°C by using a Cannon Ubbelhode capillary viscometer.

Thin solid films of TBC-1 and binary blends of TBC-1 with 2,5-PBO were spin coated onto silica or indium-tin-oxide (ITO) coated glass substrates from their solutions in trifluoroacetic acid/methane sulfonic acid (v/v, 9:1). The films were subsequently immersed in ethanol solvent containing 1 wt% triethylamine to remove the acid and dried in a vacuum chamber at 60 °C. The film thickness of all samples was typically 100 nm and was measured by an Alpha Step Profilometer (Tencor Instruments) which has a resolution of 1 nm. Optical absorption spectra were recorded with a Perkin-Elmer Model Lambda 9 UV-Vis-NIR spectrophotometer. Steady state photoluminescence (PL), photoluminescence excitation (PLE), and electric field-induced PL quenching measurements were done on a Spex Fluorolog-2 spectrofluorometer equipped with a Spex DM3000f spectroscopy computer. The thin film samples for PL or PLE studies were positioned such that the emission was detected at 22.5° from the incident radiation beam. Relative PL quantum efficiencies were obtained by integration of PL emission spectra. For PL measurement under an external field, aluminum was evaporated onto thin film samples on indium-tin-oxide (ITO) coated glass. A positive bias voltage was applied to the ITO electrode, creating an electric field across the film sandwiched between ITO and aluminum electrodes. The samples were illuminated through the ITO-glass side and the PL spectra collected under applied field. Picosecond time-resolved PL decay was performed using time-correlated single-photon-counting techniques.

Transient absorption spectroscopy was performed in a system which consisted of a Continuum PY61 Series Nd:YAG laser utilizing Kodak QS 5 as the saturable absorber to produce laser light pulses of ~25 ps FWHM. These output pulses were then amplified and the second and third harmonics generated (532 nm and 355 nm respectively). Dichroic beamsplitters in conjunction with colored glass filters were used to isolate the fundamental (1064 nm) and desired harmonic. The fundamental was directed along a variable optical delay and then focused into a 10 cm quartz cell filled with H_2O/D_2O (50:50) to generate a white light continuum probe pulse. The excitation and probe pulses (ca. 2 mm diameter) were passed approximately coaxially through the sample. The probe pulse was directed to a Spex 270 M monochromator through a Princeton Instruments fiber optic adapter and dispersed onto a Princeton Instruments dual diode array detector (DPDA 512). This allowed ~350 nm of the visible spectrum to be collected in a single experiment. A ST-121 detector controller/interface was incorporated into a 386/25 MHz PC to control the arrays and for data storage, manipulation and output. Additional experimental details, including the instruments and detailed methods used in our optical and photophysical experiments can be found elsewhere(*18,19,27(a),27(c)*).

Results and Discussions

Block Copolymer Structure and Composition. The synthetic approach used and the quantitative reaction yield ensured that the desired triblock copolymer structure was obtained. The molecular structures and composition of the copolymer TBC-1 (Figure 1) were established primarily by ^{1}H NMR and FTIR spectroscopies, thermal analysis, intrinsic viscosity, and various other spectroscopic measurements, as previously done for related polymers *(18)*. The ^{1}H NMR spectrum of the triblock copolymer in CD_3NO_2/ $AlCl_3$ was essentially a superposition of those of the two parent homopolymers (HPBT, 2,5-PBO). The proton resonances of the 2,5-PBO blocks appear as a singlet at 9.25, a doublet at 9.0, and a doublet at 8.7 ppm, which are assigned to the protons *ortho, para,* and *meta* to the oxazole nitrogen, respectively. Two of the proton resonances of the HPBT block appear at 8.45 ppm (doublet) and 8.1 ppm, which are assigned to the 1,4-phenylene protons *meta* and *para* to the hydroxy group. The 1,4-phenylene proton *ortho* to the 2-hydroxy group also appears at 8.45 ppm, thus overlapping with the resonance of the proton at 6 position. The protons of the benzobisthiazole ring system of HPBT block appear as a broad resonance peak between 9.2-9.3 ppm, thus overlapping with the proton *ortho* to the oxazole nitrogen of 2,5-PBO block. The fact that the ^{1}H NMR spectrum of the TBC-1 is a superposition of those of the HPBT and 2,5-PBO homopolymers indicates that the copolymer is a true block copolymer. The ratio of the integrated peak at 8.7 ppm (2,5-PBO block) to that at 8.45 ppm (HPBT block) was 1.95, which is consistent with the theoretical ratio of 2 for the proposed block copolymer composition (Figure 1). The FTIR spectrum of TBC-1 exhibits all the characteristic bands of the parent homopolymers. There are major differences between the spectra of HPBT and 2,5-PBO, for example, the 958-cm^{-1} band due to heteroring breathing of HPBT, is absent in the 2,5-PBO spectrum whereas the C-O-C vibration band arising from the oxazole ring at ~1060 cm^{-1} is characteristic of 2,5-PBO. The ratio of the characteristic bands (958, 1060 cm^{-1}) was consistent with the composition of copolymer. The FTIR results in conjunction with the ^{1}H NMR spectra clearly confirmed the proposed structure and composition of the triblock copolymer.

Thermogravimetric analysis of the triblock showed a single onset of decomposition in N_2 (670 °C) that was different from either homopolymer, 650 °C for 2,5-PBO and ~700 °C for HPBT. Intrinsic viscosity of TBC-1 in methanesulfonic acid (MSA) at 30 °C was 2.5 dL/g which is higher than either the starting $HOOC-B_9-COOH$ (1.0 dL/g) or a 2,5-PBO homopolymer with a degree of polymerization of 60(1.7 dL/g). All these results further confirmed the chemical linkage between the A and B blocks in the copolymer.

Exciton Localization Phenomena in Block Copolymer Thin Films. The expected exciton localization in a quantum-well heterostructure was probed in thin films of TBC-1 and its blends with 2,5-PBO by optical absorption spectroscopy. Figure 2 shows the absorption spectra of thin films of the pure TBC-1 and its binary blends with 2,5-PBO homopolymer. The optical absorption spectrum of 2,5-PBO homopolymer is also shown for comparison (The composition of TBC-1/2,5-PBO blends is specified in terms of mole % HPBT repeat units, thus a 8

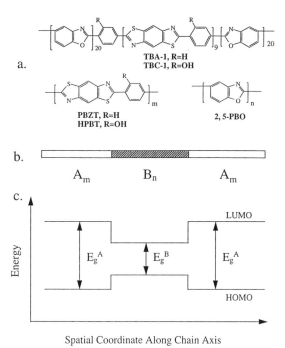

Figure 1. (a) Chemical structures of ABA block conjugated copolymers and parent homopolymers. (b) Schematic illustration of triblock copolymer chain. (c) Schematic illustration of quantum-well formation in an isolated ABA triblock chain due to electronic structure differences.

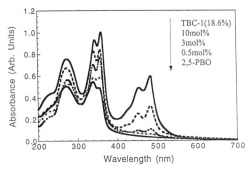

Figure 2. Optical absorption spectra of thin films of the triblock/homopolymer (TBC-1/2,5-PBO) blend system. Blend composition in mol% HPBT repeat units is shown.

mol % blend corresponds to about 1:1 triblock to 2,5-PBO homopolymer chain ratio and 1 mol % corresponds to roughly one triblock chain in fourteen 2,5-PBO chains). The highly structured absorption spectrum of TBC-1 is clearly a superposition of those of its two component homopolymers. The vibrationally resolved absorption bands in the 400-500 nm region with λ_{max} of 450 and 479 nm are characteristic of the HPBT block while the vibrationally resolved band in the 240-380 nm region with λ_{max} of 270, 340, and 355 nm is due to 2,5-PBO block. The absorption spectra of the TBC-1/2,5-PBO blends are also superpositions of those of the two homopolymers, with varying ratio of the 479-nm band intensity, as expected. Compared to the optical absorption spectra of the homopolymer thin films, the absorption spectra of the copolymer and its blends reveal bandwidth narrowing and exciton peak sharpening. For example, the lowest energy absorption peak in 2,5-PBO (355 nm) is relatively more intense in the copolymer and its blends than in the homopolymer. Similarly, the 479-nm transition which is characteristic of the HPBT chromophore is more intense in the spectra of Figure 2 than previously reported for the homopolymer (24). The observation of the absorption bands characteristic of the component homopolymers in the absorption spectrum of the block copolymer is strong evidence that excitons produced in the triblock chains are confined to two different spatial regions with different energy levels as schematically illustrated in Figure 1c. Exciton localization phenomenon revealed in the electronic absorption spectrum of the block copolymer is also an additional evidence of the block copolymer chain structure of TBC-1. However, no new optical transition that could be attributed to quantum confinement effects was observed in the copolymer or blend absorption spectra. This is due to the low sensitivity of room temperature absorption spectroscopy in contrast to photoluminescence excitation spectroscopy, which will be discussed subsequently. The sharpening of the exciton peak in the absorption spectra of the block copolymer and its blends may be an indication of reduced dimensionality of the excitons, as has been observed in inorganic semiconductor quantum-well structures (28).

Quantum Confinement Effects. Photoluminescence excitation (PLE) spectroscopy at 298K was more sensitive in revealing evidence of progressive spatial confinement in going from the pure triblock to blends of the triblock with 2,5-PBO homopolymer. Figure 3 shows the PLE spectra of the thin films of 2 mol% blend and pure TBC-1 (normalized at 479 nm). The excitation spectra were obtained by monitoring the emission at 620 nm where there is negligible emission from the 2,5-PBO chromophore. The PLE spectrum of the blend has features similar to the PLE spectrum of the pure triblock as well as a major difference. The presence of a large 340-360-nm band due to 2,5-PBO absorption shows that the 620-nm emission from the HPBT block has a significant component in excitation energy transfer from 2,5-PBO. The major difference in the spectra of Figure 3 is that a small but entirely new band centered at 540 nm (2.3 eV) appears in the PLE spectrum of the blend. The new below-gap 540-nm optical transition was dependent on the blend concentration and was only clearly seen in blends with low HPBT concentrations (<5mol%). Similar results were previously observed in other block copolymer/homopolymer blends, such as the TBA/2,5-PBO blends, where new optical transitions were seen below the band gap of PBZT(*19-22*). The new optical transition at 540 nm, which is below the optical band gap of the lowest energy component of the block copolymer, has not been anticipated in prior theoretical studies (*11-13*). Only optical transitions with energy levels intermediate between or higher than those of the two components of a block copolymer have been predicted. We propose that these below-gap excitonic states arise from spatial-confinement-induced delocalization of excitons along the length of the quantum boxes or wires (0.35 x 0.58 x l_m nm^3, l_m = 2.4-25 nm) which is many times larger than the 1.3 nm bulk exciton Bohr radius of PBZT or HPBT (*23*). A net effect of such an exciton delocalization could be the formation of below-gap states much like the well-known photogenerated bipolaron/polaron states in conjugated homopolymers (*29*). Regardless of the exact quantum mechanical origin of the observed new optical transitions in the PLE spectra of these organic low dimensional systems, they are evidence of new exciton states induced by confinement in ultrasmall volumes.

Additional evidence of quantum confinement effects in thin films of the block copolymer /homopolymer blends was obtained through steady-state and time-resolved PL measurements as well as by electric field-modulated PL spectroscopy. Figure 4a shows the PL emission spectra of thin films of TBC-1 and TBC-1/2,5-PBO blends excited at 360 nm. Excitation at 360 nm preferentially excites 2,5-PBO, as can been seen from the absorption spectra of Figure 2. Excitation of TBC-1 at 360 nm results in a broad PL emission band in the 460-640 nm region with a dominant peak at 566 nm and a shoulder at 495 nm. Similar PL emission spectra of TBC-1 were obtained when the HPBT block was directly excited at either 450 or 479 nm where the 2,5-PBO block does not absorb. The broad, featureless PL emission spectrum of the block copolymer TBC-1 is very similar to that of the HPBT homopolymer (λ_{max} =570 nm) which has been previously assigned to excimer emission (*24*). This similarity of emission spectra of HPBT and TBC-1 suggests that microphase separation has occurred in the TBC-1 block copolymer and thus that its luminescence was from excimer-forming aggregates or microdomains. Independence of the block copolymer PL emission spectrum from excitation energy and the nearly complete quenching of emission from 2,5-PBO block in the 380-420 nm region indicates that there is efficient excitation energy transfer from 2,5-PBO blocks to the middle HPBT block and also that the lower energy HPBT block acts as a potential well which traps excitons and serves as a radiative recombination center.

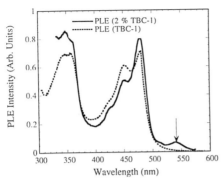

Figure 3. Room temperature (298K) PL excitation spectra (620-nm emission) of thin films of TBC-1 and the 2% blend. The PLE spectra were normalized at 479 nm. Vertical arrow marks a new optical transition.

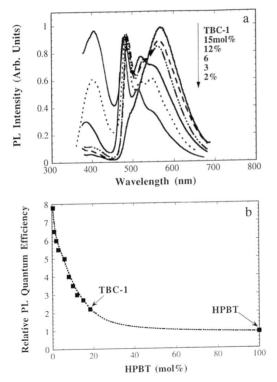

Figure 4. PL emission spectra of TBC-1/2,5-PBO blend system thin films at 298K. (a) PL emission spectra (360-nm excitation) as a function of blend composition. (b) Relative PL quantum efficiency as a function of blend composition for the blend spectra in (a) and excitation of 450 nm.

The PL emission spectra of thin films of the binary block copolymer (TBC-1) /homopolymer (2,5-PBO) blends of varying composition are also shown in Figure 4a. The blend PL spectra are dramatically different from that of the pure block copolymer in three ways: the emission band spectral position, the PL emission lineshape, and the Stokes shift of the emission maximum from the absorption maximum. As the block copolymer content of the blends, as measured by mol% HPBT, is progressively reduced to 2%, the HPBT block emission maximum is blue shifted to 479 nm. In the dilute blends (<6%), a vibrationally resolved PL emission spectrum is observed. The emission linewidth in terms of full width at half maximum (FWHM) is reduced from 144 nm (571 meV) for the 15% blend to 47 nm (280 meV) for the 2% blend. The Stokes shift of the PL emission is reduced from 413 meV for the block copolymer to only 11 meV for the 2% blend. Furthermore, the blend emission spectra shown in Figure 4a were independent of excitation wavelength. The progressive evolution of the spectrally narrowed and vibrationally resolved PL emission spectrum of the dilute (2%) blend from that of the microphase-separated copolymer reflect the increasing spatial confinement of the HPBT blocks and emergence of the isolated potential well as exciton traps and radiative recombination centers. Since the Stokes shift in conjugated polymers and molecules commonly arises from excitation relaxation, the relatively small Stokes shift in the PL emission of dilute blends indicates that excitons photogenerated in the HPBT blocks are confined to the lowest energy exciton states.

The blend PL emission spectra in Figure 4a also show the progressive increase of the 2,5-PBO emission centered at ~380-400 nm with decreasing concentration of HPBT. Although there is nearly complete quenching of emission from the 2,5-PBO chromophores in the pure triblock and in the 12-15 % blends, further dilution results in increased emission from the predominantly 2,5-PBO-containing blends because of the reduction in the number of HPBT blocks which act as luminescence quencher for excited 2,5-PBO. The highly efficient excitation energy transfer from the excited 2,5-PBO blocks to HPBT blocks is thus regulated by the composition of the blends.

The observed PL emission spectral change in the block copolymer /2,5-PBO blends was accompanied by enhancement in PL quantum efficiency with increasing HPBT block confinement as shown in Figure 4b. About a factor of 4 enhancement in PL quantum yield was observed in going from the pure TBC-1 triblock (18.4 mol% HPBT) to 0.5 % blend. The enhancement with increasing spatial confinement of HPBT blocks can be understood to arise from a decrease in the nonradiative recombination rate with HPBT chromophore dilution in the blends. This interpretation was confirmed by the results of picosecond-resolved PL decay dynamics study which showed an increase in lifetime or decay time with decrease in HPBT block concentration. The decrease of nonradiative recombination rate with decreasing concentration of HPBT blocks in the blends in turn is a consequence of the progressive elimination of the bimolecular recombination component of nonradiative decay as the distance between the HPBT blocks is increased.

The results of electric field-modulated PL spectroscopy on ca. 100-nm thin films of pure TBC-1 and its 3% blend are shown in Figure 5 for 450-nm excitation. Up to 50% quenching of the photoluminescence of the pure block copolymer is observed at an applied bias voltage of 12 (Figure 5a) which corresponds to an electric field of 1.2 x 10^6 V/cm. This large electric field-induced quenching of photoluminescence in a block conjugated copolymer is similar to previous observation on conjugated homopolymers by our laboratory (*19*) and others (*30*). The electric field-induced quenching of luminescence arises from field-induced dissociation of excitons in the materials and the extent of dissociation depends on exciton binding energy, spatial confinement, and electric field. In contrast, electric field-induced quenching of luminescence was not observed in dilute blends, such as the 3% blend shown in Figure 5b, at electric fields of up to 2.5 x 10^6 V/cm. This exceptional stability of excitons in HPBT quantum boxes under large electric fields reflects their strong three-dimensional confinement and large binding energy.

Stimulated Emission From Block Copolymer Heterostructures. Transient photoinduced absorption (PA) spectroscopy with picosecond time resolution was done on a different block copolymer /homopolymer blend system (TBA-1/2,5-PBO) which is related to the present TBC-1/2,5-PBO blend system (see Figure 1). This experiment is essentially a pump-and-probe technique in which a 12 mol% TBA-1/2,5-PBO blend thin film with optical density of 0.6 at 470 nm was photoexcited at 355 nm and then the excited state was probed with white light continuum (400 - 950 nm). Figure 6 shows transient absorption spectra at selected delays after photoexcitation. No transient species was observed at 0 ps delay. However, photoinduced bleaching in the 380 - 490 nm region with peaks at 440 and 470 nm was observed at t > 0 ps delays. Since the spectral position of photobleaching overlaps with the absorption band of PBZT blocks in TBA-1, it can be clearly identified as the bleaching of PBZT chromophores. Photobleaching of the PBZT absorption reached its maximum with an optical density of -0.12 at 470 nm and 100 ps delay. At subsequent delay times (100-200 ps, Figure 6) the photobleached signal progressively decays. Similar PA experiments on the pure homopolymer thin films (2,5-PBO thin film with an optical density of 2.0 at 355 nm and PBZT thin film with an optical density of 0.6 at 470 nm) photoexcited at 355 nm showed that photobleaching in PBZT reach a maximum at 50 ps delay with an optical density of -0.015 at 470 nm. No photoinduced absorption signal in the 400-700 nm was observed in 2,5-PBO at all delay times, although occurrence of PA below 400 nm, which was not probed, cannot be ruled out. These results show that the 12 % blend thin film which has an identical optical density as the PBZT homopolymer thin film exhibits an enhanced photobleaching by a factor of 8 compared to the homopolymer. The enhanced photobleaching in the ABA/A

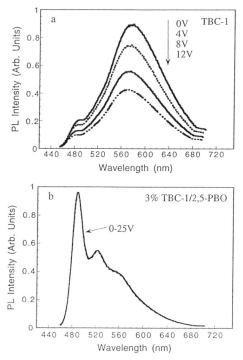

Figure 5. Electric field-modulated PL spectra of thin films of TBC-1 (a) and 3% blend (b) for 450-nm excitation at different bias voltages.

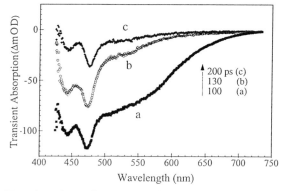

Figure 6. Transient absorption spectra of 12 mol% TBA-1/2,5-PBO blend thin film recorded at selected delay times after a 355-nm laser excitation: (a) 100, (b) 130, and (c) 200 ps.

block copolymer blend is a consequence of the efficient excitation energy transfer from A* to the middle B block.

Stimulated emission (SE) in the 490 - 700 nm region, which coincides with the emission region of PBZT (*18-20*), was also observed in the PA spectra of the 12% copolymer blend thin film as shown in Figure 6. The large SE signal at 100 ps delay in the 500-520 nm region, with an optical density of -0.08, corresponds to an optical gain of 20%. The SE signal persists to 200 ps (Figure 6). SE signal of any magnitude was not observed in either 2,5-PBO or PBZT homopolymer. Clearly, the observed efficient stimulated emission in the TBA-1/2,5-PBO blend is a direct result of excitation energy transfer from the excited 2,5-PBO to the PBZT blocks in the copolymer.

The dynamics of the PA in the block copolymer blends revealed that whereas the photobleaching occurred instantly (<1 ps) after photoexcitation, there was about 80 ps risetime for the stimulated emission. Given the high excitation density of 100 mJ/cm^2 and the large A absorption cross section in the ABA/A blends, the excited state energy level of A will be significantly populated by the pump pulse. Since excitation energy transfer occurs on the subpicosecond time scale (*31*), the observed instant photobleaching of the B can be understood. The observed risetime of SE is explained by the fact that SE occurs only when the population of B is inverted by energy transfer and accumulation of B in the excited state. Blend thin films of various TBA-1/2,5-PBO compositions were investigated under similar PA experimental conditions to gain more insight into the factors that control SE in these materials. Enhanced photobleaching was observed in all the blends. However, significant SE was only observed in blends of 3-15 mol% PBZT but not in the more dilute (< 3%) or more concentrated (> 15%) blends. These results suggest that a balance between spatial confinement and chromophore concentration is essential to achieving population inversion and efficient stimulated emission in block copolymer heterostructures.

Conclusions

A new symmetric ABA triblock conjugated copolymer, poly(2,5-benzoxazole)-*block*- poly(benzobisthiazole-2-hydroxy-1,4-phenylene)- *block* -poly(2,5-benzoxazole), has been prepared and thin films of its blends with the higher energy gap 2,5-PBO homopolymer have been shown by photoluminescence emission and excitation, electric field-modulated photoluminescence, and picosecond absorption spectroscopies to exhibit a range of quantum confinement effects at room temperature. Some of the observed quantum effects include a new optical transition in the photoluminescence excitation spectrum, enhanced luminescence efficiency, PL emission spectral narrowing, and unusual exciton stability in large applied electric fields. However, thin films of the pure block conjugated copolymer did not exhibit any quantum size effects because of microphase separation. These new HPBT-based block copolymer quantum-well

176

heterostructures thus expand the rather small number of currently known organic quantum wires and quantum boxes.

Cooperative interaction between the two electroactive and photoactive components of the block copolymer/homopolymer (ABA/A) blends is exemplified by the efficient excitation energy transfer from A to B which led to the observed enhanced photobleaching and large stimulated emission. Block conjugated copolymer/ homopolymer blends thus represent a promising novel class of solid state lasing materials.

Acknowledgments

This research was supported by the Office of Naval Research and in part by the National Science Foundation (CTS-9311741, CHE-9120001).

Literature cited

(1)   Esaki, L.; Tsu, R. *IBM J. Res. Develop.* **1970**, *14*, 61.

(2)   Dingle, R.; Wiegmann, W.; Henry, C.H. *Phys. Rev. Lett.* **1974**, *33*, 827.

(3)   Bryant, G.W. *Phys. Rev. B* **37**, 8763 (1988); Miller, D.A.B.; Weiner, J.S.; Chemla, D.S. *IEEE J. Quantum Electron.* **1986**, *QE-22*, 1816.

(4)   So, F.F.; Forrest, S.R.; Shi, Y.Q.; Steier, W.H. *Appl. Phys. Lett.* **1990**, *56*, 674.

(5)   So, F.F.; Forrest, S.R. *Phys. Rev. Lett.* **1991**, *66*, 2649.

(6)   Haskal, E.I.; Chen, Z.; Burrows, P.E.; Forrest, S.R. *Phys. Rev. B* **1995**, *51*, 4449.

(7)   Imanishi, Y.; Hattori, S.; Kakuta, A.; Numata, S. *Phys. Rev. Lett.* **1993**, *71*, 2098.

(8)   Fujitsuka, M.; Nakahara, R.; Iyoda, T.; Shimidzu, T.; Tsuchiya, H. *J. Appl. Phys.* **1993**, *74*, 1283.

(9)   Ruckh, R.; Sigmund, E.; Kollmar, C.; Sixl, H. *J. Chem. Phys.* **1986**, *85*, 2797.

(10)   Ruckh, R.; Sigmund, E.; Kollmar, C.; Sixl, H. *J. Chem. Phys.* **1987**, *87*, 5007.

(11)   Seel, M.; Liegener, C.M.; Forner, W.; Ladik, J. *Phys. Rev. B* **1988**, *37*, 956.

(12)   (a) Bakhshi, A. K.; Liegener, C.-M.; Ladik, J. *Synth. Met.* **1989**, *30*, 79; (b) Bakhshi, A. K. *J. Chem. Phys.* **1992**, *96*, 2339; (c) Bakhshi, A. K.; Pooja *J. Chem. Soc., Faraday Trans.* **1996**, *92*, 2281.

(13)   (a) Piaggi, A.; Tubino, R.; Colombo, L. *Phys. Rev. B* **1995**, *51*, 1624; (b) Musso, G. F.; Dellepiand, G.; Cuniberti, C.; Rui, M.; Borghesi, A. *Synth. Met.* **1995**, *72*, 209; (c) Quattrocchi, C.; dos Santos, D. A.; Bredas, J. L. *Synth. Met.* **1995**, *74*, 187.

(14)   (a) Jenekhe, S.A.; Chen, W.C. *Mat. Res. Soc. Proc.* **1990**, *173*, 589;

(b) Jenekhe, S.A.; Chen, W.C.; Lo, S.K.; Flom, S.R. *Appl. Phys. Lett.* **1990**, *57*, 126.

(15) Piaggi, A.; Tubino, R.; Borghesi, A.; Rossi, L.; Destri, S.; Luzzati, S.; Speroni, F. *Phys. Lett. A* **1994**, *185*, 431.

(16) Yoshimura, T.; Tatsuura, S.; Sotoyama, W. *Mater. Res. Soc. Symp. Proc.* **1992**, *247*, 829.

(17) (a) Yang, C.-J.; Jenekhe, S. A. *Chem. Mater.* **1991**, *3*, 878; (b) Yang, C.-J.; Jenekhe, S. A. *Macromolecules* **1995**, *28*, 1180; (c) Yang, C.-J.; Jenekhe, S. A. *Chem. Mater.* **1994**, *6*, 196; (d) Yang, C.-J.; Jenekhe, S. A.; Vanherzeele, H.; Meth, J. S. *Chem. Mater.* **1991**, *3*, 985; *Polym. Adv. Technol.* **1994**, *5*, 161.

(18) Chen, X.L.; Jenekhe, S.A. *Macromolecules* **1996**, *29*, 6189.

(19) Chen, X.L.; Jenekhe, S.A. *Appl. Phys. Lett.* **1997**, *70*, 487.

(20) Chen, X.L.; Jenekhe, S.A. *Synth. Met.* **1997**, *85*, 1431.

(21) Jenekhe, S. A.; Chen, X.L. In *Chemistry and Physics of Small Structures,* Vol 2, 1997 OSA Technical Digest Series, pp 24-26.

(22) Jenekhe, S. A.; Chen, X.L. *Polymer Prepr.* **1997**, *38(*1), 981.

(23) Martin, S.J.; Bradley, D.D.C.; Osaheni, J.A.; Jenekhe, S.A. *Mol. Cryst. Liq. Cryst.* **1994**, *256*, 583.

(24) Tarkka, R.M.; Zhang, X.; Jenekhe, S.A. *J. Am. Chem. Soc.* **1996**, *118*, 9438.

(25) Wolfe, J.F. In: *Encyclopedia of Polymer Science and Engineering*; Wiley: New York, 1988; vol. 11, 601-635, and references therein.

(26) Miura, Y.; Torres E.; Panetta, C.A.; *J. Org. Chem.* **1988**, *53*, 439-440.

(27) (a) Osaheni, J.A.; Jenekhe, S.A. *Chem. Mater.* **1995,** *7*, 672; (b)*ibid* **1992**, *4*, 1283; (c)Osaheni, J.A.; Jenekhe, S.A. *J. Am. Chem. Soc.* **1995**, *117*, 7389; (d) Jenekhe, S.A.; Osaheni, J.A. *Science* **1994**, *265*, 765.

(28) Yoshimura, T. *Optics Communications* **1989**, *70*, 535.

(29) J.L. Brédas, in: T.A. Skotheim, Ed., *Handbook of Conducting Polymers* (Dekker, New York, 1986) vol. 2, pp. 859-913.

(30) Deussen, M.; Scheidler, M.; Bässler, H. *Synth. Met.* **1995**, *73*, 123.

(31) (a) Kersting, R.; Mollay, B.; Rusch, M.; Wenisch, J.; Leising, G.; Kauffmann, H. F. *J. Chem. Phys.* **1997**, *106*, 2850. (b) King, B. A.; Stanley, R. J.; Boxer, S. G. *J. Phys. Chem. B* **1997**, *101*, 3644. (c) Renger, T.; Voigt, J.; May, V.; Kuhn, O. *J Phys. Chem.* **1996**, *100*, 15654.

Chapter 13

Polystyrene-*block*-poly(2-cinnamoylethyl methacrylate) Brushes

Guojun Liu

Department of Chemistry, The University of Calgary, 2500 University Drive, NW, Calgary, Alberta T2N 1N4, Canada

Polystyrene-*block*-poly(2-cinnamoylethyl methacrylate) (PS-*b*-PCEMA) micelles, formed in THF/cyclopentane (CP) at high CP contents with PCEMA as the core and PS as the shell, were adsorbed by silica. If the volume fraction of CP, f_{CP}, was between 0.67 and 0.90, the adsorbed micelles disintegrated on silica to form polymer brushes in which the PCEMA block spread on the silica surface like a melt and the PS chains stretched into the solution phase like bristles of a brush. This layered structure of PS-*b*-PCEMA brushes was verified by transmission electron microscopy and surface-enhanced Raman scattering studies. The brush surface coverage of different diblock copolymers was found to follow scaling relations predicted by Marques et al.

In a block selective solvent, a diblock copolymer forms micelles with the soluble block in contact with the solvent to stabilize the collapsed insoluble block. The polymer may also deposit out from the solution and self-assemble to form a polymeric monolayer on a substrate being contacted. If the interaction between the insoluble block and the substrate is favorable, a dense monolayer called a polymer "brush" may form in which the insoluble block spreads on the solid substrate like a melt and the soluble block stretches into the solution phase like bristles of a brush "as illustrated in Figure 1" (*1-3*).

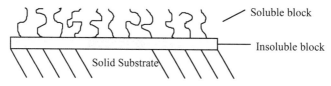

Soluble block

Insoluble block

Solid Substrate

Figure 1. Polymer brush illustrated. Reproduced from reference 19.

Polymer brushes have been traditionally used to stabilize latex and pigment particles (*4*). Due to its industrial relevance, there have been many studies of polymer brushes in the past decade (1-3). In most previous studies, dilute block copolymer solutions (< 0.1%) in which polymer chains existed as unimolecular micelles or unimers were used to coat solid substrates. In such solutions, the buoy layer chains were shown to stretch by the surface-force technique (1, 5-8). The layered structure, i.e. that the soluble buoy layer overlays the insoluble anchoring layer, of diblock brushes has been confirmed by surface tension (9) and surface-enhanced Raman scattering studies (10-12). The equilibrium surface coverage of diblocks with different block lengths under otherwise identical conditions (13-16) has been found to follow a scaling relation derived by Marques et al. (17) for unimer adsorption in the van der Waals-buoy regime.

In industrial situations, diblock adsorption occurs in concentrated solutions in which diblock copolymers exist as micelles. We were the first to study polymer brush formation under these conditions systematically. Here, I report the different scaling relations followed by polystyrene-*block*-poly(2-cinnamoylethyl methacrylate) (PS-*b*-PCEMA) adsorption from micellar solutions. Also to be reported is our experimental verification of the layered structure of the polymer brush formed under these conditions.

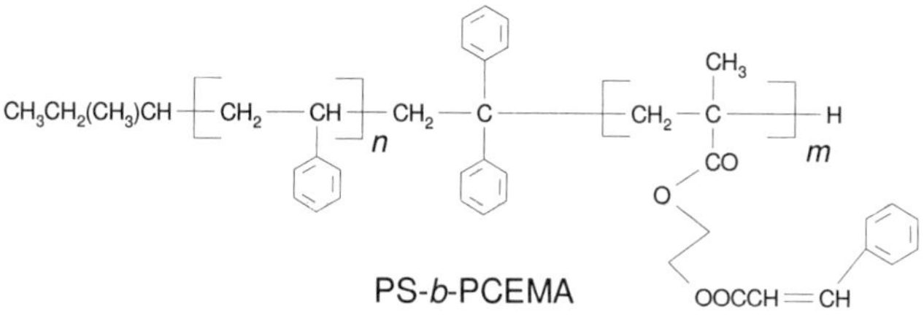

PS-*b*-PCEMA

Experimental Procedures

Polymer Synthesis and Characterization. The precursors to PS-*b*-PCEMA were synthesized by anionic polymerization. Styrene was polymerized at -78 °C in THF using *sec*-butyl lithium as the initiator (18-19). 1,1-diphenyl ethylene (DPE) and lithium chloride, both at 3.0 molar equiv. to *sec*-butyl lithium, were then added. Lithium chloride was added to improve the polydispersity of the second block. DPE reacted with polystyryl anions to produce the sterically more hindered PS-DPE anions, which initiated the polymerization of trimethylsiloxyethyl methacrylate (HEMA-TMS). The TMS protecting groups were removed by hydrochloric acid catalyzed hydrolysis of PS-*b*-P(HEMA-TMS) in THF/methanol to produce polystyrene-*block*-poly(2-hydroxylethyl methacrylate) (PS-*b*-PHEMA). To attach cinnamoyl groups, PS-*b*-PHEMA was reacted with cinnamoyl chloride in pyridine. The polymers used were characterized by GPC, ^{1}H NMR, and light scattering (LS), and their important characteristics are summarized in Table 1.

Table I. Important Characteristics of the Polymers Used.

Laboratory Code	(n/m) by NMR	$\overline{M}_w/\overline{M}_n$ by GPC	$10^{-4}\,\overline{M}_w$ by LS	$10^{-2}\,n$	$10^{-2}\,m$
Surface coverage determined in THF/CP with 86% CP					
810-59	13.8	1.17	10.0	8.1	0.59
271-22	12.0	1.10	3.4	2.71	0.22
340-34	10.1	1.14	4.4	3.4	0.34
124-13	9.9	1.13	1.62	1.24	0.13
227-23	9.7	1.30	2.97	2.27	0.23
1880-195	9.6	1.09	24.6	18.8	1.95
2190-232	9.4	1.10	28.8	21.9	2.32
204-22	9.2	1.12	2.70	2.04	0.22
300-36	8.2	1.12	4.1	3.0	0.36
1180-150	7.9	1.09	16.5	11.8	1.50
750-107	7.0	1.08	10.6	7.5	1.07
490-71	6.8	1.15	6.9	4.9	0.71
222-41	5.3	1.19	3.4	2.22	0.41
410-87	4.7	1.08	6.5	4.1	0.87
264-75	3.5	1.09	4.7	2.64	0.75
1270-360	3.5	1.08	22.6	12.7	3.6
Surface coverage determined in THF/CP with 82% CP					
75-43	1.73	1.14	1.90	0.75	0.43
90-51	1.76	1.14	2.26	0.90	0.51
110-60	1.80	1.13	2.72	1.09	0.61
179-153	1.17	1.14	5.8	1.79	1.53
237-125	1.90	1.11	5.7	2.37	1.25
300-150	1.82	1.08	6.7	2.71	1.49
428-191	2.24	1.11	9.4	4.3	1.91
777-854	0.91	1.06	30	7.8	8.5
1077-1158	0.93	1.10	41	10.8	11.6
1289-623	2.07	1.10	29.6	12.9	6.2

PS-b-PCEMA Adsorption Experiments. The block-selective solvent used for PS-b-PCEMA was cyclopentane (CP, Aldrich, 95%) which solubilized PS but not PCEMA. To adjust solvent quality for PCEMA and ensure PS-b-PCEMA micelle stability, we normally used a mixture of THF/CP, where THF was good for both PS and PCEMA.

We studied the adsorption of PS-b-PCEMA by both silica and silver and the results obtained using aerosil OX50 (spherical non-porous silica with a specific surface area of 50 m^2/g) as the adsorbent will be the focus of discussion here. To coat silica, a coating solution of a known concentration was equilibrated with a known amount of silica particles. Coated particles were separated by centrifugation, and the polymer concentration in the supernatant was determined by UV spectrophotometry. The amount of polymer adsorbed by an adsorbent was calculated from the initial and final cinnamate absorbances, A_0 and A_t, by

$$w_a = \frac{A_0 - A_t}{A_0} c_0 V_0 \qquad (1)$$

where c_0 and V_0 were the initial concentration and volume of the coating solution to which the adsorbent was added. The surface coverage, σ, was calculated using:

$$\sigma = \frac{w_a}{A_{sp} \times w_s} \qquad (2)$$

where w_s is the amount of adsorbent used, and A_{sp} is the specific surface area of the silica.

Using this method, we obtained the adsorption kinetic data of 300-36, a sample with $n = 300$ and $m = 36$, in THF/CP at $f_{CP} = 0.86$ as illustrated in Figure 2 (20). σ initially increased rapidly and then leveled off to its asymptotic value σ_{eq} with time. Similar kinetic behavior was observed for other diblocks. The adsorption equilibrium time, however, increased with the diblock molar masses.

We also established the dependence of σ_{eq} as a function of polymer concentration, c_{eq}, in equilibrium with the silica. Illustrated in Figure 3 are the adsorption isotherms of 271-22 and 300-36 at $f_{CP} = 0.86$ (20). σ_{eq} increased initially rapidly with c_{eq} and then leveled off to σ_∞, the saturated surface coverage.

Surface Enhanced Raman Scattering (SERS) Measurements. Silver foils (Aldrich, 0.025 mm thick, 99.9%) were roughened for 5-8 min in a 3 M nitric acid and 4 M sulfuric acid aqueous mixture. The foils were then rinsed in distilled water and dried with a gust of nitrogen before use. To coat the roughened silver surfaces with a thin film of SiO_2, the foils were immersed in a 1.0 mg/mL tetraethyl orthosilicate (Aldrich, 98%) solution in 95% ethanol for two hours, rinsed with the solvent, and dried in open air for one day (21). PS-b-PCEMA films on these substrates were obtained by equilibrating the substrate with a PS-b-PCEMA solution, typically between 1.0 and 5.0 mg/mL, for a period of time and then drying the withdrawn substrate surface by blowing nitrogen on it. All SERS spectra were recorded on Jarrell-Ash model 25-100 instrument.

Transmission Electron Microscopy (TEM) Studies. Silica particles were equilibrated with a 10 mg/mL 1077-1158 solution in THF/CP with $f_{CP} = 0.82$ for 3 days. The silica particles were then precipitated by centrifugation, rinsed by dispersing in fresh THF/CP with $f_{CP} = 0.82$, and re-precipitated. The rinsing procedure was repeated three times and the coated particles were irradiated for 30 min while dispersed in THF/CP with $f_{CP} = 0.82$ with light passing through a 260-nm cut-off filter from a 500-W Hg lamp. Irradiated particles were centrifuged, separated from solvent, redispersed in cyclopentane, and then sprayed onto a Formvar-covered copper TEM grid.

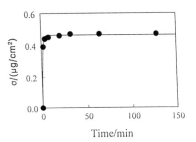

Figure 2. Increase in the amount of 300-36 adsorbed as a function of polymer and silica equilibrium time in THF/CP with $f_{CP} = 0.86$. Adapted from reference 21.

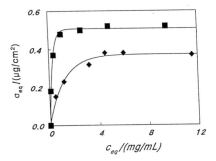

Figure 3. Upon establishing adsorption equilibrium, silica surface coverage is plotted as a function of equilibrium 271-22 (◆) and 300-36 (■) concentration at $f_{CP} = 0.86$. Reproduced from reference 21.

Results and Discussion

PS-*b*-PCEMA Adsorption in the Brush Conformation. Since the silica particles were coated in micellar solutions, it was important to examine whether the PS-*b*-PCEMA chains were adsorbed in the brush conformation (*22*). In the brush conformation, the PS chains should overlay the PCEMA block and be stretched. We could not directly demonstrate the stretching of the PS chains. Instead, we calculated from the adsorption data the surface area allowed for each PS chain and this area was found to be much smaller than that calculated from πR_G^2, where R_G is the radius of gyration the PS chain would have in the coating solvent (*18, 20*). Thus, the PS chains in the buoy layer were crowded and they should be stretched due to repulsion from neighboring chains. The layered structure of PS-*b*-PCEMA brushes was verified by our SERS and TEM experiments.

SERS operates on the principle that functional groups which are in the immediate neighborhood, e.g. <100 Å, of a curved "metal" surface may display an enhanced Raman signal (*23*). If PS-*b*-PCEMA is adsorbed in the brush conformation, the PCEMA signals should be much stronger than those of PS.

While the adsorption studies were carried out using silica, SERS studies are normally carried out on metal surfaces. To correlate the two types of data, we produced "simulated" silica surface on silver as described in the Experimental Procedure section for the SERS studies.

SERS spectra of PS, PCEMA, and 1270-360 on such simulated silica surfaces are compared in Figure 4. While the PS and PCEMA films were prepared by drying silicate-covered silver foils dipped in THF solutions of the polymers, the 1270-360 film was obtained from drying a foil soaked in a THF/CP solution with $f_{CP} = 0.86$ for several hours. PCEMA has a strong peak at 1640 cm^{-1}. Unfortunately, the strong scattering peak of PS at 1005 cm^{-1} overlaps with a peak of PCEMA. The lack of a strong characteristic PS peak made the determination of the absolute percentage of PCEMA in the vicinity of the solid substrate difficult. A measure of the relative PCEMA and PS contents near the surface was obtained using the double ratio $\left(I_{1640}/I_{1005}\right)/\left(I_{1640}/I_{1005}\right)_0$, where I_{1640}/I_{1005} represented the SERS intensity ratio measured for the peaks of a PS-*b*-PCEMA coating at 1640 and 1005 cm-1, respectively, and $\left(I_{1640}/I_{1005}\right)_0$ was the SERS intensity ratio for the peaks of a PCEMA coating at the two wavenumbers under identical sample preparation conditions. This double ratio should normalize when there are no PS chains near a solid substrate.

Illustrated in Figure 5 is the change in $\left(I_{1640}/I_{1005}\right)/\left(I_{1640}/I_{1005}\right)_0$ for 1270-360 at the concentrations of 1.0 and 5.0 mg/mL as a function of f_{CP}. The double ratio peaks between 0.67 and 0.86 and are low at high and low f_{CP}'s.

At low f_{CP}'s, PS-*b*-PCEMA was molecularly dissolved in the solvent and the polymer did not adsorb on the silica (Data to be presented later). The observed SERS signals derived from a film formed due to the evaporation of solvent from a solution layer left on the substrate after the substrate's withdrawal from a coating solution. Due to the random conformation of PS-*b*-PCEMA in the solution phase and the fast solvent evaporation rate, the polymer film derived from the solution

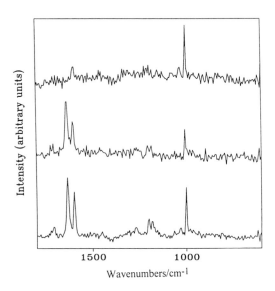

Figure 4. Comparison of SERS spectra of PS (top), PCEMA (middle), and 1270-360 (bottom) films on silver. Reproduced from reference 23.

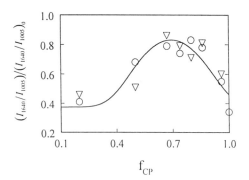

Figure 5. Change in $\left(I_{1640}/I_{1005}\right)/\left(I_{1640}/I_{1005}\right)_0$ for 1270-360 at the concentration of 1.0 (**O**) and (∇) 5.0 mg/mL as a function of f_{CP}. Adapted from reference 23.

layer should not have the layered structure for PS-*b*-PCEMA brushes and $\left(I_{1640}/I_{1005}\right)/\left(I_{1640}/I_{1005}\right)_0$ should be low.

As f_{CP} increased, PCEMA solubility decreased and PCEMA chains adsorbed on silica to form a brush layer. Upon the withdrawal of the foil from a coating solution, the brush layer structure was retained. Since the PCEMA block was spread on the simulated silica in the brush conformation, this lead to the expected $\left(I_{1640}/I_{1005}\right)/\left(I_{1640}/I_{1005}\right)_0$ increase.

The low $\left(I_{1640}/I_{1005}\right)/\left(I_{1640}/I_{1005}\right)_0$ values at high f_{CP}'s was probably caused by micellar adsorption. At high f_{CP}'s, the cores of PS-*b*-PCEMA micelles should not uptake much THF. When not plasticized by THF, the PCEMA chains should have low mobility. Stable micelles may anchor on silica surface via the PS shell layer.

The key conclusion of the SERS experiment was confirmed by our TEM results. In THF/CP with 82% CP, 1077-1158 had a surface coverage of 2.7 μg/cm². This surface coverage gives a dry PS-*b*-PCEMA film thickness of 27 nm, if the density of the PS-*b*-PCEMA film is assumed to be 1.0 g/mL. Assuming clean phase separation between PS and PCEMA, the PS layer thickness should be 7 nm, as the PS weight fraction in the diblock is 0.27.

Illustrated in Figure 6 is a TEM image of a silica particle coated with 1077-1158. Since the silica particle was not stained, the inner dark circle must correspond to the silica particle and the outer gray shell represents the polymer layer. Averaging the thickness of the brush layer over ~20 silica particles, a thickness of 29 nm was obtained with a standard deviation of 3.7 nm. The average thickness of 29 nm agrees with the expected 27 nm.

The brush-coated silica particles were then stained with OsO_4. This increased the diameter of the dark inner circle and decreased the thickness of the gray shell (Figure 7). Since PCEMA should react selectively with OsO_4, the increase in the dark inner circle diameter suggests that PCEMA was next to the silica. The remaining of a gray polymer layer after staining is in agreement with the picture that PS and PCEMA phase-separated well and PS overlaid the PCEMA layer. Averaging the PS layer thickness over 10 stained silica particles gave a PS layer thickness of 11 nm. This agrees well with the expected 7 nm, because PS chains might have partially retained the swollen configuration they assumed in cyclopentane before the particles were sprayed on copper grids for viewing by TEM. Thus, adsorbed PS-*b*-PCEMA had the brush conformation at $f_{CP} = 0.82$.

Also shown in Figure 6 is a brush-covered silica particle aggregate. The image clearly shows the formation of a brush layer at a site with a large surface curvature change. Also the contour of the substrate is approximately mapped by the brush layer.

Micelle adsorption at $f_{CP} = 0.92$ is evident from Figure 8, in which two micelles are seen fused with a silica particle. Micelles are distinguishable from coated silica particles because a silica particle appears much darker than the core of a micelle. The unambiguous evidence for micelle adsorption derived from the thickness of 20 nm for the 1077-1158 layer, prepared in THF/CP with $f_{CP} = 0.92$, in

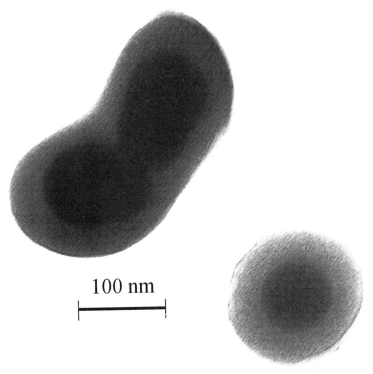

Figure 6. TEM image of single and fused silica particles coated with 1077-1158 at f_{CP} = 0.82 before OsO_4 staining. Reproduced from reference 23.

Figure 7. TEM image of a silica particle coated with 1077-1158 at f_{CP} = 0.82 after OsO_4 staining. Reproduced from reference 23.

the regions free of micelles. This thickness is considerably smaller than the brush layer prepared at $f_{CP} = 0.82$. Our later evidence will show that the surface coverage should increase with f_{CP}. Since the extra polymer adsorbed at $f_{CP} = 0.92$ did not increase the thickness of the adsorbed layer, it had to be accounted for by micelle deposition.

Brush Formation Mechanism. Although not discussed here, similar PS-b-PCEMA adsorption behaviors were observed on silver surfaces in THF/CP mixtures with different f_{CP}'s. To gain insight into the mechanism of PS-b-PCEMA brush formation on silica or silver surfaces, we monitored the variation in I_{1640}/I_{1005} for 1270-360 films prepared on silver surfaces using different silver-polymer solution equilibrium times and the results are shown in Figure 9 with two curves hand-drawn through the experimental data. The I_{1640}/I_{1005} values increased with time initially and then leveled off, the same trend as observed by Boerio et al. for PVP SERS signals in the case of PS-b-P2VP adsorption from toluene (*11-12*).

A logical explanation for this trend is that both the PS-b-PCEMA micelles and unimers were adsorbed initially. Micelle adsorption via the PS block resulted in low I_{1640}/I_{1005} values. As time progressed, unimers replaced the micelles or the micelles disintegrated on the silver surface to increase the number of PCEMA/substrate contacts and thus the I_{1640}/I_{1005} values. The I_{1640}/I_{1005} values reached their equilibrium value faster at $f_{CP} = 0.82$ than at $f_{CP} = 1.00$, because the micelles disintegrated faster in a solvent mixture containing more THF (18).

One may plausibly argue that the I_{1640}/I_{1005} value variation is correlated with the amount of polymer deposited. This argument is wrong as our adsorption kinetic data showed that the rate of polymer deposition on the silica and silver increased with f_{CP} (*18*).

Brush Thickness Control. Having established brush formation from PS-b-PCEMA between $f_{CP} = 0.67$ and $f_{CP} = 0.90$, we subsequently examined the effect of varying f_{CP}, n, and m on σ_{∞}. Plotted in Figure 10 is the variation in the σ_{∞} values of 1270-360 as a function of f_{CP}. Below a critical f_{CP}, 1270-360 adsorbed on silica negligibly. σ_{∞} then increased steadily with f_{CP}. The more abrupt increase in σ_{∞} with f_{CP} above $f_{CP} \approx 0.90$ is probably caused by micelle adsorption at high f_{CP}'s.

There are at least two distinctive cases when comparing σ_{∞} of different polymers. For large n/m, e.g. greater than 3.5, σ_{∞} did not change significantly with n and m (*20*). For samples with $n/m \approx 1$, σ_{∞} increased with n and m as illustrated in Figure 11 (*24*).

The σ_{∞} values can be converted into ρ_{∞}, polymer chains adsorbed on unit silica surface area, using:

$$\rho_{\infty} = \frac{\sigma_{\infty}}{M_w} \times N_A \tag{3}$$

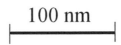

Figure 8. TEM image of a silica particle coated with 1077-1158 at $f_{CP} = 0.92$. The sample was not stained with OsO_4. Reproduced from reference 23.

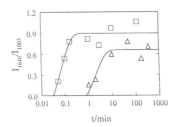

Figure 9. Variation in I_{1640}/I_{1005} for 1270-360 films prepared on silver surfaces as a function of silver-polymer solution equilibrium time t. The slower and faster rising data were obtained by soaking the silver foil in a polymer solution in CP (Δ) and THF/CP (□) with 86% CP by volume, respectively. The polymer concentration used was 3.0 mg/mL. Adapted from reference 23.

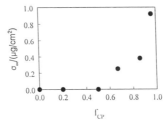

Figure 10. Variation in saturated equilibrium silica surface coverage, σ_∞, for 1370-360 as a function of f_{CP}. Adapted from reference 21.

where $\overline{M}_w$ is the weight-average molar mass of a diblock under study and N_A is Avogadro's number. Plotted in Figures 12 and 13 are the ρ_∞ values for the two sets of samples. For $n/m > 3.5$ (20) and $n/m \approx 1$ (24), ρ_∞ values were shown to obey the following scaling relations:

$$\rho_\infty \propto m^{-(12/23)} n^{-(6/23)} \tag{4}$$

and

$$\rho_\infty \propto m^{(12/25)} n^{-(6/5)} \tag{5}$$

in agreement with eqs. 26 and 28 of Marques et al. (17).

Conclusion

When deposited from micellar solutions with f_{CP} between 0.67 and 0.90, PS-b-PCEMA formed brushes on the silica particles. Brushes are formed probably due to the disintegration of adsorbed micelles. At high f_{CP}'s, PS-b-PCEMA brushes were not formed because the adsorbed micelles were too stable to disintegrate.

Depending on n/m, brush formation fell in either the van der Waals-buoy or the buoy regime as predicted by Marques et al. (17). In the van der Waals-buoy regime, σ_∞ did not vary significantly with n and m. Thick brush layers can be obtained in the buoy regime by increasing n or m.

Acknowledgment

Several post-doctoral fellows and students in my group contributed to the work reviewed. Their contributions are cordially acknowledged by citing their publications. The Natural Sciences and Engineering Research Council of Canada is thanked for its Research Grants, Equipment Grants, and an Industrial Oriented Research Grant to G. Liu as well as a Strategic Grant to V. I. Birss and G. Liu. The generous support provided by the Environmental Sciences and Technology Alliance of Canada and VX Optronics to G. Liu is also gratefully acknowledged.

Literature Cited

1) Milner, S. *Science* **1991**, *251*, 905.
2) Halperin, A.; Tirrell, M.; Lodge, T. P. *Adv. Polym. Sci.* **1992**, *100*, 31.
3) Liu, G. p. 1548 in *the Polymeric Materials Encyclopedia - Synthesis, Properties, and Applications* Salamone, J. C. Ed. CRC Press: **1996**.
4) Napper, D. H. *Polymeric Stabilization of Colloidal Dispersions* Academic Press: London, 1983.
5) Taunton, H. J.; Toprakcioglu, C.; Fetters, L. J.; Klein, J. *Macromolecules* **1990**, *23*, 571.
6) Watanabe, H.; Tirrell, M. *Macromolecules* **1993**, *26*, 6455.

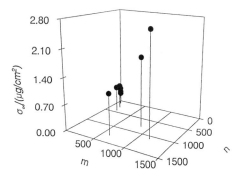

Figure 11. Variation in σ_∞ for samples with $n/m \approx 1$ as a function of n and m at $f_{CP} = 0.82$. Adapted from reference 24.

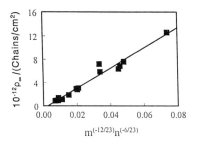

Figure 12. Plot of the number of polymer chains adsorbed on unit silica surface, ρ_∞, as a function of $m^{-(12/23)}n^{-(6/23)}$ for samples with $n/m > 3.5$ at $f_{CP} = 0.86$. Adapted from reference 21.

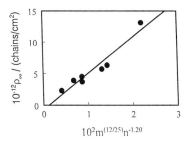

Figure 13. Plot of ρ_∞ as a function of $m^{(12/25)}n^{-1.20}$ for samples with $n/m \approx 1$ at $f_{CP} = 0.82$. Reproduced from reference 24.

Literature Cited

1) Milner, S. *Science* **1991**, *251*, 905.
2) Halperin, A.; Tirrell, M.; Lodge, T. P. *Adv. Polym. Sci.* **1992**, *100*, 31.
3) Liu, G. p. 1548 in *the Polymeric Materials Encyclopedia - Synthesis, Properties, and Applications* Salamone, J. C. Ed. CRC Press: **1996**.
4) Napper, D. H. *Polymeric Stabilization of Colloidal Dispersions* Academic Press: London, 1983.
5) Taunton, H. J.; Toprakcioglu, C.; Fetters, L. J.; Klein, J. *Macromolecules* **1990**, *23*, 571.
6) Watanabe, H.; Tirrell, M. *Macromolecules* **1993**, *26*, 6455.
7) Kumacheva, E.; Klein, J.; Pincus, P.; Fetters, L. J. *Macromolecules* **1993**, *26*, 6477.
8) Tirrell, M. in *Solvents and Self-Organization of Polymers*, NATO ASI Series E: Applied Sciences - vol 327; eds. Webber, S. E.; Munk, P.; Tuzar, Z.; Kluwer Academic Publishers: Dordrecht, 1996.
9) Munch, M. R.; Gast, A. P. *Macromolecules* **1990**, *23*, 2313.
10) Venkatachalam, R. S.; Beerio, F. J.; Roth, P. G.; Tsai, W. H. *J. Polym. Sci., Part B: Polym. Phys.* **1988**, *26*, 2477.
11) Hong, P. P.; Boerio, F. J.; Tirrell, M.; Dhoot, S.; Guenoun, P. *Macromolecules* **1993**, *26*, 3953.
12) Tsai, W. H.; Boerio, F. J.; Tirrell, M.; Parsonage, E. *Macromolecules* **1991**, *24*, 2538.
13) Bossé, F.; Schreiber, H. P.; Eisenberg, A. *Macromolecules* **1993**, *26*, 6447.
14) Parsonage, E.; Tirrell, M.; Watanabe, H.; Nuzzo, R. G. *Macromolecules* **1991**, *24*, 1987.
15) Watanabe, H.; Tirrell, M. *Macromolecules* **1993**, *26*, 6455.
16) Stouffer, J. M.; McCarthy, J. *Macromolecules* **1988**, *21*, 1204.
17) Marques, C.; Joanny, J. F.; Leibler, L. *Macromolecules* **1988**, *21*, 1051.
18) Underhill, R. S.; Ding, J.; Birss, V. I.; Liu, G. *Macromolecules* in press.
19) Tao, J.; Guo, A.; Liu, G. *Macromolecules* **1996**, *29*, 1618.
20) Liu, G.; Smith, C. K.; Hu, N.; Tao, J *Macromolecules* **1996**, *29*, 220.
21) Ding, J.; Tao, J.; Guo, A.; Stewart, S.; Hu, N.; Birss, V. I.; Liu, G. *Macromolecules* **1996**, *29*, 5398.
22) Hill, W.; Rogalla, D.; Klockow, D. *Anal. Methods Instrum.* **1993**, *1*, 89.
23) Ding, J.; Birss, V. I.; Liu, G. *Macromolecules* **1997**, *30*, 1442.
24) Tao, J.; Guo, A.; Stewart, S.; Birss, V. I.; Liu, G. *Macromolecules* in press.

Chapter 14

Langmuir and Langmuir–Blodgett–Kuhn Films of Poly(vinylidene fluoride) and Poly(vinylidene fluoride-co-trifluoroethylene) Alternated with Poly(methyl methacrylate) or Poly(octadecyl methacrylate)

Rigoberto C. Advincula,[1] Wolfgang Knoll,[1,2,4] Lev Blinov[3], and Curtis W. Frank[1,4]

[1]Center for Polymer Interfaces and Macromolecular Assemblies (CPIMA), Department of Chemical Engineering, Stanford University, Stanford, CA 94305–5025
[2]Max Planck Institute for Polymer Research, Ackermannweg 10 Mainz, Germany D-55021
[3]Institute of Crystallography, Russian Academy of Sciences, Leninsky prosp. 59, Moscow 117333, Russia

Langmuir and Langmuir-Blodgett-Kuhn (LBK) films of poly(vinylidene fluoride) (PVDF) and poly(vinylidene fluoride -co- trifluoroethylene) (co-PVDF) were investigated for their morphological and layer ordering behavior. Such materials are of interest because of their possible use as ultrathin ferroelectric switching materials. Our studies involved the use of various spectroscopic and microscopic methods, the foremost of which are surface plasmon spectroscopy (SPS) and atomic force microscopy (AFM). The multilayer film properties of co-PVDF, as alternated with poly(methyl methacrylate) (PMMA) and poly(octadecyl methacrylate) (PODMA), were also investigated in comparison to bulk properties. This was done in order to create two-dimensional structures where the domain behavior and blending characteristics of co-PVDF can be studied in restricted geometries. The co-PVDF monolayer is observed to be highly viscous with isotherm and domain formation dependent upon the spreading conditions and experimental configuration. The PVDF polymer does not form a true monolayer at all, although multilayers can be fabricated by vertical deposition as verified by ellipsometry, X-ray diffraction and SPS. The possible phase transitions of the copolymer and alternate films were investigated by applying an *in-situ* surface plasmon spectroscopy heating probe. A phase transition observed for the copolymer film at 88 °C is suggested to be related to the ferro-paraelectric transition (Curie temperature). AFM images correlated with the morphological properties and transfer behavior of the copolymer and the blend at the air-water interface.

[4]Corresponding authors.

192

Poly(vinylidine fluoride), (PVF$_2$ or PVDF), -(-CH$_2$-CF$_2$-)- and its copolymers have been widely investigated for their piezoelectric, pyroelectric and ferroelectric properties *(1)*. The highly polar nature of the C-F bond results in macroscopic polarization which can be influenced by mechanical stress, temperature and applied electric field. Copolymerization with trifluoroethylene, -(-CHF-CF$_2$-)- and tetrafluoroethylene, -(-CF$_2$-CF$_2$-)- at different percentage compositions influences the melt crystallized microstructures (α,β,γ phases) by introducing "defects" of Head - Head or Tail - Tail units *(2)*. This polymorphism has a direct effect on the observed electrical properties. Thus, for a 70/30 trifluoroethylene copolymer -(-CH$_2$-CF$_2$-)$_n$-(-CF$_2$-CHF-)$_m$-, the ferro-paraelectric transition (Curie temperature) is lowered to within the range 80-110 °C compared to the homopolymer at 170°C *(3)*. While these properties have been observed in bulk, little work has been focused on thin (1000 - 10,000 Å) and ultrathin (< 1000 Å) film configurations where phase transitions might be very different. Ferroelectric switching in such regimes is both of fundamental and applied interest *(4,5)*. The nucleation and growth mechanism of particular polymer domains is often influenced by the asymmetry of the interface *(6)*. An important question is to determine the minimum layer thickness by which the ferroelectric phenomenon can be observed, i.e. as the domain size approaches that of extended chain coil dimensions. Another issue is to determine how much of the spontaneous polarization process is influenced by the non-polar interactions (Van der Waals, steric, etc.) and dipole-dipole interactions within the ultrathin regime. While not directly attempting to correlate with ferroelectric switching measurements, the thrust of this investigation is to address the morphological and layer ordering issues with the use of the LBK technique.

Recently, Blinov and coworkers have reported the fabrication of ultrathin LBK films of the 70/30 trifluoroethylene copolymer in which they investigated ferroelectric switching behavior with spontaneous polarization *(7)*. They found that in comparison to bulk switching properties, a new phenomenon of "conductance switching" was observed, which could be related to the layered ordering and total thickness of the polymer film *(8)*. Bistable switching has been observed up to thicknesses of about 100 Å. Below this thickness only monostable switching occurred. A high pyroelectric response at room temperature was also observed, switchable in the presence of an external electric field. At the molecular level, they have attributed the reversal of the polarization vector for each layer as playing a key role in defining the ON and OFF state for conductance switching. The main assumption is that the individual monolayers are comprised of laterally extended polymer chains with a defined polymorph (β-phase) and a net average dipole preference.

In this work, we report our investigations on Langmuir and LBK films of poly(vinylidenefluoride) and its 70/30 trifluoroethylene copolymer at the air-water interface. Traditionally, the route to the β-phase, all-*trans* polymorph (required for ferroelectric properties) is by mechanical deformation of the melt-crystallized films *(1)*. In our case, the Langmuir monolayer method confines insoluble films into quasi-two dimensional structures by the application of mechanical stress from a movable barrier (e.g. a 20 Å thick film subjected to a pressure of 20 mN/m is approximately equivalent to a 100 atm pressure across the thickness of the monolayer *(9)*). It will be interesting to compare analogous conditions of isothermal high-pressure crystallization in bulk systems to monomolecular films *(10)*. In addition, the air-water interface can have an orienting effect on the dipoles (the C-F bond on PVDF, for example) and the hydrophobic-hydrophilic balance in amphiphilic systems. These investigations are necessary to define the extent of ordering with this technique and their consequence on the ferroelectric behavior *(11)*. In addition, we have investigated the miscibility behavior of these polymers with poly(methyl methacrylate) (PMMA) and poly(octadecyl methacrylate) (PODMA) as alternately-deposited ultrathin films. It is known that melting point depression and kinetic effects on crystallization are affected by their blend composition ratios *(12)*. It is also interesting to observe the influence on these properties in ultrathin configurations., i.e. thicknesses less than 1000 Å with defined stacking order (alternating layers).

Experimental

The homopolymer was obtained from Elf Atochem and one of us (LB) provided the 70/30 trifluoroethylene copolymer. Solutions of the polymers were prepared in spectrograde solvents at concentrations of 0.5 mg/ml to 1 mg/ml. The PVDF homopolymer was soluble only in polar aprotic solvents, such as N-methyl pyrrolidone (NMP) and dimethylformamide (DMF). The copolymer was soluble in a variety of organic solvents such as dimethyl sulfoxide (DMSO) and cyclohexanone. Poly(methyl methacrylate), Mw = 60,600 and poly(octadecyl methacrylate), Mw = 97,200 (Aldrich Chemicals) solutions were prepared in methylene chloride. Langmuir film measurements and LBK depositions were done on a KSV 5000 alternate trough using ultrapurified water (MilliQ quality). The temperature of the trough was at 21°C and pH = 5.6 unless otherwise specified. Brewster angle microscope (BAM) images were taken using a p-polarized He-Ne laser incident at the Brewster angle for the air-water interface *(13)*. The images were captured with a CCD camera, recorded, and digitized for analysis. Deposition was done on hydrophilic or hydrophobic glass, gold or silver-coated glass and silicon wafers. Glass and silicon wafers were treated with Nochromix/sulfuric acid mixtures and NH_4OH/H_2O_2 cleaning procedures. Hydrophobization was carried out using hexamethyldisilazane (HMDS) coupling agent with 4 hours minimum exposure in a sealed chamber at 70°C. Gold and Ag/SiOx coated substrates (400-450 Å) were prepared by metal vacuum evaporation on pre-cleaned glass. Surface plasmon spectroscopy (SPS) was done at $\theta-2\theta$ geometry using the Kretschman configuration. A polarized He-Ne light source was used (632.8 nm). The set-up and interpretation of the plasmon curves has been described previously *(14)*. The LBK films of successive layer increases were prepared on individual substrates by vertical deposition from stable monolayers. A heating stage was utilized to heat the sample *in situ* at various ranges and rates while scanning the reflectivity changes. Atomic force microscopy (AFM) images using tapping and contact mode were taken on a Digital Instruments Nanoscope III. Preliminary X-ray measurements were done on a Rigaku $\theta-\theta$ diffractometer with an 18 kW rotating anode X-ray source. Ellipsometric measurements were done on a Gaertner L116C Ellipsometer at 70° angle of incidence and 632.8 nm wavelength. A value of 1.46 was assumed as the refractive index for both SPS and ellipsometry.

Results

Langmuir Isotherms. The surface pressure-area isotherms were obtained with a constant speed compression rate of 3 Å^2/repeat unit x min, as shown in Figure 1. Typical of many polymers, the isotherm curves were monotonic, exhibiting collapse pressures near 55 mN/m. It should be noted that the term collapse refers to a slow phase transition where the film tends to "buckle-up" *(15)*. Compression rate dependence measurements at ambient temperature showed no apparent trend (increasing or decreasing area) for either polymer at a range of 0.2 to 5 Å^2/repeat unit x min. Both exhibited strong concentration dependence (variations up to 0.5 Å^2 in mean molecular area, Mma),with concentration ranges from 0.2 mg/ml to 0.8 mg/ml in NMP for the homopolymer and cyclohexanone for the copolymer. This deviation is significant considering it is almost 9 % of the extrapolated Mma occupied per repeat unit.

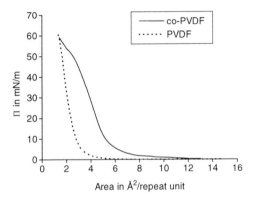

Figure 1. The surface pressure - area isotherm for PVDF and co-PVDF in dimethylformamide (DMF) and cyclohexanone, respectively, at a temperature of 21°C and pH = 5.6. Note the difference in the extrapolated mean molecular area, Mma, for the two polymers: PVDF = 2.7, co-PVDF = 5.7 Å^2/repeat unit. The molecular weight of the monomeric unit was used to define the Mma. Based on the Mma, the homopolymer does not form a usable monolayer.

The variable spreading behavior for the copolymer was also observed with different spreading solutions at 0.5 mg/ml concentration (*N,N*-dimethyl propylene urea, (DMPU), *N,N*-dimethyl acetamide, (DMAC), and DMSO) as shown in Figure 2. Hysteresis experiments (compression-expansion cycles) showed *irreversible* collapse behavior for both polymers as with subsequent cycles. This was observed at all surface pressures prior to collapse, especially for the homopolymer. For the copolymer, a reversible behavior can be obtained with isotherms using slow compression rates, e.g. 0.2 Å^2/repeat unit x min up to 25 mN/m. In addition, isobaric creep measurements showed that only the copolymer is stable for LBK deposition with change in area less than 1 Å^2/repeat unit in one hour at constant surface pressures of 5, 10, 20, 35 mN/m. For the copolymer, the extrapolated limiting area of 5.7 Å^2 was consistent with that observed by Blinov et.al. on the basis of STM and molecular modeling data (8,11,17).

However, the limiting area for the homopolymer is too small (by a factor of 2) to represent any true monolayer limiting area of an extended chain conformation. Together with the variable spreading behavior, this indicates that the homopolymer is not capable of forming a true monolayer.

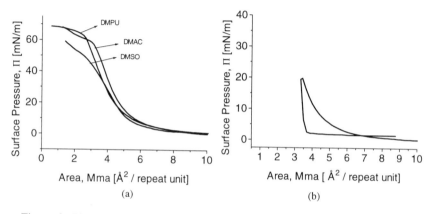

Figure 2. The spreading behavior of the copolymer for different solutions (a): N,N-dimethylpropylene urea (DMPU), dimethylacetamide (DMAC), and dimethyl sulfoxide (DMSO). Note that the copolymer is also soluble in acetone and cyclohexanone. (b) An irreversible hysteresis behavior is characteristic for both homopolymer and copolymer at higher compression rates.

Brewster Angle Microscopy. To characterize the domain attributes and polymer morphology of co-PVDF directly at the air-water interface, Brewster angle microscopy (BAM) was utilized. From the images in Figure 3, the film as spread formed precipitated domains or "flakes" with irregular shapes even at high areas and did not exhibit any flow behavior. Upon compression, these flakes were observed to coalesce at higher surface pressures where sintering of the domain boundaries was observed.

Once the fused domains were observed at about 5 Å^2/ repeat unit, this was accompanied by increasing reflectivity intensity (brightness) especially at the most focused region. No long-range ordering was observed within these domains. Variable reflection was observed though within some of the initially spread domains which may indicate a heterophase thickness profile or domain birefringence contributing to optical heterogeneity (16).

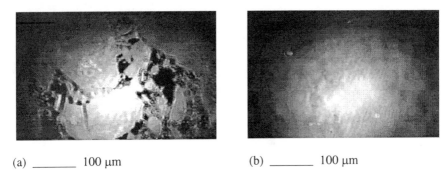

(a) _____ 100 μm   (b) _____ 100 μm

Figure 3. Brewster angle microscope (BAM) images of co-PVDF at low surface coverage (a) 8 Å²/r.u., π = 1 mN/m (left image) and at a higher surface coverage (b) 4 Å²/r.u. area, π =25 mN/m. The precipitated domains are characteristic even prior to compression (gaseous region), while a uniform film is observed at higher pressures.

LBK Films. Multilayer deposition was initially done on both hydrophilic and hydrophobic glass substrates. Even with various parameter iterations, deposition was observed only on the hydrophilic glass. The changes in the contact angle measured before and after deposition are also consistent with the surface energy changes. With subsequent layers, transfer ratios were within 1.0 ± 0.1. However, transfer was observed only on the upstroke mode indicating a z-type deposition behavior. This was maximized by programming the substrate to deposit only on the upstroke mode and to pass the monolayer-free subphase compartment on the downstroke. Depositions were done at typical rates between 1-5 mm/min with 30 minutes drying time in-between. For ellipsometric and surface plasmon spectroscopy measurement samples, polished silicon and silver-coated BK-7 glass were used as substrates. The same deposition behavior was observed.

Ellipsometry. As a preliminary step, ellipsometry was used to verify thickness increase with deposition on silicon substrates, as shown in Figure 4. The thickness obtained for these films is greater by a factor of two compared to the literature *(17)*. It should be pointed out that the method of estimation in the literature had a 20% margin of error in comparison to the direct layer-by-layer measurement in this case. The difference may also be explained by the higher surface pressure (20 mN/m) used for transfer. Nevertheless, a generally linear trend is observed with an average thickness of 17.8 Å per two layers compared to 10 Å for two layers reported in the literature.

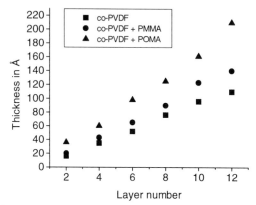

Figure 4. Ellipsometric thickness data for the LBK films. The measurements were made on films deposited on individual silicon substrates. The average two layer values can be calculated based on thickness values up to 12 layers.

Polymer Film Blends. Studies were initially made on alternating multilayers with PMMA. Since PMMA is known to be miscible with PVDF in the bulk, we were interested in determining its miscibility behavior as ultrathin films and its effect on phase transitions with the PVDF copolymer film. It is known that melting point depression and kinetic effects on crystallization are affected by the blend composition ratios (12). Nevertheless, it is interesting to observe the influence on these properties in ultrathin configurations., i.e. thicknesses less than 1000 Å. These studies first involved making alternate multilayers for both polymers. The deposition parameters for PMMA are known in the literature (surface pressure of 5-10 mN/m at 1-5 mm/min) (18). Transfer ratios were within 1.0 ± 0.1 on the upstroke, also indicating z-type deposition behavior. The PVDF copolymer was first deposited on the upstroke. After a drying time of 30 minutes, the substrate was then passed downstroke through the monolayer-free compartment. On the upstroke mode, the substrate was passed through the PMMA monolayer depositing the second layer. This cycle was repeated up to the desired number of multilayers. The deposition was monitored both by ellipsometry and surface plasmon spectroscopy. From the ellipsometry data, slightly thicker films were formed (ave. double layer thickness, co-PVDF + PMMA = 22.3 Å, Figure 4) on pairing the co-PVDF with a layer of PMMA. Surface plasmon spectroscopy data were consistent with those obtained from ellipsometric measurements (ave. double layer thickness of 22.4 Å, Figure 5). An increasing broadening of the plasmon curve could indicate increasing surface heterogeneity for the film.

Alternate deposition was also done with PODMA and the PVDF copolymer. This is an interesting system to investigate since the two polymers are known to be immiscible in the bulk (1). The deposition was done at 25 mN/m surface pressure for the PODMA with good transfer ratios. The same sequence of upstroke and downstroke modes was used as described above except for the PMMA being replaced by the PODMA monolayer. From ellipsometry (Figure 4), the observed average double layer thickness was found to be 32.7 Å. Preliminary X-ray diffraction measurements showed a first order Bragg peak at $2\theta = 3.06°$ corresponding to a d-spacing of 28.6 Å. There is a discrepancy of 4 Å between these two experimental values but no measurements are available at lower angles.

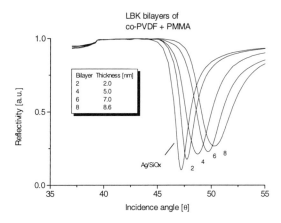

Figure 5. Surface plasmon spectroscopy data for co-PVDF+PMMA LBK Films. Fitting of the plasmon shift at an assumed refractive index value of 1.46 gave an average double layer thickness of 22.4 Å. Increasing film inhomogeneity is evident from the plasmon curve broadening.

Thermal Investigations. Temperature dependent reflectivity experiments were done on the copolymer and alternating blend LBK films *in-situ* with surface plasmon spectroscopy. The samples were air-dried for three weeks under clean room conditions prior to their use as substrates for the measurements. A heating rate of 10°C/min was used in air. Figure 6 shows the resonance plasmon curves before, during, and after heating beyond 110°C for the co-PVDF + PMMA film. A shift to lower resonance angles is observed during and after heating of the film.

By fixing the incident angle at a value just before the minimum, the change in reflectivity can be measured as a function of temperature which would be sensitive to phase transitions that affect the polymer film (Figure 7). An observed change in slope can be traced based on the intersection of the R vs. T slopes. Since the films are ordered as LBK multilayers, only one heating direction can be made (first annealing cycle destroys the layer order). Heating temperatures were carried up to 170°C then in order to measure the full extent of the linearity. For the 6, 8, and 10 layers, the change in reflectivity commences with an initial decrease followed by a less steep slope beginning at the region 75-100°C. By extrapolation of these two slopes to an intersection, a slope change can be determined. For the co-PVDF, this slope change corresponds to a value of about 88°C and for the co-PVDF + PMMA LBK film a slope change is observed at 110°C.

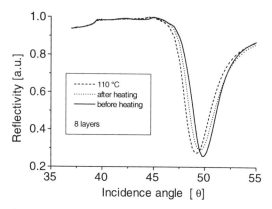

Figure 6. Surface plasmon spectroscopy data for co-PVDF + PMMA LBK film with heating at a temperature beyond 110°C. Plasmon curves measured before and after heating were taken at ambient temperature.

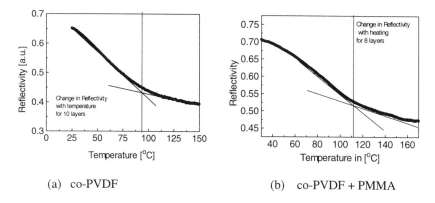

(a) co-PVDF

(b) co-PVDF + PMMA

Figure 7. Surface Plasmon Spectroscopy data for co-PVDF, co-PVDF + PMMA LBK film with heating. The change in reflectivity at a fixed angle of 48° is measured with respect to temperature. Extrapolation of the adjacent linear regions determines the slope change.

AFM Measurements. Measurements were made in both contact and tapping modes on the 6 layer co-PVDF and the 8 layer alternate, co-PVDF + PMMA LBK film (Figure 8). For the copolymer, the estimated rms roughness for a 2 x 2 μm region is at 1.1 nm. The image gives the appearance of forming globular morphology although the analysis needs to be done further using SEM and STM. No domain-like structures or long range ordering is observed even for large area sampling. For the co-PVDF + PMMA LBK film, the appearance of fractured domains is evident on all areas of the sample. The difference in bright contrast between domain structures may indicate separation between the two types of polymers. Again, no long range ordering was observed. The estimated rms roughness for a 30 x 30 μm size film is at 1.7 nm. The AFM image of the blend is also shown after heating the sample to 150°C for 20 minutes. It is evident from the image that the sharp boundaries characterized by the embrittled domains before heating seemed more diffuse. The estimated rms roughness also decreased to 1.2 nm.

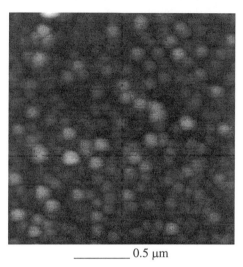

_____ 0.5 μm

Figure 8 (a). The AFM image of the co-PVDF gives the appearance of forming globular morphology at the nm scale.

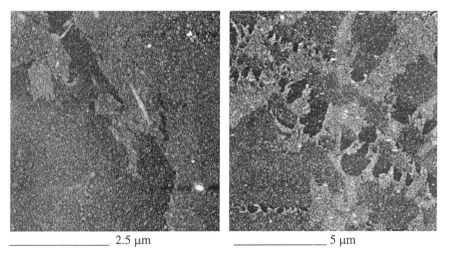

_____ 2.5 μm   _____ 5 μm

Figure 8. (b) Tapping mode AFM image of co-PVDF + PMMA LBK Film, 8 layers showing the embrittled domains as a consequence of film transfer (left) and the diffused boundaries.

Discussion

Langmuir Film and LBK Deposition. The Langmuir film of the copolymer is highly viscous with low compressibility, especially at lower surface areas. The observed spreading behavior by BAM seems to explain the variable Langmuir isotherm measurements with concentration, solvent and compression rate dependence. In this case, the domain size and distribution can be influenced by the spreading conditions and experimental configuration. However, the extent or ordering within these domains is unknown. Similar behavior has been observed with other polymeric and low molecular weight amphiphiles in monolayers which have been interpreted as being very viscous *(19)*.

The fact that the film is highly viscous did not prevent it from forming multilayers upon vertical deposition on the upstroke mode (z type). The monolayer film was sufficiently stable based on the isobaric creep measurements. For example, a highly viscous monolayer of a liquid crystalline siloxane homopolymer has been observed to form well-ordered multilayers (up to 110 layers as verified by UV-vis absorbance and X-ray diffraction) despite the fact that the homopolymer also formed precipitated domains as spread and showed similar concentration, compression rate and solvent isotherm variance *(20)*. The preference of the copolymer to deposit on hydrophilic glass is interesting in terms of compatibility with the substrate surface. A highly hydrophobic co-PVDF monolayer is expected because of the presence of fluorinated carbons on the polymer backbone; thus, a dewetting should have been observed with a hydrophilic surface. It is possible that dipole interaction is dominant at the Si-OH surface of glass with the fluoropolymer backbone. F and Si are highly electronegative and electropositive atoms, respectively.

All these observations point to the fact that the extent of ordering imposed by the Langmuir monolayer technique may be limited by the spreading enthalpy, ΔH, of the copolymer *(21)*. The poor flow behavior and precipitation upon spreading indicates a positive ΔH where there is little subphase-polymer interaction *(22)*. The polymer monolayer as spread will then have a tendency to retain the inherent order within the domains as determined by the spreading conditions, e.g. solvent, evaporation rate, concentration, etc. Compression to high surface pressures towards collapse is needed to produce any significant morphological ordering.

Blend Studies and Film Characterization. Alternating LBK multilayers of the copolymer with PMMA and PODMA allowed blending configurations not possible with the bulk. Ellipsometric and surface plasmon thickness data showed a consistent alternate layer-by-layer deposition on the substrate. In addition, the broadening of the plasmon curves (Figure 5) indicated increased inhomogeneity consistent with the observed domain features from AFM. The observed discrepancy between the ellipsometry data (32.7 Å) and X-ray diffraction (28.6 Å) for the co-PVDF + PODMA blend can only be resolved by more complementary techniques such as X-ray reflectometry. Diffraction experiments are only capable of interpreting periodic regions within the film but not the overall layer order and thickness. With the absence of any higher order Bragg peaks, it also indicates a lack of long range periodicity. It is interesting to note that homogeneous mixtures between polymethacrylate derivatives and PVDF in the bulk are found only for short alkyl chains *(23)*. It is known that compatibility between these polymers is attributed mostly to the carbonyl moiety of the methacrylate group and the C-F dipole. In this case, the ability to form well-defined alternate multilayers even with a longer alkyl chain derivative such as PODMA should allow polymer blend investigations not possible with bulk systems. A more direct interaction between the carbonyl moiety and the C-F dipole of adjacent alternate layers may contribute towards blend compatibility.

Temperature Dependent Reflectivity. For the co-PVDF + PMMA film, a shift to lower resonance angles is observed during and after heating of the film at 110°C. This is somewhat puzzling in that a shift to higher resonance angle is expected with increase in temperature assuming a greater contribution of the thermal expansivity coefficient, β, compared with the isotropic refractive index, n. It is therefore important to distinguish the actual contribution of the isotropic refractive index increment with temperature ($\Delta n/dT$) for these very thin films *(24)*. The fact that the curve is shifted after heating is already indicative of a possible structural or morphological rearrangement within the film. This is correlated by the observed changes in the AFM images before and after heating (Figure 8).

For the fixed angle reflectivity measurements on co-PVDF, the slope change does not correspond to the Tg since this should be observed at much lower temperatures (-40 °C for PVDF) *(25)*. On the other hand, the melting temperature is usually observed at higher temperatures (between 150-170 °C for PVDF) *(26)*. Since the ferro-paraelectric (Curie transition) temperature for a 70/30 trifluoroethylene copolymer is observed at 80-110 °C range, one might speculate that the observed transition is related to this. This is possible considering that the ferro-paraelectric transition is accompanied by a reorientation of the polymer backbone and C-F dipole direction. This would likely influence the molecular polarizability, α, (hence the refractive index) and the thermal expansion coefficient, β. A positive identification of layer ordering and domain characteristics that affect this transition is important as a possible direct link to the ferroelectric switching behavior.

For the co-PVDF + PMMA LBK film a slope change is observed at 110 °C. At present, this property cannot be attributed to any particular phase transition but may be the result of multiple events like Tg of the PMMA and the influence of the Curie transition temperature *(3)*. This temperature may also be related to the melting temperature depression observed for PVDF + PMMA blends in bulk *(12)*. In comparison to LBK films of just PMMA, a depression of the Tg has been observed between 8 layers of PMMA (d / layer = 9 Å) and 22 layers. The differences in the slopes of the R vs T change before and after the inflection point has been shown to be a function of parameter variations of β and $\Delta n/dT$ relative to the Claussius-Mosotti equation *(27)*. Again, the effect of isotropic refractive index increment with temperature on the film thickness also needs to be distinguished.

AFM Measurements. For the 6 layer co-PVDF (estimated rms roughness for a 2 x 2 μm region is at 1.1 nm), the image gives the appearance of forming globular morphology. This type of morphology has been observed on spin-cast PVDF in DMF at larger dimensions (μm) *(28)*. This was found to disappear with directional solidification of the melt or stretch orientation by zone drawing, for example *(29)*. In this case, it would be interesting to do *in-situ* heating and annealing while imaging the sample by AFM. The lack of any long range ordering and domains indicate that the co-PVDF film is morphologically homogeneous at the lateral scale (up to 100 μm as observed). Future experiments will focus on scanning tunneling microscopy (STM) measurements on the film as it relates to the monolayer spreading and deposition conditions.

For the co-PVDF + PMMA LBK film, the appearance of fractured domains may indicate the degree of disorder and inhomogeneity in forming the film blends. It is not clear though if this is a consequence of stress induced by the transfer conditions or the differences in viscoelastic properties between the two monolayer films. .For example, during alternate transfer between PMMA and co-PVDF, the coalesced domains may tend to break-up with the stress of pulling the substrate through the monolayer. The estimated rms roughness for a 10 x 10 μm size film is 1.7 nm. The higher rms roughness compared to the co-PVDF film seems consistent with the dimensions of a partially embrittled film which does not cover the total surface area during transfer (although the observed transfer ratio = 1). Therefore, the AFM image supports the observed broadening of the resonance plasmon curves with increasing layer thickness. The AFM image of the blend is also shown after heating the sample to 150 °C for 20 minutes. It is evident that the sharp boundaries characterized by the embrittled domains before heating seemed more diffuse. The estimated rms roughness also decreased to 1.2 nm.

Conclusion

The Langmuir and LBK film forming properties of the co-PVDF polymer are characteristic of a highly viscous film capable of forming multilayers on hydrophilic substrates. The extent of morphological order is limited by the polymer enthalpy of spreading. These surface spectroscopic and microscopic investigations should provide a framework for understanding the film forming and morphological properties of the PVDF copolymer as it relates to the nature of the ferroelectric switching mechanism in such ultrathin films.

Acknowledgments

This work was supported by the Center on Polymer Interfaces and Macromolecular Assemblies (CPIMA) sponsored by the NSF-MRSEC program.

Literature Cited

1. Lovinger, A.; In *Developments in Crystalline Polymers*, Basset, D.C.; ed.; Applied Science Publishers, 1982., Ch. 5.
2. Lando, J.; Doll,W.; *Macromol. Sci. Phys,*. **1968** ,*B2*,205.
3. Lopez Cabarcos, E.; Camara Canalda, J.; Martinez Salazar, J.; Balta Calleja F.; *Colloid and Polymer Science* **1988**, *266*, 41.

4. Scott, J. F.; Duiker, H.; Beale, P.; Pouligny, B.; Dimmler, K.; Parris, M.; Butler, D.; and Athens, S; *Physica B,* **1988** *150,* , 160.
5. Lines, M and Glass, A.; *Principles and Applications of Ferrolelectrics and Related materials,* Clarendon Press, Oxford, 1977 Ch 15.
6. Andrade, J.D.; *Polymer Surface Dynamics,* Plenum press; New York, 1988.
7. Sorokin, A.; Palto, S.; Blinov, L.; Fridkin,V. and Yudin, S.; *Mol. Mat.* **1996**, *6,* 61.
8. Palto, S.; Blinov, L.; Bune, A.; Dubovik, E.; Fridkin,V.; Petukhova, N.; Ducharme, S.; Yudin, S.; *Appl.Phys. Lett.* **1995**, *67* , 397.
9. Shaw, D.; *Introduction to Colloid and Surface Chemistry,* Butterworths; London, 1970.
10. Scheinbeim, J.; Nakafuku, C.; Newman, B.; Pae, K.; *J. Appl. Phys.,* **1979**, *50,* 439.
11. Palto, S.; Blinov, L.; Bune, A.; Dubovik, E.; Fridkin, V.; Petukhova, N.; Verkhovskaya, K.; Yudin,S.; *Ferroelectric Letters* **1995**, *19,* 65.
12. Nishi, T. and Wang,T.; *Macromolecules* **1975**, *8,* 909.
13. Goedel, W; Wu, H.; Friedenberg, M.; Fuller, G.; Foster, M.; Frank, C.W.; *Langmuir* **1994**, *10,* 4209.
14. Aust, E.; Ito, S.; Swodny, M.; Knoll, W.; *Trends Polym.Sci.***1994**, *2,* 9
15. Crisp, D..J.; *J. Colloid Sci.* **1946**,*1, 49,* 161.
16. Hoenig, D.; Moebius, D.; Overbeck, G.; *Adv. Mater.* **1992**, *4,* 419.
17. Palto, S.; Blinov, L.; Sorokin, A.; Dubovik, E.; Fridkin,V.; Petukhova, N.; Verkhovskaya, K.; Yudin, S and Zlatin, A.; *Europhys. Lett..* **1995**, *34* , 465.
18. Brinkhaus, R.; Schouten, J.; *Macromolecules* **1991**, *24,* 1487.
19. Peng, J.B.; Barnes, G.T.; *Langmuir* **1990**, *6,* 578
20. Adams, J.; Rettig, W.; Duran, R.; Naciri, J.; Shashidhar, R.; *J. Phys. Chem.* **1993**, *97,* 2021.
21. MacRitchie, F.; *Chemistry at Interfaces,* Academic Press; San Diego, 1990.
22. Advincula, R. C.; Ph.D. Thesis, University of Florida, 1994.
23. Paul, D.; Barlow, J.; Bernstein, R. and Wahfmund, D.; *Polym. Eng. Sci.* **1978**, *18,* 1225.
24. Prucker, O.; Christian, S.; Block, H.; Rühe, J.; Frank, C.W.; Knoll, W.; this *ACS Symposium Series,* Chapter XXX.
25. McBrierty, V.; Douglass, D.; and Weber, T.; *J. Polym. Sci-Polym. Phys. Ed.* **1976**, *14,* 1271.
26. Prest, W. Jr.; Luca, D.; *J. Appl. Phys.* **1975, 46,** 4136.
27. Prucker, O.; Christian, S.; Bock, H.; Rühe, J.; Frank, C.W.; Knoll, W. submitted to *Macromol. Chem. Phys.*
28. Malmonge, L. and Mattoso, L.; *Synthetic Metals* **1995**, *69,* 123.
29. Lovinger, A. and Gryte, C.; *Macromolecules* **1976**, *9,* 247.

Chapter 15

The Effect of Ion Type and Ionic Content on Templating Patterned Ionic Multilayers

Sarah L. Clark, Martha F. Montague, and Paula T. Hammond

Department of Chemical Engineering, Massachesetts Institute of Technology, Cambridge, MA 02139

The effects of NaCl, NaBr, NaI, Na_2SO_4, Li_2SO_4, LiCl, and KCl on templating the deposition of ionic multilayers of sulfonated polystyrene (SPS) and polydiallyl dimethylammonium chloride (PDAC) were studied. Self-assembled monolayers (SAMs) of alkane thiols on gold with terminal functional groups of carboxylic acid (COOH) and oligo ethylene glycol (EG) directed the adsorption of the SPS and PDAC multilayers. Ionic strengths of salt added to the polyelectrolyte solutions ranged from 0.001 M to 4.0 M. The anion had the greatest effect on bilayer thickness, with the chloride ion producing the thickest films. The cation, specifically Li, had the greatest effect on the EG surface's ability to resist polyion deposition. NaCl and LiCl proved to be the optimal salts of those studied to fabricate thick multilayers while preserving the SAMs' ability to control deposition.

Layer-by-layer assembly of ionic multilayers is a promising method for building supramolecular architectures using benchtop processing and inexpensive materials. A wide variety of applications from electroluminescent devices to biosensor arrays are currently being developed by many groups interested in the versatility of the technique *(1-13)*.

So far, all of these systems consist of uniform multilayer films on a substrate. By templating the deposition of polyelectrolyte multilayers onto chemically patterned surfaces, we are developing a method to confine these films to discrete regions on a surface, opening up new potential applications like waveguides and sensors. Previous work has included the effect of polyelectrolyte molecular weight *(14)* and sodium chloride *(15)* on the templating behavior of chemically patterned

206

surfaces of self-assembled monolayers (SAMs) and we have fabricated structures such as the one depicted in Figure 1.

To construct films for optical applications, there is a need for large numbers of bilayers to be templated by these chemically patterned surfaces. It is also desirable to reduce the number of bilayers required to achieve a certain final film thickness. We have decided to utilize the known effect of salt content to maximize film thickness of polyelectrolytes while preserving the power to template the deposition of multilayer adsorption *(1,16,17)*. In the following paper, we discuss the results of these experiments utilizing two oppositely charged polyions which have high segmental charge densities that are relatively independent of pH.

Experimental

Substrate Preparation and Patterning. N-type test grade silicon wafers (Silicon Sense) were coated with a 1000 Å gold film by thermally evaporating gold shot (99.99% purity, American Gold & Silver). A 100 Å film of chromium served as the adhesive metal between the gold film and the silicon surface.

Microcontact printing *(18)* was used to pattern two alkane thiols (R = COOH, $(OCH_2CH_2)_3OH$) onto the gold surface and create discrete regions of specific functionality. The first thiol was patterned onto the surface by stamps inked with the supernatant of a saturated solution of $HS(CH_2)_{15}COOH$ (COOH hereafter) in hexadecane. The second alkane thiol $HS(CH_2)_{11}(OCH_2CH_2)_3OH$ (EG in this paper) was deposited on the remaining bare gold regions of the surface from a 1 mM solution in absolute ethanol. The polydimethylsiloxane (PDMS) stamps used in this study printed either straight lines of 3.5 µm with spaces of 2.5 µm between the lines or the negative of a two dimensional array of 6 µm crosses.

Results from unpatterned films presented in this paper were derived from continuous films of the two alkane thiols. The COOH ink was spread on half the surface using a cotton tipped applicator. The excess material was rinsed using absolute ethanol and dried. The substrate was then dipped in the solution of EG terminated thiol to cover the remaining half of the surface. Each SAM surface was between 1 and 3 cm^2.

Polydiallyldimethylammonium chloride (PDAC) (~150,000 Mw, Aldrich, low molecular weight PDAC 20% solution in water) and sodium polystyrene sulfonate (SPS) (35,000 Mw, Polysciences, Inc., Mw/Mn = 1.10, molecular weight standard) were each dissolved in Milli-Q water (Millipore) to prepare the aqueous solutions required to fabricate ionic multilayers. SPS was prepared in 10 and 20 mM concentrations on a repeat unit basis. PDAC was used at a 20 mM repeat unit concentration. Sodium chloride, sodium bromide, sodium iodide, sodium sulfate, and potassium chloride were used as received from Mallinckrodt. Lithium chloride and lithium sulfate were bought from Aldrich. These salts were added to the polyelectrolyte solutions and were grouped by ionic content. The salt concentrations ranged from 0.001 M to 4 M. Solutions were filtered with a 0.22 µm MILLEX-GS Filter Unit (Millipore) to remove any particulates.

Polyion Multilayer Formation. The SAM functionalized surfaces were dipped in the PDAC solutions for 20 minutes to form an initial PDAC layer on the ionized COOH region. After the substrate was removed from the PDAC solution, it was rinsed with Milli-Q water to remove any excess material. The substrate was then placed in the solution of the polyanion, SPS, for 20 minutes to form the next ionic layer. The total thickness of material deposited on the surface could be carefully controlled by fixing the number of dipping steps used in the overall fabrication procedure. After a full bilayer was deposited, an ultrasonic cleaning step was performed for 4-5 minutes in an ultrasonic cleaning bath (Bransonic).

Characterization of Samples. Ellipsometry was used to measure the film thickness of the multilayer films. A Digital Instruments Dimension 3000 Atomic Force Microscope (AFM) was used in tapping mode with a standard etched silicon tip to collect data on the surface roughness, the height of patterned features, and the overall appearance of patterned surfaces.

Safety Considerations. This process is relatively benign and few precautions beyond standard labwear (gloves, lab coat, and safety glasses) were worn to avoid skin contact when working with the alkane thiol and polyelectrolyte solutions with added ionic content. The exceptions were the SPS and PDAC solutions containing LiCl and Li_2SO_4. Both are possible teratogens with Li_2SO_4 having "target organ: nerves" listed on the label. Extra precautions included donning two pairs of gloves, working in a hood, and bottling all rinse water for separate waste disposal.

Results and Discussion

Patterned Samples. Varying the ionic content of NaCl in solutions of SPS and PDAC affected the final structure of the patterned multilayers by altering the polymer-surface interactions and the roughness and thickness of the films *(15)*. The surface properties of the COOH and EG terminated SAMs altered depending on the NaCl concentration and the presence of a nitrogen drying step, changing the templating behavior of ionic multilayers. Figure 2 depicts three AFM micrographs of patterned surfaces. The crosses are regions of EG, a "resist" for polyelectrolyte adsorption at low ionic contents. The surrounding matrix is the COOH surface which promotes polyion adsorption at low NaCl concentrations. NaCl concentrations ranged from 0.001 M in micrograph 2a to 1.0 M in micrograph 2c.

The simplest way to approach these results is to discuss what is happening to each surface in turn. At low ionic content the polycations adsorb readily to the COOH region from aqueous solution. As the concentration of salt is increased, the charges on the polyion in solution become screened, reducing the attractive electrostatic interaction that leads to adsorption. The intramolecular repulsion between like-charged repeat units on the polyelectrolyte is also screened, changing the conformation of individual polyelectrolytes in solution from an extended chain to a coil. The comparisons between Figure 2a and b show rougher multilayer surfaces (the light colored regions) as the ionic content is increased from 0.001 M to 0.1 M. By examining the vertical scale of the two images, it is also possible to observe an increase in multilayer height. Further increases in ionic content serve to screen attractive ionic interactions between the solvated polyion and the surface until no polyelectrolyte deposition is observed in the patterned film of Figure 2c. In contrast, the EG surface changes from an area that resists polyion adsorption at low ionic contents in Figure 2a and 2b to a region that promotes adsorption at high ionic contents in Figure 2c. There appears to be an optimum point at 0.1 M NaCl where EG is the most effective resist and the EG region imaged in Figure 2b is relatively free of polyelectrolyte deposition. This is caused by an enhancement of the EG structure at the surface as small ions stabilize the hydrated brush structure of the EG repeat units. At high ionic content the salt ions, along with the nitrogen drying step, dehydrate the EG brushes and allow macroscopic deposition of the polyelectrolytes. These two simultaneous behaviors work in conjunction to reverse the templating of the ionic multilayers in a process we have termed "reverse deposition". When the drying step is removed from the fabrication process, the EG surface remains hydrated at all ionic contents while the multilayer adsorption on the ionized COOH surface is again reduced due to charge screening. Reverse deposition, which requires the EG region and COOH region to swap roles in templating ionic multilayers, is only observed with dried films adsorbed at high ($\geq$ 1.0 M) NaCl concentrations.

Salt Studies. After observing the effects of NaCl in templating patterned ionic multilayers, a series of experiments was performed with six other salts to determine if some of the effects were unique to NaCl or could be generalized to the ionic content of any salt. The experiments consisted of depositing 5 bilayers of PDAC / SPS on unpatterned surfaces of COOH and EG terminated SAMs. The ionic content in each pair of PDAC and SPS solutions was varied according to Table 1.

Table I. Experimental Design for Salt Series Studies

Salt Type	Ionic Strengths (M)
NaCl	0.001, 0.01, 0.04, 0.1, 0.4, 1.0, 2.0, 4.0
NaBr	0.01, 0.1, 0.4, 1.0
NaI	0.01 (PDAC was insoluble in solutions of higher ionic content)
Na_2SO_4	0.01, 0.1, 0.4, 1.0
Li_2SO_4	0.003, 0.01, 0.03, 0.1, 0.3, 1.0
LiCl	0.01, 0.1, 0.4, 1.0
KCl	0.01, 0.1, 0.4, 1.0

The final film thickness was measured by ellipsometry. Samples were not dried with nitrogen while fabricating the thin films, but were thoroughly rinsed in-between each

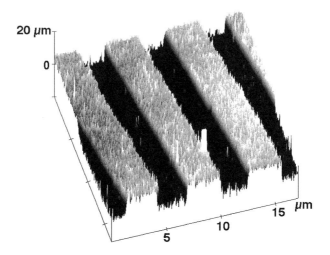

Figure 1. AFM micrograph of a typical patterned ionic multilayer. Features depicted have a width of 3.5 μm.

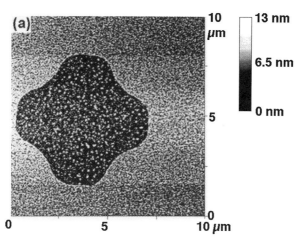

Figure 2. AFM micrographs of 13 bilayer patterned multilayers of SPS and PDAC. Lighter regions are the multilayer film. NaCl concentrations used are a) 0.001 M, b) 0.1 M, and c) 1.0 M.

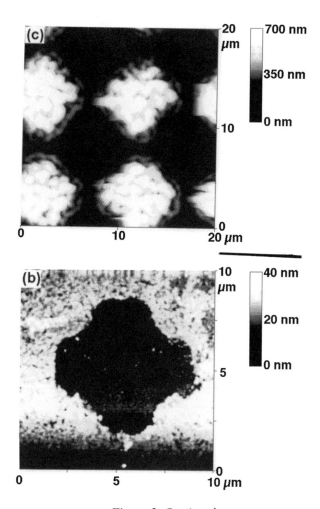

Figure 2. *Continued.*

adsorption cycle to remove excess adsorbed polymer and residual ions. Previous studies of similar layer-by-layer systems indicate that the salt ions are not detectable by XPS, and it is generally thought that most ions are removed during the rinsing step of the process (19). The polyion solutions used in these studies were below the critical concentration c* for SPS polyanion solutions at all ionic contents except zero (no added electrolyte). For PDAC polycation solutions, the solutions were also found to be below c*, although the c* limit was approached for PDAC solutions at lower ionic content. In general, we can assume that the data discussed here are primarily for systems in the dilute solution regime. This point is further confirmed by the correlation of ionic strength to film thickness discussed below.

All sets of multilayers exhibited similar behavior as the ionic content was increased, excluding NaI. NaI proved to have a great capacity to precipitate PDAC out of solution at any concentration higher than 0.01 M, allowing us only one data point in the desired range of salt concentrations. Figure 3 is a typical plot of total film thickness versus ionic content for NaCl. Circles represent the total thickness of films deposited on COOH SAMs and triangles correspond to the total film thickness of EG regions. The typical thickness of COOH and EG SAMs are around 10-15 Å and are part of the ellipsometric thickness plotted in this paper. Three basic features of this plot were common to all the resulting data from the study. The first characteristic is the film thickness of 5 bilayers on COOH at low ionic strengths ($\leq$ 0.01 M) which did not vary with the type of salt used. All plots also indicated a gradual increase in the bilayer thickness as the conformation of the polyelectrolyte was altered by screening the intramolecular repulsion of the repeat units, resulting in the adsorption of more loops and tails, rather than as a flat macromolecular configuration. This increased thickness occurred at salt contents above 0.01 M and continued to a maximum ionic content that shifted depending on the salt ions present. The observed increase in film thickness is approximately proportional to the square root of the ionic strength for all systems described here, which is consistent with theoretical predictions of the change in polyelectrolyte radius of gyration with added ionic content. These findings provide confirmation that polyelectrolyte adsorption for these systems may be modeled as adsorption of isolated macromolecules. This relationship was also observed in other experimental studies (20,21) of polyelectrolyte adsorption. These results are counter to the square dependence of thickness on ionic strength observed by Lvov et al in the adsorption of similar polyion multilayered systems (19).

After the maximum was reached, the total bilayer thickness on the COOH samples decreased sharply to the expected thickness of the bare COOH SAM monolayer. The driving force for polyelectrolyte deposition was screened by the high ionic content present in the aqueous solutions. This screening-reduced polyelectrolyte adsorption has been observed by other researchers for a single layer of polyelectrolyte adsorbed on an oppositely charged substrate (22,23). There is a balance between the increased adsorption due to charge screening, the decreased adsorption due to loss of the electrostatic driving force and competition between polyions and small ion for the surface. The maximum observed thickness is the point at which the latter driving forces become greater than the former. The maximum film thickness for NaCl is labeled in Figure 3 for an ionic content of 0.4 M; this number is of the same order of magnitude as the 0.2 M concentration recently described by Hoogeven et al for adsorption of polycation onto TiO_2 (22). All subsequent data on film thickness for bilayer height (COOH region) and resist height (EG region) is compiled from examining this characteristic maximum for each salt and the ionic concentration at which it occurred. The single data point measured for NaI was assumed to be its maximum. In contrast to the behaviors observed for the COOH region, the EG region showed relatively little variation with the ionic content and salt type. Excluding the sample series for NaBr, NaI, and KCl, the EG film thickness remained between 10-20 Å, indicating the SAM surface was hydrated and resisted the adsorption of polyelectrolytes at all ionic concentrations. Ionic strength was chosen as a parameter to compare data from across the different salt systems since it accounts for the valency of ions as well as their respective concentrations (24). Ionic strength is derived from electrostatic theory and is defined in Equation 1,

$$I = \frac{1}{2} \sum_i c_i z_i^2 \qquad (1)$$

where c_i and z_i represent the molar concentration and valency, respectively, of each ionic species, i, present in a salt. Ionic strength is in molar units.

Figure 4 plots the maximum film thickness of the bilayers deposited on the COOH region (circles) and the total film thickness of the EG region (triangles) for each of the salts as a function of ionic strength. The lines fit to each set of film thickness are included to guide the eye through the data and show that bilayer height increases with ionic strength regardless of the salt type while the resist height decreases only slightly with ionic strength. This is not unexpected if one assumes a purely electrostatic argument. The higher ionic strength screens charges on the repeat units of the polyelectrolyte, allowing it to change from an extended conformation to a more coiled one. The coiled polyelectrolyte layers are thicker and rougher than the flatter, smoother films formed at lower ionic strengths. Since certain salts like LiCl, KCl, and NaCl do not reach a maximum film thickness until high ionic contents, it follows that the multilayers formed under these conditions are much thicker than the multilayers formed from salts which peak at a lower ionic content. One of the motivations for undertaking this study was to find a system that deposited the thickest bilayers possible while still being "template-able", reducing the total number of dipping cycles required to form films of a certain thickness. Examining the data for the resist height, the trend seems to indicate that a higher ionic content leads to a cleaner resist surface. This fits with previous observations of chemically patterned surfaces that show a moderate amount of salt in solution reinforces the EG surface's ability to reject polyelectrolyte adsorption and produces the best patterned structures.

An important observation obtained from Figure 4 is the striking correlation between ionic strength of the maximum film thickness and anion type. All salts with

214

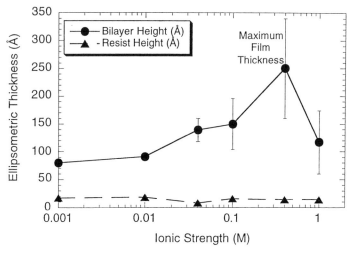

Figure 3. Plot of the thickness of 5 unpatterned bilayers of SPS/PDAC on COOH surfaces (Bilayer Height) and EG surfaces (Resist Height) versus the ionic strength of NaCl.

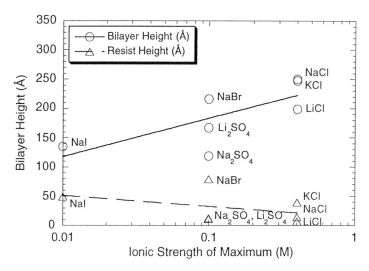

Figure 4. Plot of maximum film thickness for seven salts as a funtion of ionic strength.

a chlorine anion peaked at 0.4 M while the sulfate and bromine containing salts peaked at 0.1 M. The assumption that NaI maximum film thickness was achieved at 0.01 M is not inconsistent with theories proposed by other researchers. One such theory was derived by van de Steeg (23) when she modeled the adsorption of a single polyelectrolyte layer on an oppositely charged surface in the presence of salt. She was able to predict the form of the experimental data in Figure 3 and predicted the maximum film thickness would shift to lower ionic contents and lower film thickness as the salt had an increased "non-electrostatic" interaction with the charged surface. Ordering the anions based on the Hofmeister series for surface affinity, Figure 5 shows a decrease in the ionic strength of the maximum film thickness as surface affinity increases for all the anions except sulfate.

Figures 6 and 7 show histograms of the average maximum film thickness of the bilayer (COOH region) and resist (EG region) heights versus increasing radius of the anion and cation present in the salt. Bilayer height increases as the size of the cation increases while the opposite trend is observed as the size of the anion increases. The trend for anions is explained in terms of the Hofmeister series. If chlorine anions are able to achieve maximums at higher ionic strengths because of a weaker surface affinity, it follows that salts containing chlorine would have thicker bilayers. As was previously explained, the polyelectrolytes deposit in thicker bilayers at higher ionic contents. Examining Figure 4 again shows there is very little correlation between cation type and maximum film thickness. The trend in Figure 7 is more likely a consequence of the types of anions associated with each cation in the course of the study. Potassium comes out with the highest bilayer thickness only because it is associated with the chlorine anion. Both anions and cations interact with the ionized surface and polyions in solution, but the cation can also associate with the oxygen in the three ethylene glycol repeat units of the EG terminated SAM surface (25,26). A hydrated EG brush leads to good adsorption resistance and an EG brush supported by a cation that "fits" into the brush's hydrated structure, giving it extra stability, would be a more effective resist. In Figure 7 the smallest cation, lithium, gives an average resist height of around 15 Å, a film thickness that is consistent with the thickness of a bare EG terminated SAM surface. Sodium and potassium show an increase in this film thickness, corresponding to a decrease in the resist's ability to prevent adsorption.

Conclusions

Our results indicate salts containing a chlorine ion can achieve the thickest films since they allow polyelectrolyte multilayers to be constructed at the highest ionic strengths. Minimizing the amount of deposition on the EG surfaces requires a cation that bolsters the hydrated brush structure of the EG functionality. The lithium and sodium cations have shown an ability to improve resist characteristics due to their smaller ionic radii. Combining these observations, we conclude that NaCl and LiCl are the best choices for templating future ionic multilayer experiments.

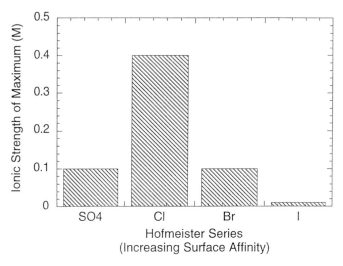

Figure 5. Histogram of the maximum ionic strength for each of four anions ordered according to the Hofmeister series for surface affinity.

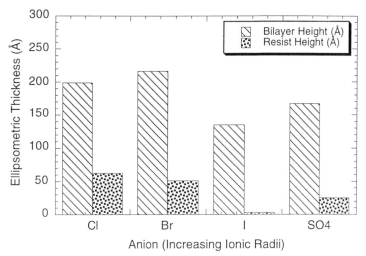

Figure 6. Histogram of the average maximum film thickness versus the size of each anion.

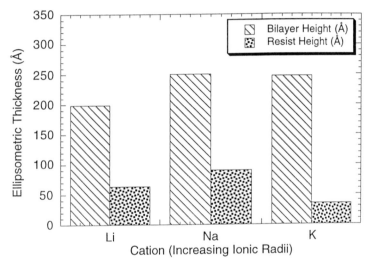

Figure 7. Histogram of the average maximum film thickness versus the size of each cation.

218

Acknowledgments

The authors thankfully recognize the research support of the Office of Naval Research, grant number N00014-96-1-0789. Other services used in the course of the study were funded by the MRSEC program of the National Science Foundation, DMR-9400334.

Literature Cited

1)Decher, G.; Hong, J.-D. *Makromol. Chem., Macromol. Symp.* **1991**, *46*, 321-327.
2)Decher, G.; Hong, J-D *Ber. Bunsenges. Phys. Chem.* **1991**, *95*, 1430-1434.
3)Ferreira, M.; Cheung, J.H.; Rubner, M.F. *Thin Solid Films* **1994**, *244*, 806-809.
4)Ferreira, M.; Rubner, M. F. *Macromolecules* **1995**, *28*, 7107.
5)Fou, A. C.; Rubner, M. F. *Macromolecules* **1995**, *28*, 7115.
6)Dressick, W.J.; Calvert, J.M. *Jap. Jour. of App. Phys., Part 1* **1993**, *32*, 5829.
7)Keller, S.W.; Kim, H-N; Mallouk, T.E. *J. Am. Chem. Soc.* **1994**, *11*, 8817.
8)Kleinfeld, E.R.; Ferguson, G.S. *Science* **1994**, *265*, 370.
9)Service, R.F. *Self-Assembly Comes Together*; Service, R.F., Ed., 1994; Vol. 265, pp 316-318.
10)Lee, H.; Kepley, L.J.; Hong, H.-G.; Akhter, S.; Mallouk, T.E. *J. Phys. Chem.* **1988**, *92*, 2597.
11)Katz, H.E.; Scheller, G.; Putvinski, T.M.; Schilling, M.L.; Wilson, W.L.; Chidsey, C.E.D. *Science* **1991**, *254*, 1485-1487.
12)Tsukruk, V.V.; Rinderspacher, F.; Bliznyuk, V.N. *Langmuir* **1997**, *13*, 2171-2176.
13)Cheung, J. H.; Stockton, W. B.; Rubner, M. F. *Macromolecules* **1997**, *30*, 2712.
14)Clark, S.L.; Montague, M.; Hammond, P.T. *Supramolecular Science* **1997**, *4*, 141-146.
15)Clark, S.L.; Montague, M.F.; Hammond, P.T *Macromolecules* **1997**, submitted.
16)Lvov, Y.; Decher, G.; Mohwald, H. *Langmuir* **1993**, *9*, 481-486.
17)Fou, A.C.; Onitsuka, O.; Ferreira, M.S.; Howie, D. ; Rubner, M.F. *Self-Assembled Multilayers of Electroactive Polymers: From Highly Conducting Tranparent Thin Films to Light Emitting Diodes*; Fou, A.C.; Onitsuka, O.; Ferreira, M.S.; Howie, D. ; Rubner, M.F., Ed.: Anaheim, 1995.
18)Kumar, A.; Biebuyck, H. A.; Whitesides, G.M. *Langmuir* **1994**, *10*, 1498-1511.
19)Lvov, Y. M.; Decher, G. *Crystallography Reports* **1994**, *39*, 628.
20)Baur, J. W. *Fabrication and Structural Studies of Sequentially Adsorbed Polyelectrolyte Multilayers*; Baur, J. W., Ed.; Massachusetts Institute of Technology: Cambridge, 1997.
21)Böhmer, M. R.; Cohen-Stuart, M. A.; Fleer, G. J. *Macromolecules* **1990**, *23*, 2301-2309.
22)Hoogeveen, N.G.; Cohen Stuart, M.A.; Fleer, Gerald J. *Journal of Colloid and Interface Science* **1996**, *182*, 133-145.

23)van de Steeg, H. G. M.; Cohen Stuart, M. A.; de Keizer, A.; Bijsterbosch, B. H. *Langmuir* **1992**, *8*, 2538-2546.

24)Dahlgren, M.A.G. *Langmuir* **1994**, *10*, 1580-1583.

25)Bailey, F.E.; Koleske, J.V. *Poly(ethylene oxide)*; Academic Press: New York, 1976.

26)Dennison, K.A. *Radiation Crosslinked Poly(ethylene oxide) Hydrogel Membranes*; Dennison, K.A., Ed.; Massachusetts Institute of Technology: Cambridge, 1986, pp 367.

Chapter 16

Organized Multilayer Films of Charged Organic Latexes

V. N. Bliznyuk[1,3], A. Campbell[2], and V. V. Tsukruk[1,4]

[1]College of Engineering and Applied Sciences, Western Michigan University, Kalamazoo, MI 49008
[2]Materials Directorate, Wright Laboratory, WL/MLPJ, Wright-Patterson AFB, OH 45433

We study several multilayer latex-latex films fabricated by electrostatic self-assembly and composed of charged polystyrene (PS) latexes by combined scanning probe microscopy and X-ray reflectivity. PS nanoparticles of 20 - 200 nm in diameter possess carboxy, sulphate, and amidine surface groups. Most of latexes used are able to form monolayers with short-range local ordering of nanoparticles within the monolayer. Layer-by-layer deposition of latexes with alternating charges results in steady increase of film thickness that is consistent with dense centered cubic packing of nanospheres up to the first five layers. Virtually linear growth of film thickness is observed for further deposition (up to 50 layers) with average increment of 7.3 nm per layer for 20 nm nanoparticles due to the incomplete monolayer formation. During thermal treatment in a wide range of temperatures, we observe that strong tethering of charged nanoparticles to surfaces prevents their surface diffusion and rearrangements required for formation of perfect lateral ordering.

Supramolecular engineering is a modern approach to fabrication of organized films from specially designed functional polymers for electronics and optics applications (1). A principal limitation of the known molecular multilayer assemblies (Langmuir-Blodgett and self-assembled films) is the nanometer modulation of their properties (e. g., density or refractive index). Scaling this molecular organization to an "optical" range (200-1000 nm) is a challenge for organic solid state chemistry. Usage of pre-formed polymer nano-phases such as latex and dendrimer nanoparticles can provide required routes towards fabrication of such organized systems.

It is speculated that nano-particle arrays can be used as narrow-band optical diffraction filters (rugate filters), photonic-band-gap structures ("photonic crystals") and nonlinear optical switches (2-4). Two dimensional (2D) arrays of such particles on solid substrates can be employed in some modern technical applications such as data storage as well as in optical and microelectronics devices (5,6). Knowledge of the arrangement of particles on the substrate gives insight into fundamental forces and mechanisms involved in formation of nanostructures and has practical importance for engineering of materials with advanced interfacial properties.

[3]Also: Institute of Semiconductor Physics, National Academy of Science, Kiev, Ukraine
[4]Current address: Physics Department, University of California, Santa Cruz, CA 95064.

Among ways proposed for latex particles thin film organization are: solution casting, dip-coating, and alternated layer-by-layer deposition *(7-13)*. In Iler's pioneering work, he proposed electrostatic layer-by-layer deposition as a tool for fabrication of organized films *(7)*. Coulombic interactions between oppositely charged latexes and micro-fibrils adsorbed from the solution are used as a driving force for multilayer organization. Presently, this method has been further developed and applied for soft-chain polymers, conductive polymers, liquid crystals, dendrimers, organic-inorganic composites, and biomolecules *(13-16)*.

Most of the research in this field was done on the macroscopic rather than the microscopic (submicron) level. Recently, the interest in these systems has been renewed due to both developments of more sensitive experimental tools and measurement devices as well as promising new applications.

This publication presents our results on fabrication of alternating layered films from latex nanoparticles by electrostatic self-assembly and forced solution removal techniques (self-assembling and dip-coating) (Figure 1). Various types of negatively and positively charged latexes of 20 - 200 nm in diameter were used to build films up to 50 layers thickness. These films were examined under both scanning probe microscopy (SPM) and X-ray reflectivity.

Experimental

Colloidal particles - polystyrene nanospheres of submicrometer diameter, and negative or positive surface net charge produced by surface carboxyl (CML), sulfate (SL), and amidine (AL) groups (Interfacial Dynamics Corporation, Portland, OR) are used in the present study (Table, Scheme 2). These latexes are quasi-monodisperse with standard size deviations from 1.5 to 23%. We use 0.4 - 8 wt.% latex solutions in Milli-Q water and adsorption times of 10 - 20 min. for deposition. The initial resistivity of the purified water was 18 MΩ.cm

Negatively charged bare silicon wafers treated with chromic sulfuric acid and the positively charged surface of 3-aminopropyltriethoxysilane self-assembled monolayer (SAM) *(17)* are used as substrates in the beginning of the adsorption process. Due to recharging of the surface after deposition of the monolayer, the next latex layer (with the charge opposite to the previous one) is adsorbed in the same route. An idealized multilayer assembly constructed after one deposition cycle (i.e. bilayer) is shown in Scheme 1. By varying the pH of the colloids and thus changing the hydroxylation-hydrogenation equilibrium, the surface charge of the substrate can, in most cases, be adjusted opposite to that of the colloid solution. We have used 0.1 N aqueous solution of HCl (Aldrich) or 0.1N aqueous solution of NaOH (Aldrich) for this purpose.

Several latex films were also prepared by spin- or dip-coating methods for comparison purposes. In the former case a standard spin-coating apparatus (EC101DT Photo Resist Spinner, Headway Research, Inc., Garland, TX) was applied. Influences of the solution concentration (0.4 to 8 wt.%) and angular rotation speed (1000 to 3000 rpm) on thickness and quality of the films were tested. In the latter case an R&K (Wiesbaden, Germany) Langmuir trough dipper mechanism was used with a controlled substrate withdrawal speed of 6- 20 μm/s. Concentration of the solution was varied in 0.5-2 wt.% range.

SPM imaging, surface microroughness and topography parameters measurements of the dried samples were performed with a Nanoscope III, Dimension 3000 scanning probe microscope (Digital Instruments, Santa Barbara, CA) using the tapping mode in air. In such a regime forces applied to particles are minimized and latex arrays can be imaged without altering the position of the particles on the surface. To avoid contamination the SPM tip it was modified with hydrophobic (CH$_3$

222

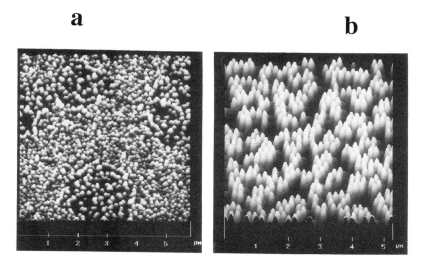

Figure 1. SPM images (3 x 3 mm scale) of typical latex SAMs: 20-nm latex (CML20) on NH$_2$-terminated SiO$_2$ (a); and 200-nm latex monolayer (AL190) on hydrophilic SiO$_2$ (b).

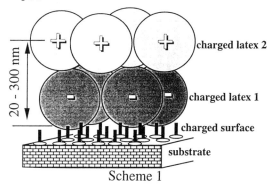

Scheme 1

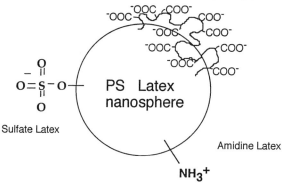

Scheme 2

Table. Nanoparticle characteristics and notations.

Particle Type	Mean Particle Diameter (D), nm	Standard Deviation of D, %	Surface Charge Type and Density	Type of the Surface	Notation
Amidine- modified PS latex	15	25	positive 0.64 μC/cm^2	hydrophobic	AL15
Amidine-modified PS latex	20	23.3	positive 1.7 μC/cm^2	hydrophobic	AL20
Carboxyl-modified PS latex	20	15.3	negative 17.7 μC/cm^2	hydrophilic	CML20
Sulfate-modified PS latex	40	13.9	negative 2.5 μC/cm^2	hydrophobic	SL40
Amidine-modified PS latex	190	1.5	positive 8.3 μC/cm^2	hydrophobic	AL190
Carboxyl-modified PS latex	190	2.7	negative 221 μC/cm^2	hydrophilic	CML190
Sulfate-modified PS latex	200	4.6	positive 1.1 μC/cm^2	hydrophobic	SL200
Copper phthalocyanine tetrasulfonic acid	0.5	0	negative 40 μC/cm^2	N/A	CuPcTc

terminated) or appropriately charged (SO_3 or NH_2) SAMs *(17)*. X-ray reflectivity measurements were performed on a Siemens D-5000 diffractometer equipped with a reflectometry stage *(18)*. Optical absorption spectra were measured utilizing a Perkin-Elmer Lambda 9 spectrophotometer.

Results and Discussion

Self-assembling and layer-by-layer deposition. Examples of SPM images of monolayer films of different latexes are shown in Figure 1. We observed formation of a latex monolayer on an oppositely charged substrate such as AL latexes on SiO_2, and CML or SL latexes on NH_3^+ terminated substrate. The surface coverage achieved was 80% for 20-nm particles. Adsorption time had no influence on the film quality for time interval between 5 minutes to several hours. Surface roughness of 9.5 nm (measured for 5 μm x 5 μm area) corresponded to the expected value for packed nanospheres. However, adsorption of AL latexes on positively charged SAM or CML latexes on negatively charged SiO_2 was negligible at the given conditions. Monolayers of large latexes possess lower degree of surface coverage in comparison to smaller particles (surface coverage parameter is calculated here as a ratio of occupied area to total available space on the surface). If the small latex monolayers have the coverage close to 100% with randomly occurring round voids of micrometer scale diameter, the large latex monolayers, on the contrary, consist mainly of relatively small islands with randomly occurring crystalline domains of 1- 10 μm size. The average coverage for these films does not exceed 60%.

Formation of the second layer on top of the monolayer is demonstrated in Figure 2. SPM images reveal that multilayer films of 20-nm latexes possess liquid type lateral packing of the particles with a positional correlation expanded only over the nearest neighbors, which can be caused by relatively wide particle size distribution (Table). We explain this (and also difficulties with a more regular multilayer growth) by inhomogeneity of particle sizes. Much more uniform 200-nm latexes display better local order. As can be seen in Figure 3, they have small domains of perfect structural ordering. Fourier analysis of the image 3b reveals an appearance of correlated regions of hexagonal packing with an average lateral dimension of 3- 5 coordination spheres and interplanar distance of 180 nm (corresponding to average latex diameter).

Typical multilayer growth pattern (thickness versus number of deposited layers) is shown in Figure 4 for CML20/AL20 alternating deposition. As the number of layers increases, the total thickness increases too but at decreasing increments. The first layer has a thickness of 20 nm (particle diameter) and the second layer is only 17 nm thick. These values suggest hexagonal packing of spheres in a bilayer. The average increment for the first five layers is about 15 nm which corresponds to centered cubic packing of spheres. However, for n > 5 the increment decreases to 7.3 nm per layer due to incomplete filling of preceding layers during film growth. Surface roughness gradually increases for the first five layers and reaches a constant value of 22 ± 2 nm for thicker films. Apparently, this suggests that some equilibrium growth process is established at this stage. We can speculate that roughly two layers are "under construction" during each deposition cycle. X-ray reflectivity curves for multilayer films are diffuse and do not exhibit Bragg reflections associated with internal periodicity. Obviously, this is due to homogeneous density distribution along the surface normal caused by overlapping of nanoparticles belonging to adjacent layers and exceeding roughness of the film surface.

Lateral ordering in uncharged latex films can be dramatically improved by thermal treatment near the glass transition temperature, which provides additional surface mobility and stimulates particle aggregation *(9,11)*. However, contrary to

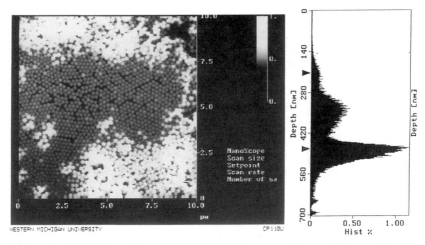

Figure 2. SAM bilayer film of SL200/AL190 latex particles on NH$_2$-terminated SiO$_2$

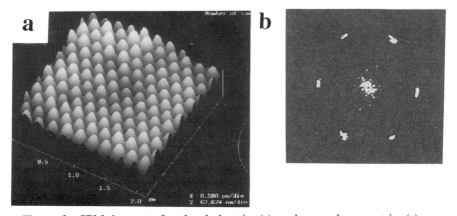

Figure 3. SPM images of ordered domain (a) and amorphous matrix (c) representative places of SL200 latex SAM. corresponding fourier transforms are shown in (b) and (d).

Continued on next page.

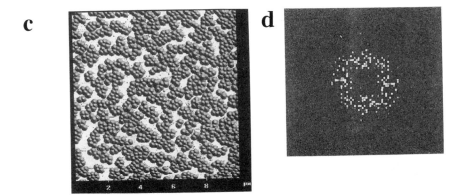

Figure 3. *Continued.*

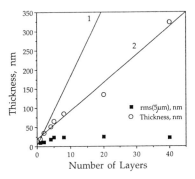

Figure 4. Thickness and surface roughness of multilayer CML20/AL20 films depending on a number of deposition cycles (number of layers). Straight lines 1 and 2 represent calculations for centered cubic packing and linear fit of experimental data, respectively.

neutral latexes, clusters of charged polystyrene (PS) nanoparticles possess unusual thermal stability (in terms of their shape and position) for temperatures up to 90°C (Figure 5). Furthermore, temperature higher than glass transition (110°C) can lead to local "melting" and spreading of particle clusters over the substrate surface and formation of locally smooth and flat areas (Figure 5b-c). Three-hour thermal treatment at 150°C resulted in further film spreading and formation of a virtually complete thin film with random holes and no visible tracks of particle packing (Figure 5d). Average thickness of the initial film gradually decreases from 190 nm to 56 nm to compensate for an increase of surface coverage. Strong tethering of charged nanoparticles to surfaces prevents their surface diffusion and rearrangements required for formation of perfect lateral ordering at intermediate temperatures.

Spin-coating and dip-coating. A very important factor in promoting homogeneous film formation is the development of the water surface tension and resulting negative capillary pressure in the system. Under "in situ" adsorption conditions, a uniform distribution of particles (with short-range order at high surface coverage) over the substrate is usually demonstrated *(19)*. As has been found in several experiments, already adsorbed colloidal particles can rearrange upon drying into a microstructure of two-dimensional, close-packed clusters *(20)*. The structural rearrangement is attributed to attractive capillary forces between particles in near contact during film evaporation *(9, 10, 21)*. The capillary pressure increases with decreasing amounts of water in the interstitial regions of the spherical particle monolayer (having a specific surface area at least tripled in comparison to a flat and plain surface) and can reach a very high level even with low amounts of water *(22)*. Interparticle interactions are described by electrostatic repulsion and steric attraction (excluded-area effect), which usually leads to monolayer or submonolayer coverage.

A special mechanism of 2D crystallization involving capillary forces for nucleus formation (for particles partially immersed in a liquid layer) and consequent crystal growth through convective particle flux caused by water evaporation from already ordered film has been proposed by Dimitrov and Nagayama *(10, 21)*. Such a concept can be applied for a family of deposition techniques (so called forced deposition methods). It should be pointed out that as a drying procedure is common for any of the "wet" methods of colloidal arrays preparation, consideration of the capillary forces is crucial importance for understanding their equilibrium structure formation.

More uniform but with less coverage films are formed at conditions where latex particle surface charges are better screened by counterions (pH 8 in case of amidine-modified latex particles shown in Figure 6). On the other hand, by increasing the surface charge to its maximum value (complete dissociation of surface groups), better coverage at the cost of worse in-plane ordering can be developed. The stability of the 2D nucleation process as well as the quality of final film are governed in addition by the velocity of the substrate removal (which controls the evaporative flow from the meniscus area).

Demonstrated in Figure 6 a, ordered domains of almost perfect hexagonal structure are formed for appropriate pH values, concentration and speed of substrate removal. These condensed domains are found to occupy as large as 70% of the surface and have average dimensions of several tens of μm. Measurements of the film thickness demonstrate that the most stable is a three-layer particle assembly in these domains. The rest of the film is usually in sub-monolayer or at least highly defect monolayer state (Figure 6 b).

Network-like pattern of particles is observed under conditions favoring particle aggregation. In contrast, faceted clusters are observed at low coverage. A

a

b

c

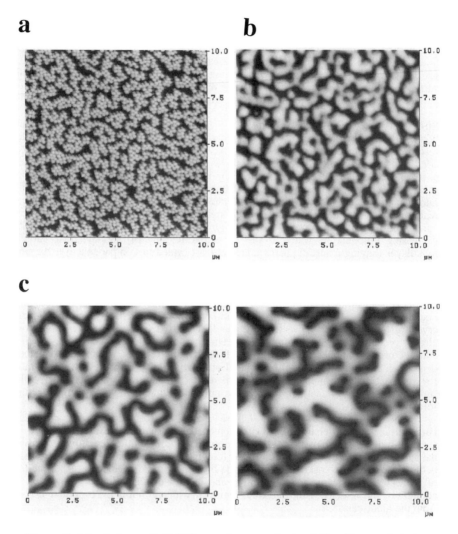

Figure 5. Morphology of AL190 monolayer on hydrophilic silicon substrate (10x10 μm scale, Z- range 0 -1 μm) for as prepared sample (a), after thermal treatment at 110°C (b), 130°C (c) and 150°C (d).

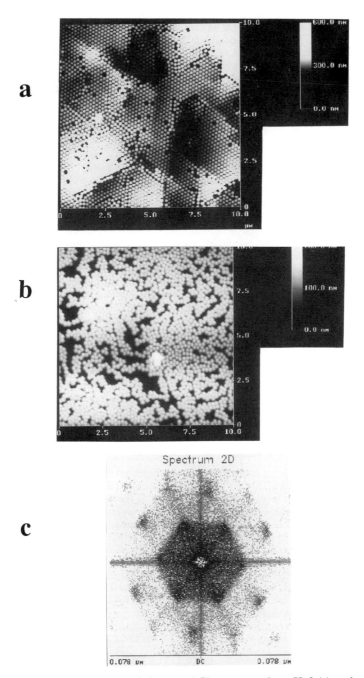

Figure 6. SPM images of dip-coated films prepared at pH 6 (a) and pH 2 (c). Fourier transform of image (a) is presented in (b). Speed of NH2-terminated substrate removal was 6 μm/s

characteristic faceted shape is typical for small clusters which depends on the number of particles in cluster. The nearest-neighbor distance of latex particles in a cluster depends on the number, n, of particles within the cluster. Some constant (equilibrium or "bulk") state is attained for n > 7.

Dense multilayer films are also produced through the spin-coating technique (Figure 7). The coverage can be varied at will in this case and can be higher than in self-assembling or dip-coating modes. The thickness follows linear dependence versus colloidal solution concentration. Only amorphous type of ordering has been observed. A simple explanation for such structural peculiarities could be in the highly non-equilibrium conditions of water evaporation from the film.

Further prospects. Improvement of the multilayer film quality is an important issue and should be a goal of further studies on charged latexes. Hopefully, some fine balance of inter-particle interactions through surface charge density and particle size would give such an opportunity. Unfortunately, small (several nanometers in diameter) particles have a wide size distribution which could be responsible for accumulation of defects in a multilayer limit. More uniform micrometer-scale latexes need to be optimized for stable charge adsorption (in sense of homogeneity of the surface charge distribution and charge-interaction strength - surface charge density /particle mass ratio). Alternation of latexes with polyionomers, or low molecular mass contra-pairs (23) which allow for curing of the film surface during its growth, could be considered as a good approach to a submicrone-scale multilayer assembling (Figure 8). Complexes of latexes and polyionomers, latexes and surfactants, and pre-arranged substrates (mixed SAMs) with a proper arrangement of "reactive" sites should be proven also as a route for stabilizing of latex monolayers.

New applications of latex multilayers can be implicated by using various types of latexes. Some specific species can have optically active chromophores that can be used as a control for refractive index variation along the film normal. Other prospective types of latexes are those with functional surface groups or special latexes capable of changing their surface properties under external stimuli (zwitterionic polymers). Hybrid inorganic-core/organic-shell latexes (24) are also attractive as a model system for third-order nonlinear optical films (25).

Conclusions

In conclusion, we have studied several multilayer latex-latex films composed of charged PS nanoparticles of 20 - 200 nm in diameter and fabricated by electrostatic self-assembly, spin- or dip-coating techniques. Most of the latex nanoparticles form monolayers with short-range local ordering. Layer-by-layer deposition results in a steady increase of film thickness that is consistent with dense centered cubic packing of nanospheres of up to five initial layers. Linear growth is observed for further deposition (up to 50 layers) with an average much smaller increment per added layer (7.3 nm) for 20 nm nanoparticles.

Strong tethering of charged nanoparticles to surfaces prevents surface diffusion and rearrangement required for formation of perfect lateral ordering. Improvement of the multilayer film quality is an important issue and should be a future goal for studies on charged latexes. Theoretical consideration of charged polymer adsorption shows that strong inter-particle interactions can suppress formation of complete monolayers (26). This would result in incomplete recharging of the surface, and disrupt a regular layer-by-layer film growth.

Force assisted methods of latex particles deposition generally produce better film quality due to additional capillary forces which favor lateral diffusion. These methods, however, need additional optimization for the case of multilayer deposition.

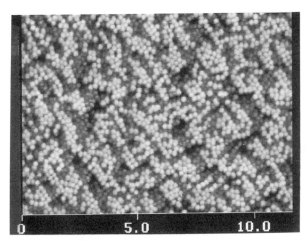

Figure 7. Spin-coated bilayer film of AL190 on hydrophilic silicon (0.4 % wt. solution in water, 3000 rpm, scale in µm).

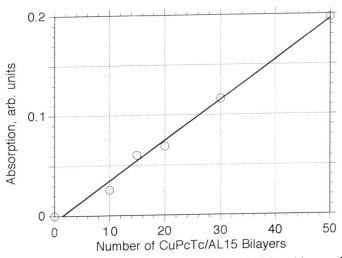

Figure 8. Optical absorption versus number of deposited latex/ low molecular mass compound (AL15/CuPcTc) bilayers.

232

Acknowledgments

This work is supported by AFOSR, Contract F49620-93-C-0063 and The National Science Foundation, CMS-94-09431 Grant. We thank S. Mirmiran and J. K. Kalsi for technical assistance.

Literature Cited

1. Tsukruk, V. V. *Prog. Polym. Sci.* **1997**, *22*, 247.
2. Weissman, J.M.; Sunkara, H.B.; Tse, A.; Asher, S.A. *Science* **1996**, *274*, 959.
3. Joannopoulos, J.D.; Villeneuve, P.R.; Fan, S. *Nature* **1997**, *386*, 143.
4. Rosenberg, A.; Tonucci, R.J.; Bolden, E.A. *Appl. Phys. Lett.* **1996**, *69*, 2638.
5. Deckman, H.W.; Dunsmuir, J.H. *Appl. Phys. Lett.* **1982**, *41*, 377.
6. Deckman, H.W.; Dunsmuir, J.H. *J. Vac. Sci. Technol. B.* **1983**, *1*, 1109.
7. Iler, R.K. *J. Colloid and Interface Sci.* **1966**, *21*, 569.
8. Denkov, N.D.; Velev, O.D.; Kralchevsky, P.A.; Ivanov, I.B.; Yoshimura, H.; Nagayama, K. *Langmuir* **1992**, *8*, 3183; Gao, M.; Gao, M.; Zhang, X.; Yang, Y.; Yang, B.; Shen, J. *J. Chem. Soc., Chem. Commun.* **1994**, 2777; Kotov, N.A.; Dekany, I.; Fendler, J.H. *J. Phys. Chem.* **1995**, *99*, 13065
9. Dimitrov A.S.; Dushkin, C.D.; Yoshimura, H.; Nagayama, K. *Langmuir* **1994**, *10*, 432; Feldheim, D.L.; Grabar, K.C.; Natan, M.J.; Mallouk, T.E. *J. Am. Chem. Soc.* **1996**, *118*, 7640;
10. Dimitrov, A.S. and Nagayama, K. *Langmuir* **1996**, *12*, 1303.
11. Li, J. and Meier, D.J. *Polymer Preprints* **1996**, *37* , 591.
12. Johnson, C.A. and Lenhoff, A.M. *J. Colloid and Interface Sci.* **1996**, *179*, 587.
13. Deher, G. and Honig, J.-D. *Makromol. Chem., Macromol. Symp.* **1991**, *46*, 321; Lvov, Yu.M. and Deher, G. *Crystallography Reports* **1994**, *39*, 628; Ferreira, M.; Cheung, J. H. and Rubner, M. *Thin Solid Films* **1994**, *244*, 806; Schmitt, J.; Decher, G.; Dressik, W.J.; Brandow, S.L.; Geer, R.E.; Schashidhar, R.; Calvert, J.M. *Adv. Mater.* **1997**, *9*, 61; Ariga, K.; Lvov, Yu.; Onda, M.; Ichinose, I.; Kunitake, T. *Chem. Lett.* **1997**, 125.
14. Fendler, J. H. and Meldrum, F.C. *Adv. Mater.* **1995**, *7*, 607.
15. Tsukruk, V.V. andWendorff, J.H. *Trends in Polymer Science* **1995**, *3*, 82.
16. Watanabe, S. and Regen, S.L. *J. Amer. Chem. Soc.* **1994**, *116*, 8855; Tsukruk, V.V.; Bliznyuk, V.N. and Rinderspacher, F. *Polymer Preprints* **1996**, *37* , 571.
17. Tsukruk, V.V.; Bliznyuk, V.N.; Wu, J.; Visser, D.W. *Polymer Preprints* **1996**, *37*, 575.
18. Tsukruk, V.V.; Rinderspacher, F.; Bliznyuk, V.N. *Langmuir* **1997**, *13*, 2171.
19. Johnson, C.A.; Lenhoff, A.M. *J. Coll. Inter. Sci.* **1996**, *179*, 587.
20. Böhmer, M. *Langmuir* **1996**, *12*, 5747.
21. Enoch, T.; Carr, A.; Nurse, P. *Nature* **1993**, *361*, 26.
22. Lin, F.; Meier, D.J. *Langmuir* **1995**, *11*, 2726.
23. Zhang, X.; Gao, M.; Kong, X.; Sun, Y.; Shen, J. *J. Chem. Soc., Chem. Commun.* **1994**, 1055-1056; Cooper, T.M.; Campbell, A.L.; Crane, R.L. *Langmuir* **1995**, *11*, 2713; Sun, Y.; Zhang, X.; Sun, C.; Wang, Z.; Shen, J.; Wang, D.; Li, T. *J. Chem. Soc., Chem. Commun.* **1996**, 2379-2380.
24. Liz-Marzan, L.M.; Giersig, M. and Mulvaney, P. *Langmuir* **1996**, *12*, 4329.
25. Zyss, J. *Molecular Nonlinear Optics: Materials, Physics, And Devices.* Acad. Press, N.Y., **1994**.
26. Slomkovski, S.; Miksa, B.; Trznadel, M.; Kowalczyk, D.; Wang,F.W. *Polymer Preprints* **1996**, *37*, 747.

Chapter 17

Glass Transition in Ultrathin Polymer Films

Oswald Prucker[1], Stefan Christian[2], Harald Bock[2], Jürgen Rühe[2],
Curtis W. Frank[1,3], and Wolfgang Knoll[1-3]

[1]Center on Polymer Interfaces and Macromolecular Assemblies (CPIMA), and
Department of Chemical Engineering, Stanford University,
Stanford, CA 94305–5025
[2]Max-Planck-Institute for Polymer Research, Ackermannweg 10, D-55128
Mainz, Germany

The glass transition temperatures (T_g) of thin polymer films have been
determined using evanescent wave optics. The polymer-substrate
interactions as well as the molecular architecture of the polymer layers
were systematically varied. A very pronounced T_g depression is
observed for polymethylmethacrylate (PMMA) films thinner than 50 -
100 nm on hydrophobic surfaces. The T_g of the thinnest samples (few
nm) is typically up to 25°C lower than that of the bulk material. This
effect appears to be independent of the internal architecture because
spin-cast, end-grafted and LBK films exhibit identical results.
However, no such depression can be found for assemblies where each
repeat unit of the polymer can interact specifically (e.g. by hydrogen
bonding) with the surface. Regardless of their thickness, polyhydroxy-
styrene films on silane monolayers with terminal ester groups always
show bulk-like T_g values. Identical results were obtained from PMMA
layers directly deposited on SiO_x surfaces.

Thin films of polymers are subject to high thermal and mechanical stress when used
as protective coatings or lubricants (1,2). As a consequence, there has been
considerable interest in the determination of the physical and dynamical properties of
those films especially if their thickness is only a small multiple of the coil dimensions
of the film-forming polymer. It was found (3-5), for example, that the degree of
crystallization of poly(di-n-hexylsilane) decreases significantly if the polymeric
material is constrained to thin layers (<30 nm). Stamm et al. (6) have shown that the
packing in liquid crystalline side chain polymers is strongly altered in layers thinner
than ca. 50 nm. There is also some evidence that the restricted geometry may
influence the glass transition temperature (7-12) (T_g) of amorphous polymer films

[3]Corresponding authors.

and their thermal expansion coefficients *(11-13)*. Occasionally, a T_g depression can be observed, again for thicknesses below ca. 50 - 100 nm *(7-9)*. These findings are often interpreted on the basis of chain mobility near the interfaces. Conformational features such as coil deformations in polymer 'brushes' or 'pancakes' *(14)* are also believed to be important in this context. However, due to the still limited database formed from the existing studies, these explanations remain rather qualitative and speculative. Additionally, it is sometimes rather difficult to obtain reliable results. For example, Reiter *(15-17)* has shown that many polymers tend to dewet from their substrate if heated above their T_g.

In our studies we attempt to broaden the knowledge about the influence of confined geometries on the T_g of ultrathin polymer films. We do this by systematically varying several parameters that might influence the thermal properties of these assemblies, namely surface/polymer interactions and conformational features. Variation of the first parameter was mainly achieved by investigating spin-cast films of polymethylmethacrylate (PMMA) on either bare SiO_x substrates or on those previously made hydrophobic by gas phase deposition of hexamethyldisilazane (HMDS). In the first case, hydrogen bonding between the surface silanol groups and the carbonyl moieties of the polymer is possible, whereas this interaction is prohibited in the latter case by capping the surface silanols with trimethylsilyl groups. Besides these systems, we also studied films of end-grafted PMMA prepared according to a procedure developed by Rühe et al. (see Figure 1) *(18-20)*. These assemblies are characterized by one strong interaction - the covalent bond - between one chain end and the surface. Further interactions between segments along the chain and the surface are unlikely in this case, as will be explained in detail in the Discussion section.

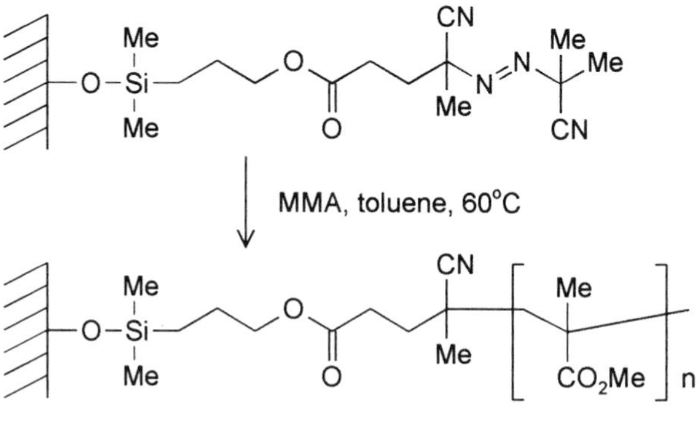

Figure 1. Generation of films of covalently attached PMMA through *grafting from* polymerization *(18,19)*.

Besides the differences in the polymer/surface interaction of these assemblies, they also exhibit different conformational features. The coils of spin-cast films - especially for the case of low thicknesses - usually lie more or less flat on the surface in a so-called 'pancake' conformation. In contrast, it is known for the grafted films used in this study that the anchoring density is comparably high (average distance between two anchors: 7 - 2.5 nm) as are the molecular weights ($> 10^5$ g mol^{-1}) *(18,19)*. Hence, the respective coils are significantly stretched perpendicular to the surface ('brush' conformation). Additionally, we also investigated multilayers of PMMA (solely on hydrophobic substrates) prepared by the Langmuir-Blodgett-Kuhn technique. These assemblies should exhibit some degree of chain orientation as the coils are forced into very thin (ca. 1 nm) separate layers.

Experimental

Sample preparation. Films prepared by spin casting and the LBK technique have been deposited by literature procedures *(21)* on glass/silver/SiO$_x$ substrates (see below) previously made hydrophobic by depositing hexamethyldisilazane (HMDS) from the gas phase. In some case, the polymers were spun directly onto unmodified substrates. All materials and solvents (HPLC grade) were used as received. Table I gives the molecular weights and polydispersities of the polymers used for the preparation of the various films.

Table I. Summary of molecular weights and polydispersities of the polymers used for the samples prepared by spin coating and the LBK technique; P4HS = poly(4-hydroxystyrene).

Polymer	M_n [g mol^{-1}]	M_w/M_n	Film type
PMMA	104,000	1.10	spin coating
PMMA	188,000	1.06	LBK, spin coating
P4HS	32,000	n\a	spin coating

Layers of covalently attached PMMA were created according to the procedures described in references 18 and 19. The layers prepared by this method are characterized by high graft densities (average distance between two anchoring sites: 2.5 - 7 nm) and high molecular weights ($M_n > 10^5$ g mol^{-1}).

Prior to any measurement, the spun cast and grafted layers were annealed at 150°C and 10^{-2} mbar for at least 12 h. After that the samples were slowly cooled back to room temperature (cooling rate: ca. 1°C min^{-1}) and either used directly or stored under nitrogen atmosphere. Such a procedure is not suitable for LBK films as it would most likely destroy their unique multilayer structure. These samples were therefore only annealed *in vacuo* at room temperature.

Evanescent Wave Optics. The use of evanescent wave optics for the characterization of ultrathin organic films has already been discussed in great detail *(22)*. Thus, only a short summary will be given here.

Surface plasmons are transverse magnetic waves that travel along an interface between a metal and a dielectric. Their field amplitudes decay exponentially, both into the metal and perpendicular to the surface. For a flat surface the dispersion relation is given by

$$k_{sp}^0 = \frac{\omega}{c} \sqrt{\frac{\varepsilon_m(\omega)\varepsilon_d(\omega)}{\varepsilon_m(\omega) + \varepsilon_d(\omega)}} \tag{1}$$

with k_{sp}^0 being the plasmon wave vector, ω the wave frequency, c the speed of light, and $\varepsilon_m(\omega)$ and $\varepsilon_d(\omega)$ the complex dielectric constants of the metal and the dielectric, respectively. It can be seen from this equation that photons cannot directly excite plasmons as their momentum,

$$k_{ph}^0 = \frac{\omega}{c} \sqrt{\varepsilon_d(\omega)} \sin\theta \tag{2}$$

is too small, at any incidence angle θ, to match both energy and momentum between them and the surface plasmons. One way, however, to allow for resonant coupling and to introduce surface plasmons is to use a prism in an attenuated total internal reflection (ATR) setup, known as the Kretschmann configuration *(23)* (Figure 2). Using a prism, the momentum of the photons is increased by n_0 (refractive index of the prism) and resonant coupling occurs at a well-defined angle of incidence θ_0.

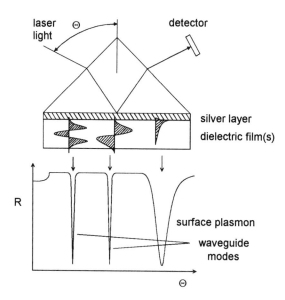

Figure 2. ATR setup for surface plasmon and waveguide spectroscopy in the Kretschmann configuration. For details see text.

A thin dielectric layer, e.g. an organic thin film, deposited on top of the metal film causes an increase in the plasmon wave vector and consequently shifts the resonance to a somewhat higher angle. Using Fresnel's equations, one can then calculate the optical thickness of this coating and, if the refractive index of the thin film is available, one can also determine the film geometrical thickness h.

If rather thick films (for organic or polymeric layers typically thicker than ca. 100 nm) are deposited onto the metal, one can actually introduce guided optical modes into the film. These new resonances can also be seen in an angular scan as very narrow dips in the reflectivity curve. Figure 2 again illustrates this phenomenon. The number of modes that can be excited depends solely on the sample thickness. Furthermore, the excitation of these modes is no longer limited to p-polarized photons but can also be achieved by using s-polarized light. A general advantage of OWS arises as soon as the films are thick enough to excite at least two optical waveguide modes (i.e. one s- and one p-mode). In this case one can actually distinguish between the refractive index and the geometrical thickness of the coating.

In practice, the polymer layers were not deposited directly on the metal surface, but on a SiO_x layer that was evaporated onto the silver film. These silicon oxide layers resemble closely the surface chemistry of other more common SiO_2 surfaces, such as those of silica gel or the oxide layers of silicon wafers. All these layers were deposited on a glass slide which was then brought into contact with the coupling prism by using an index matching fluid. This setup was mounted onto a programmable heating stage (resistance heater, temperature control: $\pm$ 0.1°C) and then to a two circuit goniometer (θ-2θ). Laser light (HeNe, 632.8 nm) was reflected from the sample through the prism. The reflected light intensity was detected by a photodiode.

T_g **measurements.** For OWS two methods have been used to study the glass transition of the various polymer films. First, complete angular scans were taken at several temperatures around the expected T_g. The reflectivity curves obtained were analyzed using the Fresnel equations to determine the film thickness and the refractive index of the sample as a function of temperature. The glass transition temperature was then identified from discontinuities in the $n(T)$ and $h(T)$ plots. The second way to determine T_g was to only measure the reflectivity at a fixed angle somewhat smaller than the resonance angle of (one of) the waveguide modes during heating and cooling ramps (kinetic mode). As the observed mode shifts to a different resonance angle due to thermal expansion and 'optical dilution', the reflectivity will either increase or decrease. The glass transition temperature can then be determined from breaks in these curves, as will be described in more detail below. This technique was also used for the SPS investigation of ultrathin samples. In all cases, complete angular scans were taken at the respective start and end temperature of such a kinetic scan. These curves were closely inspected in order to insure that the observed waveguide mode or plasmon only shifted to a somewhat different resonance angle but remained unaltered in its shape (broadness, depth of minimum). This situation was achieved by extensive pre-annealing of all samples *in vacuo* at 140°C. We also attached a small cell to the sample through which a weak flow of dry nitrogen was maintained during the measurements in order to minimize humidity effects .

Results

Optical waveguide spectroscopy (OWS) was utilized to determine the glass transition temperature of thick films of PMMA prepared by spin casting and the *grafting from* technique. As could be expected, the values of T_g obtained from these measurements were practically identical for all samples. Thus, only one example will be described that illustrates the results for all systems.

As already mentioned, OWS offers the unique possibility to distinguish between the refractive index n and the geometrical thickness h of thin organic coatings thicker than ca. 200 nm. Hence, one can also determine the temperature dependence of these two film characteristics, i.e. the refractive index increment (dn/dT) and the thermal expansivity, defined here as $h^{-1}(dh/dT)$. To do so, we have measured angular scans of different samples at various temperatures below and above the bulk glass transition temperature of PMMA. The reflectivity curves obtained at 70 and 150°C from a 579 nm thick sample of grafted PMMA are given in Figure 3. It can be clearly seen that both modes shift to somewhat different resonance angles as the temperature of the sample is varied. The low angle mode (first order) shifts to a slightly higher angle, whereas the high angle mode (zero order) moves in the opposite direction.

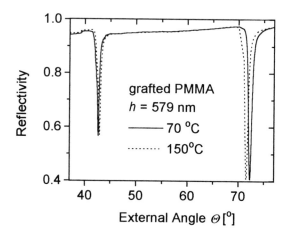

Figure 3. Reflectivity curves of a 579 nm thick sample at 70 and 150°C.

This behavior clearly represents the two effects of the temperature variation on the film. Firstly, the film expands as it is heated. This would cause a shift of the modes towards larger angles. The expansion, however, is accompanied by a decrease of the density of film and, consequently, by a decrease of the refractive index n. This effect causes both a shift of the resonances towards smaller angles and also a decrease of the 'distance' between the modes. The observed behavior reflects the combination of these two effects. This is further demonstrated by the results of a Fresnel analysis of these two curves. From these calculations, we found that the thickness of this particular PMMA film increased from 579 nm at 70°C to 591 nm at 150°C, whereas the refractive index decreased from 1.475 to 1.464. These values were derived by assuming that the optical thicknesses of the silver and SiO_x layers below the polymer film remained constant within this temperature range (n and d of these layers were determined by measuring the plasmon resonance of the unmodified substrate). This assumption seems to be reasonable for the SiO_x layer because the thermal expansivity of silicon oxides is typically two orders of magnitude smaller than that of polymers. The same can be expected for the silver layer and is actually proven by the reflectivity curves depicted in Figure 3. There, one can see that the position of the edge of total reflection at ca. 39.4° as well as the steepness of the decay at angles below that critical value is essentially the same at both temperatures. Especially the latter feature is very sensitive to the thickness and refractive index of the silver layer and any change would certainly cause a dramatic change in the shape of the reflectivity curve in this angular range. Additionally, any change of the optical properties of the silver film would alter the shape of the resonance dips, which is also not observed for the samples examined in this study.

Further measurements were recorded in 10°C steps between these two limits, and both n and h were determined from the angular scans. These values are plotted as a function of temperature in Figure 4. It is clear from this plot that the film thickness increases steadily as the sample is heated. Consequently, the density and, therefore, the refractive index of the film-forming material decreases. Both relationships exhibit two linear regions with a 'break' at ca. 120°C. This discontinuity is attributed to the glass transition of this particular sample.

From the slopes of the respective linear regions in the $n(T)$ and $h(T)$ plots we also determined the temperature dependence of n and d, i.e. the refractive index increment and the (linear) thermal expansivity $h^{-1}(dh/dT)$. The values are summarized in Table II together with the respective literature values for the bulk material *(24)*. The values obtained in this study do not vary significantly from the well-known thermal properties of bulk PMMA.

Table II. Thermal expansivity $h^{-1}(dh/dT)$ and refractive index increment dn/dT above and below T_g; values in brackets are taken from reference 24.

	$T < T_g$	$T > T_g$
$h^{-1}(dh/dT)[10^{-4}\ K^{-1}]$	1.6 (2.2)	4.2 (5.7)
$dn/dT\ [10^{-4}\ K^{-1}]$	-0.6 (-1.1)	-1.4 (-2.1)

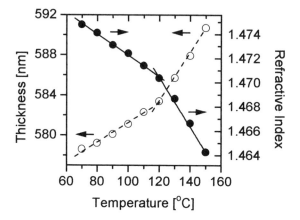

Figure 4. Thickness and refractive index of a 579 nm thick grafted PMMA film (cf. Figure 3) at various temperatures for the determination of T_g (details: see text).

Although the above mentioned procedures have been shown to be suitable for the thermal characterization of polymer films, they cannot be used generally as they are limited to relatively thick polymer films ($h >$ ca. 200 nm). They are also rather time consuming. An alternative procedure is to not take complete angular scans at different temperatures but to only measure the reflectivity at a certain angle somewhat smaller than (one of) the waveguide resonances (i.e. in the low angle slope of the dip) during a heating or cooling ramp. As can be seen from the two curves displayed in Figure 3, the resonance angles of the different modes can either shift to higher or lower angles upon heating. This behavior reflects directly the variations in both the film thickness h and the refractive index n and is indeed the reason why these values can be derived from such measurements. Consequently, the reflectivity recorded at an angle smaller than the resonance angle would either increase or decrease during a temperature ramp as the resonances shift to higher or lower angles, respectively. However, due to the rather abrupt change of the thermal expansivity and the refractive index increment, one can again expect a break or discontinuity in the $R(T)$ curve similar to the results presented in Figure 4.

A typical result of such an experiment is shown in Figure 5. This curve was obtained from the same sample as described above during a heating ramp from 80 - 140°C at a rate of 5°C min^{-1}. The reflectivity R was recorded at 42.2° with the resonance angle being 42.63° at the start temperature. It was already shown in Figure 3 that this particular waveguide mode shifts to higher angles upon heating. This behavior is reflected in the continuous increase in R with T. The curve obtained can be clearly divided into two linear regions with a break between them at ca. 112°C. In the light of the results presented above, the temperature at which this break occurs can be identified as the T_g of this film.

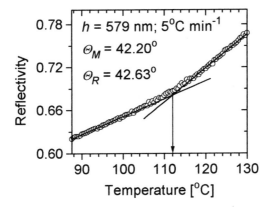

Figure 5. Reflectivity at Θ_M as a function of temperature during a heating ramp for a grafted PMMA film of 579 nm.

Samples thinner than ca. 200 nm cannot be investigated by OWS and consequently are not available for the independent determination of n and d. However, one can still monitor changes of the reflectivity near a surface plasmon mode during a heating cycle in the same way as described above for a waveguide mode. Again, the resulting $R(T)$ curve should resemble the temperature dependence of the optical thickness ($n \cdot d$), and an abrupt change in refractive index increment and thermal expansivity indicative of the glass transition should show up as a break separating two linear relationships. Figure 6 shows two examples of $R(T)$ curves obtained from ultrathin films of grafted PMMA (thicknesses: 7.5 and 50 nm) by using this technique.

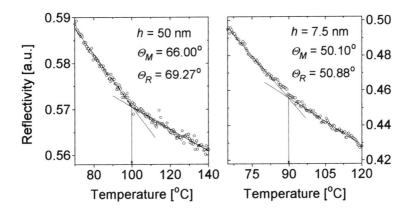

Figure 6. Reflectivity at Θ_M as a function of temperature during heating ramps for grafted PMMA films of 50 nm and 7.5 nm thickness.

Both plots show the results of heating cycles (second run) from 60 to 150°C again at a rate of 5°C min⁻¹. Due to the rather small thickness of the films, the angular shifts of the respective plasmons are very small and, therefore, the changes in reflectivity are also not very pronounced. Nevertheless, for each curve one can again identify two linear regions with a break separating them. These discontinuities, however, are now found at substantially lower temperatures, i.e. at 100°C for the 50 nm thick sample and at ca. 90°C for the 7.5 nm thick film. This observation indicates a clear depression of the glass transition temperature of the thin samples in comparison to thick, 'bulk-like' films.

However, a note of caution should be made since thickness *and* refractive index (or their respective increments) cannot be determined independently from SPS data. Thus, a direct comparison to the results of the investigation of the thick samples is not possible, and the discontinuity could also be caused by secondary effects (e.g. changes in the moisture content of the films). Despite this uncertainty, several observations indeed support our interpretation of these curves:

a) the discontinuity is always found at lower temperatures if detected from a cooling instead of a heating ramp;

b) we qualitatively detect the correct heating rate dependency: at lower heating rates it appears at lower temperatures;

c) preliminary measurements on spin-coated films of poly(ethylmethacrylate) (bulk $T_g = 65$°C) exhibit the same effect at much lower temperatures, i.e. around 40-50°C and, therefore, exactly in the range below the bulk T_g of this polymer. One example for this system is depicted in Figure 7. In contrast to the results obtained from the PMMA layers, the reflectivity increases with temperature. This opposite behavior is due to different sets of thermal expansivity and refractive index increment values for this particular polymer. Again, a clear break can be observed at ca. 42°C, which corresponds well to a 20°C depression of T_g for a ca. 20 nm thick film.

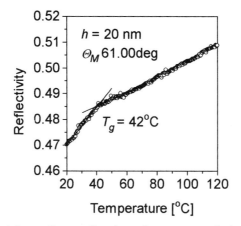

Figure 7. Reflectivity at Θ_M as a function of temperature during heating ramps for spin-cast film of PEMA on a hydrophobic surface.

In addition to the experimental evidence that supports our interpretation of these reflectivity curves, we also performed simulations of such curves on the basis of Fresnel's equations and different sets of values for the thermal expansivity and the refractive index increment (below and above T_g). These sets were either taken from the literature or calculated by using one of these values and deriving the other one by using the Clausius Mossoti equation that describes the relationship between a material's density and its refractive index. A detailed description of the results of these calculations will be published elsewhere *(21)*. The major conclusion of this study may, however, be summarized as follows: Due to the compensating effects of the thermal expansion of the film and the corresponding "dilution" of its molecular polarizibility (hence, its decreasing refractive index) even very small deviations from the bulk values for this material or the theoretically derived values may lead to a complete turn-around of the slopes in these $R(T)$ curves. Most of these curve shapes have been observed for the different assemblies showing that the thermal expansion coefficient and the correlated refractive index increment indeed vary somewhat from sample to sample, as one could expect for these delicate ultrathin samples. However, in almost every case we were able to identify two linear regions separated by a break in the way shown for several samples throughout this report.

As already mentioned, one objective of these studies was to investigate the influence of conformational features on the glass transition of constrained film geometries. The majority of measurements so far were performed on two different systems, i.e. on films spun onto non-interactive hydrophobic surfaces (SPS substrates treated with hexamethyldisilazane, water contact angles > 105°) and grafted PMMA films prepared by the method depicted in Figure 1. We also investigated three samples that were fabricated by the LBK technique, again using hydrophobic substrates. The results of the T_g measurements performed on these systems are summarized in Figure 8 together with the results obtained from the spun cast and the grafted layers.

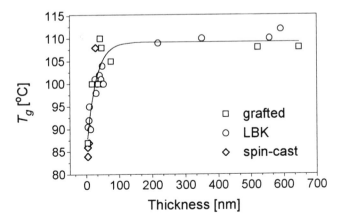

Figure 8. Glass transition temperature T_g (as determined by using SPS and OWS) as a function of film thickness for PMMA films of various molecular architecture.

It can be seen from this plot that the glass transition temperature of all three systems is clearly a function of the film thickness for layers thinner than ca. 50 - 100 nm. All thicker films show a thickness independent 'bulk-like' T_g. For thinner films, however, substantially lower T_g values are observed. This T_g depression appears to be quite independent of the actual molecular architecture of the respective sample, i.e. from the presence or absence of a covalent bond between one chain end and the surface and from differences in the conformation of the polymer chains.

A clearly different behavior, however, is observed for polymer layers deposited onto 'interacting' surfaces. Preliminary studies have been performed for two examples of such assemblies. In one case, PMMA was simply spun onto the bare SiO_x surfaces of SPS substrates. Here, hydrogen bonding between the silanol groups of the surface and the carbonyl moieties of the polymer is possible. In a second system, the reverse situation was established by depositing poly(3-methyl-4-hydroxystyrene) onto self-assembled monolayers of methyl 16-trichlorosilyl hexadecanoate, $Cl_3Si(CH_2)_{15}CO_2Me$. Figures 9 and 10 show results from T_g measurements on ca. 20 nm thick films of these compositions and one reflectivity curve obtained from a ca. 200 nm thick sample of the latter system. All of these samples seem to exhibit thickness independent 'bulk-like' T_g's of ca. 110°C and 135°C, respectively.

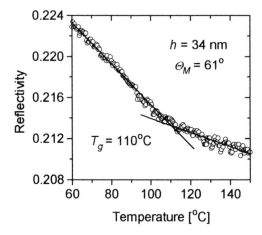

Figure 9. Reflectivity at Θ_M as a function of temperature during a heating ramp (5°C min^{-1}) for spin-cast film of PMMA on an unmodified SiO_x surface.

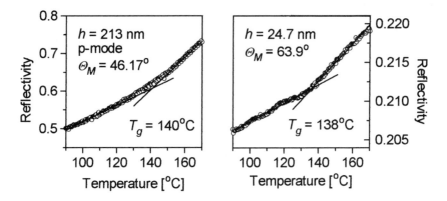

Figure 10. Reflectivity at Θ_M as a function of temperature during heating ramps (5°C min^{-1}) for spin-cast films of poly(3-methyl-4-hydroxystyrene) on a self-assembled monolayer of $Cl_3Si(CH_2)_{15}CO_2Me$.

Discussion

Before actually discussing the findings of this study, we want to point out that all polymers used (or created in the case of the grafted layers) for this investigation had molecular weights greater than 10^5 g mol^{-1}. This is significantly higher than the critical value for entanglement, which for PMMA is between 12,000 and 30,000 g mol^{-1} *(25)*. We therefore conclude that any deviation from the bulk T_g value (ca. 110°C as measured by OWS - see Figure 8 - and confirmed by DSC) cannot be attributed to molecular weight effects but rather reflects the specific geometry of these ultrathin assemblies.

As shown in Figure 8, we can indeed confirm similar findings by Jones *(7)* that for the case of films deposited on non-interacting surfaces a very pronounced T_g depression can be found for films thinner than approximately 50 - 100 nm. The data presented in this plot can also be described by the function that was used by Jones et al. *(7)*:

$$T_g(d) = T_g^\infty\left[1-\left(\frac{A}{d}\right)^\delta\right] \tag{3}$$

The values for the characteristic length $A = 0.35$ nm and the exponent $\delta = 0.8$ that we find are significantly different from the reported ones *(7)*, however. Thus, it remains unclear what meaning these parameters might have. Rather than defining a new empirical law, we prefer to compare our findings to the well-known bulk situation. One major parameter that describes the conformation of polymer coils both in solution as well as in solids and melts is the radius of gyration $<R_g>$. In various investigations *(26)* for various polymers it has been shown by small angle neutron scattering (SANS) that $<R_g>$ of a given polymer is essentially identical in the melt, the amorphous solid and in Θ solution. For PMMA in particular, Kirste et al. *(27)* have obtained the following relation between the molecular weight M_w and $<R_g>$:

$$\left\langle R_g \right\rangle = \left(0.096\, M_w^{0.98} \right)^{0.5} \quad \text{(in Å)} \tag{4}$$

In our case we have used PMMA with molecular weights between 10^5 and 10^6 g mol^{-1}. According to equation 4, this translates to radii of gyration of 10 - 30 nm. Theoretical predictions by Binder *(28)* and Theodorou *(29)* suggest that interfacial layers of the thickness of (small multiples of) $<R_g>$ may show enhanced chain mobility and, hence, suppressed glass transition temperatures. Our result of a noticeable T_g depression for layers thinner than approximately 50 -100 nm is indeed consistent with these predictions.

The T_g values obtained from the samples of grafted PMMA do not show any differences as compared to the layers spun onto hydrophobic substrates. This observation may at first be surprising as one might expect an influence of the covalent anchoring of chains to the surfaces. It is, however, known that the *grafting from* technique used to create these samples leads to very high molecular weight polymers (10^5 to 10^6 g mol^{-1}). Although the covalent bond will certainly reduce the mobility of the first few polymer segments, this influence will soon level off and the diffusion of segments further away from the tether will remain unaltered. Furthermore, any significant influence from the decomposed initiator molecules on the surface that do not carry a polymer chain can be disregarded. Azo compounds like the one used for these layers form ketenimine and succino nitrile moieties *(30)*. Although a surface carrying these structures is certainly not hydrophobic, there is also no specific interaction between these groups and the PMMA chains possible. In addition, one could also argue that extensive interaction between chain segments further away from the tether and groups on the surface is prohibited by the specific brush-like conformation of these tethered chains. In such a brush the coils are elongated perpendicular to the surface, which clearly minimizes the number of chain segments *per coil* that are close enough to the surface to interact with functional groups there. Hence, the similarity of the thermal behavior of the grafted and spun films in the absence of specific interactions is not surprising.

Less obvious, however, is the apparently similar behavior of the films prepared by the LBK technique. One might argue that the conformational differences between a polymer brush and a 'pancake' arrangement obtained by spin coating are not too significant. The system created by the LBK technique, however, should lead to a flat chain conformation that is significantly different from a three dimensional coil. Indeed, from SPS and x-ray reflectivity measurements we know that the average thickness per PMMA layer is of the order of 1 nm *(21)*. This value is much smaller than the dimensions of the unperturbed chain as demonstrated above. However, a note of caution must be made as a final conclusion for the behavior of these systems would certainly be premature on the basis of results obtained from only four samples. Also, in addition to a broader database it will be necessary to study these films with complementary techniques such as x-ray or neutron reflectivity in order to learn more about the actual architecture of these assemblies.

Based on the findings presented so far it is seems to be a reasonable assumption that the observed T_g depression reflects the influence of the two interfaces, i.e. the solid wall and the superstrate air. Indeed, preliminary results from systems that do allow for specific interaction between the solid wall and all segments of the deposited polymer chains that are close enough to the surface show a completely different picture. For systems where hydrogen bonds can be formed between appropriate donor and acceptor groups of the surface and the polymers, we cannot detect any pronounced dependence of T_g on the film thickness. Although the available set of data is up to now somewhat limited, all samples studied so far show glass transition temperatures close to the bulk values regardless of the thickness of the polymer film. This result is essentially in agreement with data published by Jones et al. *(7)* obtained from a comparable system (PMMA on a hydrophilic, interacting native silicon oxide surface). They have detected a slight T_g elevation of the order of 6°C for ultrathin films. However, given the fact that their curve shows about the same scattering as our T_g versus thickness plot (see Figure 8) the results of both groups appear to be consistent.

Related studies by van Zanten and Wallace on systems where even stronger interaction between chain segments and surface sites are possible (polystyrene on hydrogen-terminated silicon *(11)* and poly-2-vinylpyridine on SiO_x surfaces *(12)*) show that in these a very pronounced T_g elevation can be found. One might now argue that a even stronger, i.e. *covalent* interaction between film and surface is present in the case of the grafted layers investigated in the present study. These films however show the opposite behavior. It can therefore be concluded that the actual chain segment mobility near the solid wall is determined not only by the strength of the respective interaction but also by the number of interaction sites *per chain*. The chains of the grafted layers used in this study are connected to the surface by only a single tether, and further interactions between other surface groups and chain segments are unlikely due to the chemical nature of these groups as well as for conformational reasons (see above).

Summary

The first major goal of this study was to make evanescent wave optical techniques, namely optical waveguide spectroscopy and surface plasmon spectroscopy, available for the thermal characterization of ultrathin polymer films. The results presented here clearly confirm that these techniques are indeed suitable for the detection of the glass transition temperature of ultrathin polymer films. The T_g of films of different polymers with thicknesses varying from several nanometers up to the micron range could be identified from discontinuities in the $R(T)$ curves recorded during temperature ramps. Additionally, for films thicker than ca. 200 nm we were also able to independently determine the temperature dependence of both the film thickness and the refractive index of the layers. The values obtained from these studies essentially match the well-known bulk values of the polymer mostly used throughout these studies, i.e. PMMA. Up to now, however, the overall shape of curves obtained from thin polymer films cannot be sufficiently explained due to the fact that surface plasmon spectroscopy does not allow for the independent determination of the temperature dependence of film thickness and refractive index.

As a result of our investigation we could confirm a pronounced T_g depression for ultrathin polymer films of d < 50 - 100 nm. This effect can be found in very different assemblies as shown by identical findings of PMMA films deposited by different techniques (spin casting, end-grafting, LBK) but always onto a non-interacting surface. For substrates that enable specific interactions between the polymer segments and surface sites we could not detect the above-mentioned T_g depression. Regardless of the sample thickness, we always obtained the bulk values for the glass transition temperature of the respective polymer. This finding implies that it is indeed the presence of the two surfaces that leads to the increased chain mobility in very thin polymer films. However, to further elucidate this situation more work is needed. Questions about the free volume as well as about the role of defects (e.g. pinholes or voids) and the influence of the environment (e.g. moisture) need to be addressed. Thus, methods complementary to the optical techniques employed for this study will have to be applied, such as x-ray and neutron reflectometry.

Acknowledgments

OP wants to thank the *Deutsche Forschungsgemeinschaft* (DFG) for a fellowship ("Forschungsstipendium"). HB acknowledges support from the *Volkswagen Stiftung*. This work was supported in part by the *Center on Polymer Interfaces and Macromolecular Assemblies (CPIMA)* through the NSF-MRSEC program.

Literature Cited

1. Homola, A.; Mate, M.; Street, G.B.; *MRS Bull.* **1990**, *15*, 45.
2. V. Novotny, J.D. Swalen, J.P. Rabe, *Langmuir* **1989**, *5*, 485.
3. Frank, C.W.; Rao, V.; Despotopoulou, M.M.; Pease, R.F.W.; Hinsburg, W.D.; Miller, R.D.; Rabolt, J.F.; *Science* **1996**, *273*, 912.
4. Despotopoulou, M.; Miller, R.D.; Rabolt, J.F.; Frank, C.W.; *J. Polym. Sci., Polym. Phys.* **1996**, *34*, 2335.
5. Despotopoulou, M.; Frank, C.W.; Miller, R.D.; Rabolt, J.F.; *Macromolecules* **1996**, *29*, 5797.
6. Mensinger, H.; Stamm, M.; Boeffel, C.; *J. Chem. Phys.* **1992**, *96*, 3138.
7. Keddie, J.L.; Jones, R.A.L.; Cory, R.A.; *Faraday Discuss.***1994**, *98*, 219.
8. Keddie, J.L.; Jones, R.A.L.; Cory, R.A.; *Europhys. Letters* **1994**, *27*, 59.
9. Keddie, J.L.; Jones, R.A.L.; *Israel J. Chem.*. **1995**, *35*, 21.
10. Forrest, J.A.; Dalnoki-Veress, K.; Stevens, J.R.; Dutcher, J.R.; *Phys. Rev. Lett.* **1996**, *77*, 2002 and Erratum, ibid. p. 4108.
11. Wallace, W.E.; van Zanten, J.H.; Wu, W.; *Phys. Rev. E* **1995**, *52*, R3329.
12. van Zanten, J.H.; Wallace, W.E.; Wu, W.; *Phys. Rev. E* **1996**, *53*, R2053.
13. Orts, W.J.; van Zanten, J.H.; Wu, W.; Satija, S.; *Phys. Rev. Lett.* **1993**, *71*, 867.
14. de Gennes, P.G.; *J. Phys.* **1976**, *37*, 1443.
15. Reiter, G.; *Langmuir* **1993**, *9*, 1344.
16. Reiter, G.; *Macromolecules* **1994**, *27*, 3046.

17. Reiter, G.; *Europhys. Lett.* **1993**, *23*, 579.
18. Rühe, J.; *Nachr. Chem. Tech. Lab.* **1994**, *42*, 1237.
19. Prucker, O.; Rühe, J.; *submitted to Macromolecules.*
20. Schimmel, M.; Rühe, J.; *personal communication.*
21. Prucker, O.; Christian, S.; Bock, H.; Rühe, J.; Frank, C.W.; Knoll, W.; *submitted to Macromol. Chem. Phys.*
22. Knoll. W, *MRS Bull.* **1991**, *16*, 29.
23. Kretschmann, E.; *Opt. Commun.* **1972**, *6*, 185.
24. Wunderlich, W.; In *Polymer Handbook*; Brandrup, J.; Immergut, E.H.; Eds., J. Wiley & Sons, New York, 1989, p V-77.
25. Fox, T.G.; Loshaek, S.; *J. Polym. Sci.* **1955**, *15*, 371.
26. Abe, A.; In *Comprehensive Polymer Science*; Allen, G.; Bevington, J.C., Eds.; Pergamon Press, Oxford, 1989, Vol. 2, p. 66 and references therein.
27. Kirste, R.G.; Kruse, W.A.; Ibel, K.; *Polymer* **1975**, *16*, 120.
28. Baschnagel, J.; Binder, K.; *Macromolecules* **1995**, *28*, 6808.
29. Mansfield, K.F.; Theodorou, D.N.; *Macromolecules* **1991**, *24*, 6283.
30. Engel, P.S.; *Chem. Rev.* **1980**, *80*, 99.

PHOTONICS

Chapter 18

Design and Synthesis of a Perfluoroalkyldicyanovinyl-Based Nonlinear Optical Material for Electrooptic Applications

Fang Wang[1], Aaron W. Harper[1], Mingqian He[1], Albert S. Ren[1], Larry R. Dalton[1,3], Sean M. Garner[2], Araz Yacoubian[2], Antao Chen[2], and William H. Steier[2]

[1]Departments of Chemistry and of Materials Science and Engineering, University of Southern California, Los Angeles, CA 90089–1661
[2]Center for Photonic Technology, Department of Electrical Engineering–Electrophysics, University of Southern California, Los Angeles, CA 90089–0483

A novel perfluoroalkyldicyanovinyl-based second-order nonlinear optical material is synthesized and characterized. The new material has excellent processibility, large electro-optic coefficient (r_{33}=28pm/V at 1.06 µm for 32 wt% loading), and low optical loss (0.77 dB/cm) at 1.3µm.

The use of nonlinear optical polymers for electro-optic applications has received intensive research effort in the last decade. Recently, much focus has been directed toward the realization of bulk nonlinearities by the translation (*i.e.*, $<\cos^3\theta>$) of microscopic nonlinearity (*i.e.*, high β) to macroscopic nonlinearity (*i.e.*, high r_{33}). Many meaningful real world applications of electro-optic polymers require electro-optic coefficients greater than 20 pm/V at 1.3 µm. To date, very few electro-optic polymeric materials have achieved this requirement(*1-4*). Adequate poling efficiencies of these high-β systems have been sacrificed at the expense of optimizing the molecular electro-optic response of the individual chromophores. This has recently been attributed to increased chromophore-chromophore electrostatic interactions for these molecular-response-optimized chromophores, as compared with conventional chromophores, which affects the poling efficiencies of electric-field poling processes(*5,6*).

Optical losses create another major hurdle for many chromophore-containing polymer systems. Material-related optical losses include intrinsic (absorptive) optical losses and scattering losses. Some chromophore systems reported using cyanovinyl-type acceptors (*i.e.*, dicyanovinyl, tricyanovinyl, and TCNQ derivatives) do possess device-quality bulk nonlinearities, but still are well below the values expected from their molecular electro-optic responses. Furthermore, the general poor solubility of cyanovinyl-type chromophores makes the exploitation of these materials rather difficult. The low solubility of these chromophores facilitates phase-separation phenomena in polymer matrices, even for covalently-attached chromophores. Domains of aggregated chromophores may result in severe scattering losses. Also, most of these systems suffered large intrinsic optical losses at typical operating wavelengths, due to the long-wavelength absorption of the chromophores

---

[3]Corresponding author.

(λ_{max}=650~850 nm). Tails of these absorption bands may extend into the near infrared causing unacceptably large intrinsic optical losses at technologically important wavelengths such as 1.3 μm.

Auxiliary stabilities (e.g., thermal, chemical) of the NLO chromophore are additional criteria for device quality material. A usable chromophore must tolerate device operating temperatures on the order of 120°C for extended periods of time, and in some instances short excursions to *circa* 250°C (for some device fabrication photobleaching, electrochemical degradation, and oxidation. Unfortunately, the general architecture and composition of cyanovinyl-type chromophores render these types of chromophores susceptible to the abovementioned modes of degradation.

We report here the synthesis and properties of a new perfluoroalkyldicyanovinyl-based NLO chromophore. This chromophore was designed to overcome the problems mentioned above. Results indicate that the new chromophore gives good poling efficiency, excellent solubility, high chromophore thermostability ($T_d > 300°C$ in PMMA), low optical loss at 1.3 μm and a high electro-optical coefficient.

Experimental

The synthesis of the acceptor reagent 1,1-dicyano-2-chloro-2-(pentadecafluoroheptyl) ethylene was performed as follows. A solution of 6.60g (0.1 mole) malononitrile in ethanol was added to one equivalent of potassium ethoxide in ethanol. Methyl perfluorooctanoate was added dropwise and the resulting solution was allowed to stir overnight. The reaction mixture was evaporated to dryness under reduced pressure to give the potassium salt of β, β–dicyano-vinyl-α-(perfluoroheptyl)vinyl alcohol in nearly quantitative yield. In a simple distillation setup, 50.0g (0.1 mole) of the anhydrous salt was mixed with phosphorus pentachloride (0.1 mole) under argon and then heated slowly. Phosphorus oxychloride was collected from 90°C to 116°C. When no further distillation occurred, heating was discontinued. Vacuum distillation of the mixture left in the distillation flask gave 1,1-dicyano-2-chloro-2-(pentadecafluoroheptyl)ethylene as a volatile crystalline solid which was driven into a cooled receiver using a heat gun (10.7g). Further purification can be achieved through sublimation *in vacuo*. ^{13}C NMR (ppm, in chloroform): 119.3, 116.4, 112.1, 111.2, 110.7, 110.1, 108.8, 107.7, 107.2, 106.3, 105.8. ^{19}F NMR (ppm, in chloroform): -81.2, -108.5, -120.3, -121.7, -122.3, -123.2, -126.6.

The donor-bridge 4-[*N,N*-di(2-acetoxyethyl)amino]phenylene-2-thiophene was prepared as follows: 8.46g (71.6 mmol) sodium *t*-butoxide in 25 mL THF was added dropwise to a mixture of 17.5g (59.7 mmol) 4-[*N,N*-di(2-acetoxyethyl)amino]-benzaldehyde and 15.39g (65.7 mmol) diethyl 2-thiophenemethylphosphonate in 30 mL THF at 0°C. The reaction mixture was stirred overnight in an unattended ice-bath, then poured into 800 mL cold water. The aqueous mixture was extracted with methylene chloride and dried over MgSO$_4$. Evaporation of the solvent after filtration gave the desired donor-bridge. The acetyl groups were hydrolyzed by the conditions of the Horner-Emmons reaction. Reprotection was carried out in acetic anhydride at 45°C for three hours. Column chromatography over silica gel, eluting with 30% ethyl acetate in hexanes afforded 14.5g pure product. ^{1}H NMR (ppm, in CDCl$_3$): 7.34 (d, 2H), 7.12 (dd,1H), 7.04 (d, 1H), 6.98 (d, 1H), 6.95 (d, 1H), 6.83 (d, 1H), 6.73 (d, 2H), 4.24 9t, 4H), 3.63 (t, 4H), 2.04 (s, 6H). ^{13}C NMR (ppm, in chloroform): 170.9, 146.7, 143.6, 128.2, 127.7, 127.5, 126.1, 124.8, 123.2, 118.2, 112.2, 61.3, 49.7, 20.9.

The final chromophore was prepared by the following procedure. A solution of 2.28 mmol of the donor-bridge in 1 mL DMF was added dropwise to a small flask containing 2.33 mmol of 1,1-dicyano-2-chloro-2-(pentadecafluoroheptyl)ethylene in 0.5 mL DMF. The mixture was stirred at room-temperature for 5 hours. 30 mL of diethyl ether was added and the resulting mixture was washed with three 30 mL portions of deionized water, dried over $MgSO_4$, and then evaporated under reduced pressure. Analytically pure chromophore was obtained by repeated column chromatography in 30.3% yield. Melting point: 102-103 °C. ^1H NMR (ppm, in chloroform): 7.34 (d, 2H), 7.12 (dd,1H), 7.04 (d, 1H), 6.98 (d, 1H), 6.95 (d, 1H), 6.83 (d, 1H), 6.73 (d, 2H), 4.24 9t, 4H), 3.63 (t, 4H), 2.04 (s, 6H). ^{13}C NMR (ppm, in chloroform): 171.0, 157.7, 148.6, 147.7, 139.3, 136.2, 129.9, 129.4, 127.3, 126.5, 124.3, 119.2, 115.5, 114.9, 114.1, 112.2, 111.6, 61.2, 49.6, 29.8, 20.9. ^{19}F NMR (ppm, in chloroform): -80.2, -104.5, -118.8, -120.8, -121.4, -122.3, -126.0. Elemental analysis: Found, C 45.66; H 2.77; N 5.19. Theoritical, C 45.54; H 2.71; N 5.14.

The chromophore thermostability in PMMA was measured by a Perkin-Elmer DSC7, and absorption maxima were determined by a Perkin-Elmer Lambda 4C UV/Vis spectrophotometer. Polymer composites of this chromophore were prepared by co-dissolving the chromophore with PMMA in 1,2-dichloroethane. Excellent quality thin films were spin cast onto ITO-coated glass substrates. After drying of the films *in vacuo*, they were poled at the temperature which corresponded to the largest thermally-dependent second-harmonic signal. Electro-optic coefficients were determined by the attenuated total reflectance (ATR) technique (7). Optical loss was measured using the "dipping technique" at 1.3μm (8).

Results and Discussion

The synthetic route and the chemical structure of the chromophore are shown in Scheme I. The synthesis of the acceptor reagent 1,1-dicyano-2-chloro-2-(pentadecafluoroheptyl) ethylene and the donor-bridge 4-[*N,N*-di-(2-acetoxyethyl)amino]-phenylene-2-thiophene follows the literature procedure with slight modification (2,9). The yield of the final product is not optimized.

Scheme I

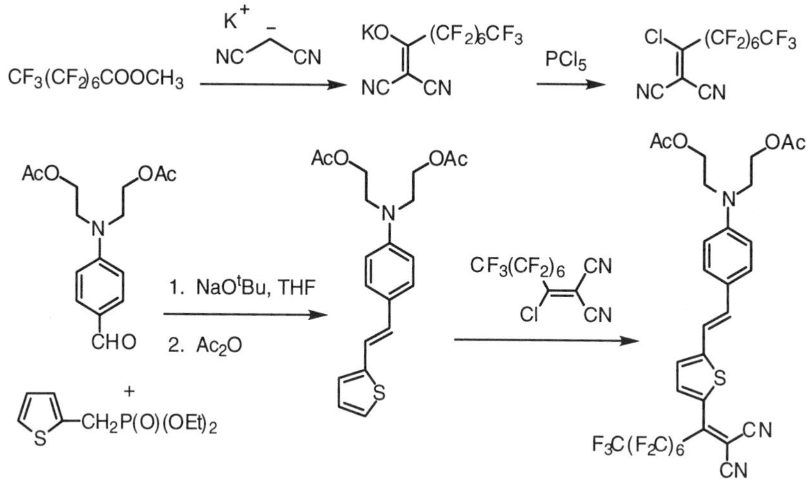

APTh-FDCV-1

The chromophore 1-{4-[*N,N*-di(2-acetoxyethyl)amino]phenylene-2-thien-5-yl}-2,2-dicyano-1-(pentadecafluoroheptyl)ethylene (denoted APTh-FDCV-1) was very soluble in common organic. It was doped into the thermoplastic host polymer PMMA at various loading densities. Optical quality films can be easily obtained with this composite polymer material.

The electro-optic coefficient, r_{33}, of the corona-poled polymer films were measured by a modified ATR method. At 31.8% by weight chromophore loading density, an electro-optic coefficient of $r_{33} = 28$ pm/V was achieved after poling. For comparison, an electro-optic coefficient of $r_{33} = 6$ pm/V was obtained for a Disperse Red-containing polymer at the same number density. The high r_{33} value of this APTh-FDCV-1 doped PMMA composite is mainly due to the improved poling efficiency, relative to other high-β containing materials. The poling efficiency of the 31.8% material is calculated based on the r_{33} value, and the μ and β_0 values of the chromophore. The dipole moment, determined by the concentration-dependent dielectric constant method, was 9.0 Debye. Nonlinearity-transparency correlation (*10*) estimates that the static hyperpolarizability of this chromophore is $\beta_0 = 350$ x 10^{-30} esu. These values lead to a poling efficiency of $<\cos^3\theta> = 0.3$, which is quite high for materials containing high-β chromophores. In most high-β containing materials, strong chromophore-chromophore electrostatic interactions effectively compete with the dipole-electric field interaction, resulting in inefficient chromophore parallel alignment. The strong electrostatic interactions of high β chromophores greatly increases the tendency of antiparallel alignment between the chromophores and causes aggregation (both transient and permanent) of chromophores, which in turn lowers the poling efficiency (*6,7*). The relatively high poling efficiency of APTh-FDCV-1/PMMA suggests that the perfluoroalkyl chain in the chromophore alleviated the aggregation problem encountered by other high β chromophore systems. However, an electro-optic coefficient of 41.8 wt% loading of the same chromphore in PMMA dropped to 22 pm/V, which may indicate that electrostatic interaction between chromophores at this loading density is of sufficient magnitude to attenuate the poling efficiency. An alternative or complementary reason for the attenuation of the electro-optic activity involves temporal decay of the induced polar order of the materials. The optimized poling temperature, which indicates the softness of the polymer system, for 41.8 wt% loading is only 40°C in contrast to 60°C for 31.8 wt% loading. This may also cause the lower r_{33} because of a more rapid relaxation of the aligned chromophores after the electric field is removed, for the composite with a higher loading density. Covalently attaching the chromophore into the polymer matrix should significantly reduce the rate of relaxation of the polar order, which translates to a higher possible electro-optic coefficient for higher loading densities.

Device-quality materials require the optical loss parameter to be very low. Due to intrinsic and scattering optical losses, many high-β chromophore systems have optical losses greater than 5 dB/cm, which is unacceptable acceptable for commercial device performance. The intrinsic optical loss can be eliminated by choosing chromophores that do not have any absorption at the operating wavelength range. As shown in Figure 1, the absorption maximum of a 20 wt% APTh-FDCV-1/PMMA thin film is 595 nm. The absorption maximum of APTh-FDCV-1 is about 50 nm lower than the corresponding tricyanovinyl chromophore (*1*). This may sacrifice the β value of the chromophore at the microscopic level, but it greatly reduces the optical losses and improves the overall properties of the chromophore. The absorption peak is symmetrical and the tail falls to zero absorption (as determined by absorption spectroscopy) well below 1.3 μm. This is an indication of less intrinsic optical loss due to the charge-transfer band of the chromphore. Indeed, an optical loss measurement on a film of 31.8 wt% loading in PMMA gave a loss value less than 1 dB/cm, as shown in Figure 2. A neat PMMA film itself possesses optical

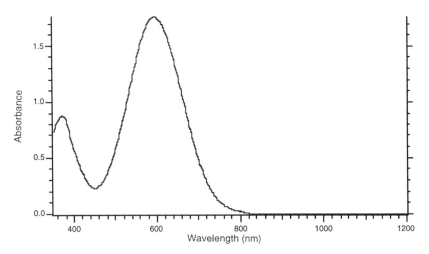

Figure 1. Absorption spectrum of a film of 22 wt% APTh-FDCV-1 in PMMA.

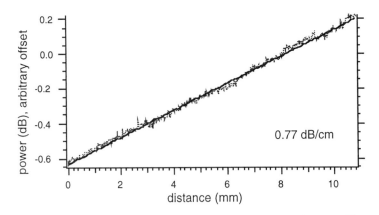

Figure 2. Optical loss of a 31.8 wt % APTh-FDCV-1/PMMA film.

losses of 0.5 - 0.75 dB/cm. Therefore, incorporation of the chromophore contributed little to the total optical loss of the film. The improved solubility of this chromophore presumably accounts for the reduction of the scattering optical loss. It is also conceivable that the oleophobic nature of the perfluoroalkyl chains on the chromophores helps to reduce the tendency of antiparallel alignment of the chromophores (parallel alignment of these chromophores maximizes both "phase separation" of the perfluoroalkyl regions and π-stacking interactions dictated by intermolecular electrostatic interactions of the chromophores' π-bridges).

Conclusions

A high-β chromophore based on a new perfluoroalkyldicyanovinyl acceptor has been synthesized. An electro-optic coefficient of 28 pm/V for 31.8 wt% loading in PMMA was achieved after electric field poling. The chromophore has excellent solubility and thermostability characteristics. The optical loss of this new material was measured at 1.3 μm, and was found to be 0.77 dB/cm. These results indicate that polymers incorporating the APTh-FDCV-1 chromophore are very promising candidates for electro-optic device applications. We are currently covalently attaching this chromophore (and variations) into a variety of polymer systems.

Acknowledgment. We wish to acknowledge Professor William P. Weber for many helpful discussions, and for generous use of his DSC instrument. Financial support for this research was provided by the Air Force Office of Scientific Research (under Contracts F-49620-94-1-0323, F-49620-94-1-0201, and F-49620-94-1-0312).

Literature Cited

1. Drost, K.J.; Rao, P.V.; Jen, A.K.-Y. *J. Chem. Soc., Chem. Commun.* **1994**, 369.
2. Boldt, P.; Bourhill, G.; Brauchle, C.; Jim, Y.; Kammler, R.; Muller, C.; Rase, J.; Wichern, J. *Chem. Commun.* **1996**, 793.
3. Sun, S.-S.; Harper, A.W.; Zhang, C.; Dalton, L.R.; Garner, S.M.; Chen, A.; Yacoubian, A.; Steier, W.H. *Mat. Res. Soc. Symp. Proc.* **1996**, *413*, 263.
4. Zhu, J.; He, M.; Harper, A.W.; Sun, S.-S.; Dalton, L.R.; Garner, S.M.; Steier, W.H. *Polym. Prepr.* **1997**, *38* (1), 973.
5. Harper, A.W.; Sun, S.-S.; Dalton, L.R.; Robinson, B.H.; Garner, S.M.; Yacoubian, A.; Chen, A.; Steier, W.H. *J. Am. Chem. Soc., submitted.*
6. Dalton, L.R.; Harper, A.W.; Robinson, B.H. *Proc. Natl. Acad. Sci. USA, in press.*
7. Dentan, *et al.* *Opt. Commun.* **1989**, *69*, 379.
8. Chia-chi Teng *Applied Optics* **1993**, *32* (7), 1051.
9. Freed, B.K.; Middleton, W.J. *J. Fluorine Chem.* **1990**, *47*, 219.
10. Harper, A.W.; Dalton, L.R. *to be published.*

Chapter 19

Progress Toward the Translation of Large Microscopic Nonlinearities to Large Macroscopic Nonlinearities in High-μβ Materials

Mingqian He[1], Jingsong Zhu[1], Aaron W. Harper[1], Sam-Shajing Sun[1], Larry R. Dalton[1,2,4], Sean M. Garner[3], Antao Chen[3], and William H. Steier[3]

[1]Department of Chemistry, Loker Hydrocarbon Research Institute, University of Southern California, Los Angeles, CA 90089–1661
[2]Department of Materials Science and Engineering and [3]Center for Photonic Technology, Department of Electrical Engineering, University of Southern California, Los Angeles, CA 90089–0483

By synthetic modification of a high μβ chromophore to defeat electrostatic intermolecular interactions, the electro-optic coefficients were doubled (to 42pm/V) in comparison to the unmodified analog. We believe that increasing the hydrodynamic volume of a chromophore is a general method to circumvent chromophore-chromophore interactions, which can lead to large macroscopic nonlinearities.

In the past several years, tremendous progress has been made in the design and preparation of chromophores with large molecular hyperpolarizibilities for use in electro-optic applications.[1-5] In order to achieve large second-order nonlinearities in bulk materials, it is a requirement to have the chromophores loaded in the polymer matrix in a noncentrosymmetric fashion. Unfortunately, several problems have hindered the incorporation of so-called high-μβ chromophores into polymer systems. First, most of the high-μβ chromophores possess little to modest solubility in appropriate organic solvents, which makes it rather difficult to synthesize polymers with required chromophore loading densities. Second, high-μβ chromophores usually have large dipole moments and linear polarizabilities, and so tend to spontaneously form aggregates with antiparallel arrangement of chromophores. Strong electrostatic interactions that govern this packing phenomenon results in low achievable poling efficiencies for this class of chromophores. Previously, we have investigated the chromophore APT-BDMI(E-(4-(N,N-di(2-acetoxyethyl)amino)-phenyl)ene-2-thien-5-ylidene 1,3-bis (dicyanomethylidene)indane),[6] the structure of which is shown in Figure 1.This chromophore has a μβ value 6144×10^{-48} esu (at 1900 nm), from which an

---

[4]Corresponding author.

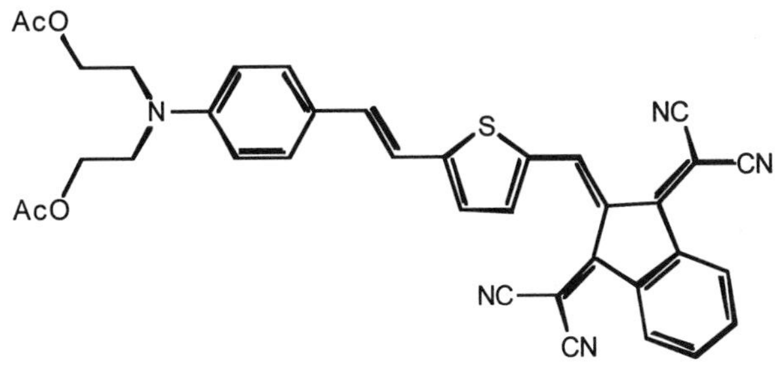

Figure 1. Structure of APT-BDMI

electro-optic coefficient, r_{33}, greater than one hundred pm/V may be expected. Unfortunately, to date, the largest value we have been able to achieve with this chromophore is 21 pm/V (at 1064 nm). Recently, we have shown that this lack of expected bulk nonlinear optical activity for APT-BDMI, and other high-$\mu\beta$ chromophores, is due primarily to intermolecular electrostatic interactions, which compete with the poling process with regard to chromophore orientation.[7,8,9] Due to these interactions, only a few scattered reports of large bulk electro-optic coefficients have appeared in the literature (even these reported values fall short of the expected values, given the chromophores' molecular hyperpolarizabilities and loading densities).[1,5] If these problems (solubility and electrostatic interactions) can be negated, materials possessing device-quality nonlinearities (i.e., electro-optic coefficient, r_{33}) on the order of a few hundred pm/V can be anticipated. Furthermore, a systematic, unambiguous study of the incorporation of high-$\mu\beta$ chromophores in polymer matrices has yet to be published.

We wish to report significant progress in this regard on a polymer system that incorporates the high-$\mu\beta$ chromophore E-(4-(N,N-di(2-t-butyl dimethyl siloxyethyl)amino)-phenyl)ene-3,4-dibutyl-2-thien-5-ylidene1,3-bis(dicyano methylidene)indane, denoted APT-BDMI-1. This is essentially APT-BDMI that has been sterically modified to address the above considerations. By structural modification of this chromophore to defeat deleterious electrostatic interactions and to increase the chromophore solubility, we put two TBDMS (t-butyldimethylsilyl) groups at the donor end and two butyl groups on the thiophene ring of APT-BDMI. For comparable number loading densities, we have obtained a significant increase in optical nonlinearities in comparison to the unmodified chromophore. Electro-optic coefficients have been determined for this modified chromophore doped into PMMA (poly(methylmethacrylate)) at various number loading densities. The largest electro-optic coefficient obtained was 42 pm/V (at 1064 nm, 56 wt% in PMMA). Also, r_{33} has been observed to go through a maximum as a function of loading density, which was expected from extended London theory treatment of the APT-BDMI system. These results suggest a rational synthetic method to realize large bulk nonlinearities from materials containing high-$\mu\beta$ chromophores.

Experimental

The synthesis of the modified APT-BDMI-1 chromophore was performed by the following reactions. The key intermediate 2-bromo-3,4-dibutyl-5-thienylmethyltributyl-phosphonium bromide was prepared either by the method of Tour and Wu,[10] or by that of Spangler and He.[11] The synthesis of another key intermediate,4-di(2-(1,1,2,2-tetramethyl-1-silapropoxy)ethyl)aminobenzaldehyde, was followed by a method which has been developed in our group.[12] The synthesis of other compounds are described below. All of the ^{1}H-NMR data were obtained from CDCl$_3$, on a Bruker 250 FT-NMR. Absorption maxima were taken from spectra obtained on a Cary spectrophotometer.

trans-1-{4-[N,N-di(2-(1,1,2,2-tetramethyl-1-silapropoxy)ethyl)amino] phenyl}-2-(2-bromo-3, 4-dibutylthienyl-5-yl)ethene

NaOEt (83 mL, 1.0 M in ethanol) was added dropwise to a solution containing 2-bromo-3,4-dibutyl-5-thienylmethyltributyl phosphonium bromide (75 mL, 0.635 M in ethanol, 0.0476 mol) and 4-di (2-(1,1,2,2-tetramethyl-1-silapropoxy)ethyl)amino-benzaldehyde (21.96 g, 0.05 mol) in 100 mL ethanol. Upon completion of addition, the mixture was allowed to reflux for 48 hours. After removal of the solvent, the mixture was extracted with diethyl ether (3 x 100 mL). The combined organic portions were washed with brine (3 x 100 mL) and dried with MgSO$_4$. The crude product was purified via column chromatography over silica, eluting with 5% ethyl acetate in hexane to give pure *trans*-1-{4-[*N, N*-di(2-(1,1,2,2-tetramethyl-1-silapropoxy)ethyl)amino]phenyl}-2-(2-bromo-3, 4-dibutylthien-5-yl)ethene (23.98 g, 67.4%) as a viscous oil. ^{1}HNMR; δ_H, 7.342-6.645 (m, 6 H, Ar-H and vinyl H), 3.768 (t, 4 H, J = 6 Hz, CH$_2$), 3.533 (t, 4 H, J = 6 Hz, CH$_2$), 2.607 (t, 2H, J = 8 Hz, CH$_2$), 2.499 (t, 2H, J = 7.75 Hz, CH$_2$), 1.446-1.265 (m, 8 H, CH$_2$), 0.961 (t, 6 H, J = 7 Hz, CH$_3$ overlap), 0.898 (s, 18 H, CH$_3$), 0.044 (s, 12 H, CH$_3$).

3,4-dibutyl-5-(2-(4-(di-(2-(1,1,2,2-tetramethyl-1-silapropoxy)ethyl) amino)vinyl) thiophene-2-carbaldehyde

t-Butyllithium (27 mL, 1.7M, 0.0462 mol) was added dropwise to a solution of *trans*-1-{4-[*N,N*-di(2-(1,1,2,2-tetramethyl-1-silapropoxy) ethyl)amino]phenyl}-2-(2-bromo-3,4-dibutylthien-5-yl) ethene (14.5 g, 0.021 mol) in THF (300mL) at -78°C. After completion of addition, the temperature of the mixture was allowed to gradually rise to -20°C, during which the solution turn dark blue and remained unchanged. DMF (4 mL) was then added, and the mixture was stirred for 3 hours. Dilute HCl (20 mL, 5%) was added dropwise to terminate the reaction. The solution was neutralized with aqueous NaHCO$_3$ (5%) and extracted with diethyl ether (3 x 100 mL). The combined organic fractions were washed with brine (3 x 200 mL) and dried with MgSO$_4$. After removal of solvent, the crude product was purified by column chromatography over silica, eluting with 10% ethyl acetate in hexane to give pure 3,4-dibutyl-5-(2-(4-(di(2-(1,1,2,2-tetramethyl-1-silapropoxy)ethyl)amino)vinyl) thiophene-2-carbaldehyde as an oil (12.37 g, 92%). ^{1}HNMR:δ_H 9.982 (s, 1H, CHO), 7.342-6.650 (m, 6 H, Ar-H and vinyl H), 3.768 (t, 4 H, J = 7 Hz, CH$_2$), 3.533 (t, 4 H, J = 7 Hz, CH$_2$), 2.607 (t, 2 H, J = 8 Hz, CH$_2$), 2.499 (t, 2 H, J = 7.75 Hz, CH$_2$), 1.446-1.265 (m, 8 H, CH$_2$), 0.961 (t, 6 H, J = 7 Hz, CH$_3$ overlap), 0.898 (s, 18 H, CH$_3$), 0.044 (s, 12 H, CH$_3$).

3,4-dibutyl-5-(2-(4-(di(2-(1,1,2,2-tetramethyl-1-silapropoxy)ethyl)amino)vinyl) thien-2-ylidene 1,3-bis(dicyanomethylidene)indane

   3,4-dibutyl-5(2(4-(di-(2-(1,1,2,2-tetramethyl-1silapropoxy)ethyl)amino)vinyl)thiophene-2-carbaldehyde (1.54 g, 2.34 mmol) and 1, 3-bis(dicyanomethylidene)indane (0.8 g, 3.3 mmol) were mixed in acetic anhydride (20 mL). The temperature of the solution was maintained at 60°C overnight, and a dark blue solution resulted. Acetic acid and residual acetic anhydride were evaporated *in vacuo*. The crude product was redissolved in THF and purified by column chromatography over silica, eluting with 40% THF in ethyl acetate to give pure 3,4-dibutyl-5-(2-(4-(di(2-(1,1,2,2-tetramethyl-1-silapropoxy)ethyl)amino)vinyl)thien-2-ylidene1,3-bis(dicyanomethylidene)indane as a solid(0.8 g, 72%), melting at 81-83°C. ^{1}HNMR: δ 8.814 (s, 1 H, vinyl), 8.521 (dd, 2 H, J$_{ab}$ = 9 Hz, J$_{ac}$ = 3 Hz, Ar-H), 7.681 (dd, 2 H, J$_{ab}$ = 9 Hz, J$_{ac}$ = 3 Hz, Ar-H),), 7.422 (d, 2 H, J = 8.5 Hz, Ar-H), 7.094 (d, 1 H, J = 14.5 Hz, vinyl), 7.036 (d, 1 H, J = 14.5 Hz, vinyl), 3.767 (t, 4 H, J = 6, CH$_2$), 3.565 (t, 4 H, J =6 Hz, CH$_2$), 2.773 (t, 2 H, J = 7.25 Hz, CH$_2$), 2.621 (t, 2 H, J = 7.25 Hz, CH$_2$),), 1.539-0.908 (m, 14 H, alkyl H),), 0.855 (s, 18 H, CH$_3$), 0.04 (s, 12 H, SiMe). Elemental analysis: theoretical; C, 70.78; H, 7.65; N, 7.79; found; C, 70.82; H, 7.63; N, 7.79.

Results and Discussion

   The detailed synthetic route of APT-BDMI-1 are summarized in scheme I and scheme II. Polymer composites of chromophore APT-BDMI-1 were prepared by co-dissolving with PMMA in 1,2-dichloroethane. The weight fractions of the APT-BDMI-1 composites were chosen so that they possessed identical chromophore number loading densities of 20%, 30%, 40%, and 50% (by weight) of APT-BDMI in PMMA. APT-BDMI-1 possessed excellent solubility in PMMA, even at very high loading density. Excellent quality thin films were made by spin casting polymer solution onto ITO-coated glass substrates. In all cases there was no observable phase separation or chromophore crystallization as determined by polarized microscopy. The absence of microscopic phase separation was confirmed via the method of Greiner, *et al.,* which used differential scanning calorimeter to determine guest miscibility with polymer hosts.[13] Following this method, we did not observe any phase separation for all of our composite systems. After drying *in vacuo*, each film was subject to the temperature and electric field potential that corresponded to the optimum second-harmonic signal and electro-optic activity. Electro-optic coefficients, refractive indices and film thickness were determined by the attenuated total reflectance (ATR)[14] technique ($\lambda = 1064$ nm). Film thickness (*ca.*2µm) were confirmed by a Sloan Detak IIA surface profiler.

Electro-optic data on the polymer composites containing APT-BDMI-1 loaded in different number densities which corresponded to the respective weight percentages of unmodified chromophore APT-BDMI in PMMA are listed in Table I.

All of the ATR measurements were repeated several times to ensure precision. The values obtained are very encouraging. With both alkyl and TBDMS groups on the chromphore, a higher poling efficiency was achieved, relative to the unmodified chromophore. Similar results were obtained for electro-optic activity of the materials. Comparing to the unmodified chromophore at the same number loading densities, each material containing the modified chromophore possessed a significantly larger electro-optic coefficient, as can be seen in Table I. We have determined from molecular modeling studies that the APT-BDMI and APT-BDMI-1 chromophores have little difference in dipole moment and optimized structure geometry. One can expect, then, that the dipole moment-hyperpolarizability product of the modified chromophore should be similar to that of the unmodified analog. The unmodified and modified chromophores have different head groups (attached to the di-(oxyethyl)amino donor) and there are some reports that acetoxethyl groups at the donor nitrogen could affect chromophore hyperpolarizability,[15] and hence affect the absorption maxima. We have used the di-(acetoxyethyl)amino functionality as the donor moiety for a great many of our chromophores and have never observed absorption spectral differences from their dialkylamino donor counterparts. In fact, in most cases, identical absorption spectral profiles are obtained, regardless of the group attached to the di-(oxyethyl)amino donor. One exception to this trend is when TBDMS moieties are attached to the head groups. Although the absorption maximum of the modified chromophore APT-BDMI-1 (760 nm in PMMA) is red shifted with respect to the unmodified analog (700 nm in PMMA), we have reason to believe that this red-shift of the absorption maximum does not result in significant resonance enhancement of the r_{33} values.

We prepared and evaluated two related chromophores, identified as chromophore 2 and chromophore 3 (Table II), which contain only the TBDMS group and butyl groups, respectively. Chromophore 2 has a loading density 38.6% by weight, which corresponding to 30% by weight of APT-BDMI. The material containing chromophore 2 yielded an electro-optic coefficient of 40 pm/V. After the loading excess 38.6%, r_{33} value begin to decrease. This chromophore has an absorption maximum of 712 nm, which is only 12 nm red shifted from APT-BDMI. The low-energy-side absorption tails of the two chromophores fall to zero well below 1064 nm; such situations do not lead to significant resonance enhancement. Two scenarios, then, can explain this increase in r_{33}. In one case, on the premise that $\mu\beta$ of the modified and unmodified chromophores are similar, our calculations indicated that the unmodified chromophore should have 46% greater polar alignment to account for the increased electro-optic activity. Because the poling electric field *axially* aligns domains of chromophore aggregates as well as *polar* aligns individual chromophores, the order parameters obtained from the electrochromism approach gives *axial*, but not necessarily *polar*, order.[16] Hence, the conventional method to determine ordering (observing attenuation of absorption upon electric field exposure) does not reliably or accurately measure polar order for materials containing high-$\mu\beta$ chromophores.[8] We are currently exploring the dependence of axial and polar ordering in materials containing these (and other) chromophores, and will report our results shortly. In the other case, assuming similar poling efficiencies, the $\mu\beta$ of the derivatized chromophore must be 46% greater than the underivatized analog to account for the increase in electro-optic activity, which is unlikely for the reason discussed earlier. We believe that the large head groups (TBDMS) and butyl groups on the APT-BDMI-1 chromophore greatly reduced the chromophore-chromophore electrostatic intermolecular interactions, which led to an increase in the fraction of polar-aligned chromophores in the composites, over the unmodified APT-BDMI composites of similar number loading densities. We are currently measuring the dipole moment and hyperpolarizability of APT-BDMI-1 for direct comparison to APT-BDMI.

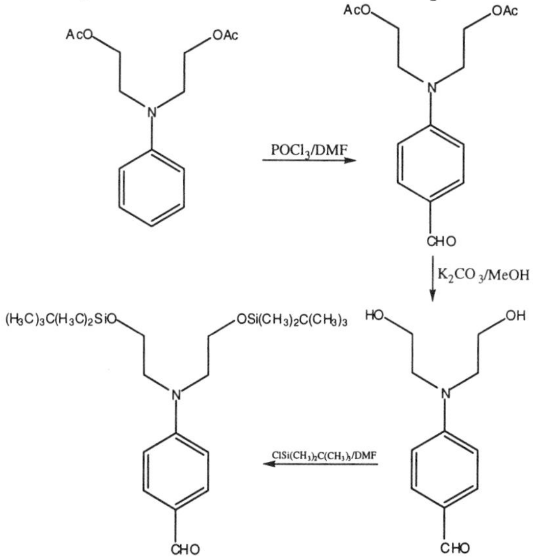

Scheme I. Synthesis of TBDMS-derivatized donor intermediate.

In order to explore the roles of the various substituents in the modified chromophore on increasing the effective electro-optic coefficients, we prepared composites of various loading densities of derivatives (with various combinations of substituents) that corresponded to that of APT-BDMI examined earlier.[6] As can be seen, adding only the butyl groups (chromophore 3) increased the r_{33} from 21 pm/V to 33 pm/V. Addition of only the TBDMS groups (chromophore 2)

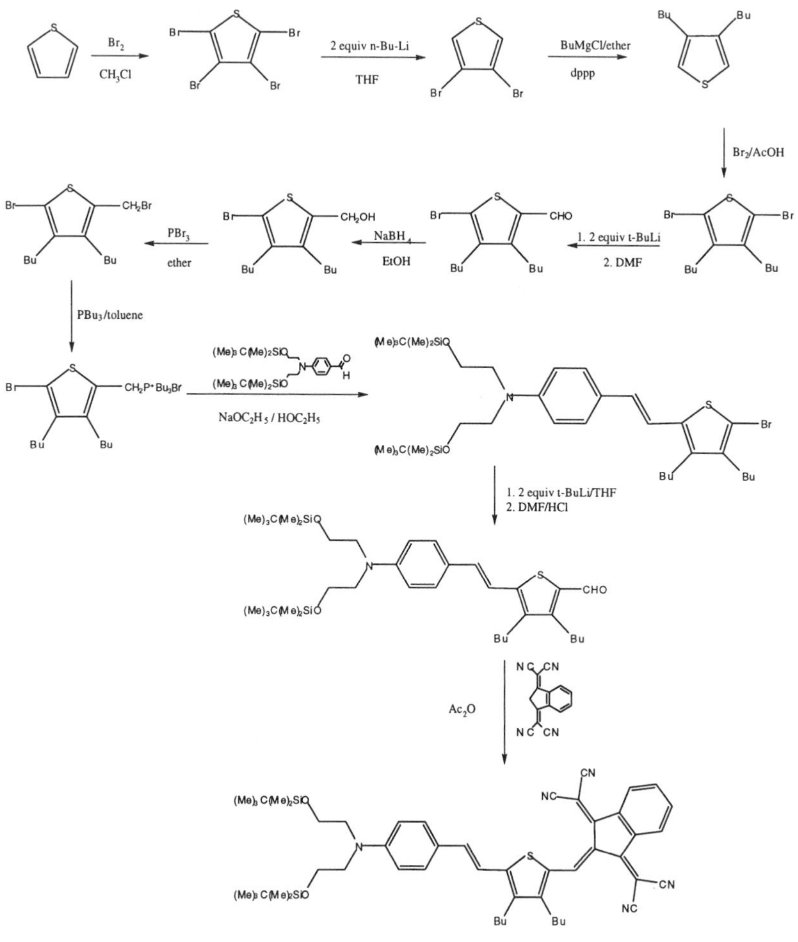

APT-BDMI-1
Scheme II. Synthesis of modified chromophore APT-BDMI-1.

increased the electro-optic coefficient to 40 pm/V. Derivitizing the parent chromophore with both butyl and TBDMS groups, resulted in an increase in electro-optic coefficient from 21 pm/V to 38 pm/V. This observation means that the modifications are not additive (else 52 pm/V would have been obtained). In the cases of the parent chromophore and modified chromophores 2 and 3, the maximum loading densities essentially corresponded to the values given in Table I. Any increase in loading density led to a reduction in the electro-optic coefficient. The APT-BDMI-1 chromophore, however, allowed for a further increase in loading density which was accompanied by an increase in electro-optic activity. The optimum material possessed an electro-optic coefficient of 42 pm/V, at a loading density of 56% by weight.

Table I. Comparison of APT-BDMI/PMMA and APT-BDMI-1/PMMA Composite Systems

APT-BDMI		APT-BDMI-1	
Loading Density	r_{33}	Loading Density	r_{33}
20 wt%	17 pm/V	28.2 wt%	34 pm/V
30 wt%	21 pm/V	42.5 wt%	38 pm/V
40 wt%	21pm/ V	56.4 wt%	42 pm/V
50 wt%	20pm/V	70.5 wt%	34 pm/V

Table II. Chromophore Derivatives versus r_{33}

Compounds	Loading density	λ_{max}	r_{33}
1	56.4* wt%	760 nm	42 pm/V
	42.5 wt%	760 nm	38 pm/V
2	38.6 wt%	712 nm	40 pm/V
3	35.6 wt%	721 nm	33 pm/V
4	30 wt%	700 nm	21 pm/V

Note: All entries have the same chromophore *number* densities except *.

Conclusion

Our results demonstrate that increasing the hydrodynamic volume of the APT-BDMI chromophore (in particular, in the direction normal to the plane of the π-system) by incorporating alkyl and/or alkylsilyl groups can not only dramatically increase solubility, but can also reduce chromophore-chromophore electrostatic intermolecular interactions.

Reduction of these interactions result in improved poling efficiencies, which in our case can double the electro-optic coefficient over the unmodified chromophore at an identical loading density. Results suggest that further refinement of the chromophore structure to increase the minimum chromophore separation distance should yield even higher electro-optic coefficients. We are presently pursuing the covalent incorporation of these modified chromophores into polymer networks to enhance the thermal and temporal stability of electro-optic activity of these materials.

Acknowledgments

This project is supported by the Air Force of Scientific Research (Contracts F-49620-94-1-0323, F-49620-94-1-0201, and F-49620-94-1-0312).

References

1.Ahlheim, M.; Barzoukas, M.; Bedworth, P.V.; Blanchard-Desce, M.; Fort, A.; Hu, Z-Y.; Marder, S.R.; Perry, J.W.; Runser, C.; Staehelin, M.; Zysset, B. *Science* **271**, 335(1996).

2.Marder, S.R.; Cheng, L-T.; Tiemann, B.G.; Friedli, A.C.; Blanchard-Desce, M.; Perry, J.W.; Skinhoj, J. *Science* **263**, 511(1994).

3.Gilm our, S.; Montgometry, R.A.; Cheng, L.T.; J, A.K.-Y.; Cai, Y.; Perry, J.W.; Dalton, L.R. *Chem.Mater.* **6**, 1603(1994).

4.Alex K.-Y. Jen; Cai Y.; Bedworth P. V.; Marder S. R.; *Adv. Mater.* **9**, 133(1997)

5. Boldt, P.; Bourhill, G.; Brauchle, C.; Jim, Y.; Kammler, R.; Muller, C.; Rase, J.; Wichern, J.; *Chem. Commun.*, 793(1996).

6.Sun, S.-S.; Harper, A.W.; Zhang, C.; Dalton, L.R.; Garner, S.M.; Chen, A.;Yacoubian, A.; Steier, W.H. *Mat. Res. Soc. Symp. Proc.* 413, 263(1996).

7.Harper, A.W.; Sun, S.-S.; Dalton, L.R.; Bobinson, B.H.; Garner, S.M.; Yacoubian, A.; Chen, A.; Steier, W.H. *J.Am.Chem. Soc.*, submitted.

8.Dalton, L.R.; Harper, A.W.; Robinson, B.H. *Proc. Natl. Acad. Sci. USA* **94**, 4842(1997).

9. Harper, A. W.; Sun, S-S.; He, M.; Zhu, J.; Dalton, L. R.; *J. Opt. Soc. Am. B. Opt. Phys.*, in press.

10. Tour, J. M; Wu, R. *Macromol.*, **25**, 1901(1992).

11.Spangler, C. W. and He, Mingqian *J. Chem. Soc. Perkin Trans. 1*, 715(1995).

12.to be published.

13.Greiner A.; Bolle B.; Hesemann P.; Oberski J.; Sander R.; *Macromol. Chem. Phys.*, **197**, 113-134(1996).

14.Dentan, et al. *Opt. Commun.* **69**, 397(1989).

15.Morley, J.O.; Hutchings, M.G.; Zyss, J.; Ledoux, I. *J. Chem. Soc. Perkin. Trans. 2*, 1139(1997)

16.Singer, K.D.; Kuzyk, M.G.; Sohn, J.E. *J. Opt. Soc. Am. B* **4**, 968(1987).

Chapter 20

Nonlinear Optical Films from Pairwise-Deposited Semiionomeric Syndioregic Polymers

M. J. Roberts[1], J. D. Stenger-Smith[1], P. Zarras[1], R. A. Hollins[1], M. Nadler[1], A. P. Chafin[1], K. J. Wynne[2], and G. A. Lindsay[1]

[1]NAWC, Research and Technology Group, Code 4B2200D, 1 Administration Circle, China Lake, CA 93555–6100
[2]Office of Naval Research, Arlington, VA 22217–5000

Polar multilayer films of syndioregic nonlinear optical polymers were made using Langmuir-Blodgett-Kuhn (LBK) deposition of a polymeric salt formed at the water surface from two complementary polymers (a polycation insoluble in water and a water-soluble polyanion). Noncentrosymmetric order in the deposited films is maintained primarily by ionic and hydrogen bonding.

An important advantage of using LBK technique to produce all-polymeric nonlinear optical films is it allows polymers to be processed near room temperature thus avoiding the disordering and degrading effects seen in high temperature electric field poling. In addition, the LBK technique offers control over final film thickness to within one monolayer and materials may be precisely located within the film to control properties for purposes such as phase matching of the fundamental and second harmonic waveguide modes.

A well-known limitation, the long-standing problem of low thermal structural stability of LBK films, may be solved by using high Tg polymers. However, a serious limitation of LBK technique remains; namely, the long processing time required to build up films of sufficient thickness (>0.5 micrometers) for waveguiding. In principle, the pairwise deposition technique reported here will increase the rate of film thickness growth.

This paper relates generally to organic polymeric thin films for photonic applications with specific focus on the process for making second-order nonlinear optical polymeric (NLOP) films, including a brief overview. Further discussions involving this technology may be found in previous reports (1,2,3). The photonic applications field has evolved rapidly over the past ten years. One class of materials within this field, NLOP films, has potential for creating breakthroughs that will enable production of low cost integrated devices for the telecommunication and data-communication industries. A key component of these integrated devices are electro-optic (EO) waveguides made from NLOP films. EO waveguides may be used to switch optical signals from one path to another and also to modulate the phase or amplitude of an optical signal (*4*).

The molecular origin of optical nonlinearity derives from the electrical polarization of the chromophore as it interacts with electromagnetic radiation. The molecular structure of the chromophore and its orientation govern the nonlinear optical properties of the system. In order for films to have a large quadratic NLO coefficient, they must contain a high concentration of asymmetric, highly polarizable chromophores arranged in a polarized orientation. The polymer's chemical structure determines the processability and temporal stability of the final product. In the past few years, several types of polymers have been developed which are effective in EO modulation of optical signals (*5*).

Two primary techniques are used to produce films thick enough for waveguiding with polar order, namely, elevated temperature electric-field poling and room temperature Langmuir-Blodgett-Kuhn (LBK) processing (*5*). Other techniques which show promise include photopoling of azo-containing chromophores (*6*), alternating polyelectrolyte deposition (*7*), and self-assembly with covalent linking (*8*).

Electric-Field Poling. It is beyond the scope of this paper to describe in detail the electric field poling method but the interested reader may find detailed articles on the technique (*9*). There are several difficulties associated with electric-field poling. First, since the polymer utilized must be heated to high temperatures, thermal disordering of the chromophores works against the torque of the electric field resulting in less than optimal average alignment. Thus the chromophore orientation is not well controlled, that is, there is an assortment of chromophore orientation angles which averages out to a perpendicular arrangement. In addition, polymers containing formal charges are very difficult to pole with an electric field because the charges tend to migrate through the polymer resulting in dielectric breakdown.

Langmuir-Blodgett-Kuhn (LBK) Processing. In conventional LBK film processing, an organic compound is adsorbed as an insoluble monolayer on a liquid surface, e.g. water, ethylene glycol or other aqueous solutions, in a trough. A solid substrate is dipped through the gas-liquid interface depositing a single molecular layer on the substrate. Thicker films comprised of multilayers are built up by repeatedly dipping the substrate into and/or out of the trough (*10*).

In contrast to electric-field poling, LBK processing is usually carried out near ambient temperature. The chromophore orientation is well controlled and the chromophore orientation angle may be specifically defined (*11*). Furthermore, unlike electric field poling, formal ionic charges on the polymer need not hinder the ordering process.

In conventional LBK processing, polymers are designed with hydrophilic and hydrophobic groups which cause the polymer to adsorb at the gas-liquid interface in a preferred conformation. Polymer amphilicity may be utilized to effect orientation of chromophores at an angle to the plane of the gas-liquid interface. However, LBK films have suffered from structural instability due to the presence of alkyl and fluoroalkyl hydrophobic chains which melt at low temperatures (*12-15*). These functional groups also dilute the concentration of chromophores which can result in a lowering of the NLO coefficient of the waveguide. The dilution of the chromophore concentration may be desirable in that the optical loss due to absorbance may be reduced. Thermal structural stability of LBK films may be improved by interlayer and/or intralayer covalent bonding (i.e. crosslinking). One recent report suggests that polymers with high Tg may be processed by the LBK technique (Roberts, M. J.; Lindsay, G. A.; Hollins, R. A.; Stenger-Smith, J. D.; Chafin, A. P.; Yee, R.; Gratz, R.G. submitted to *Thin Solid Films*) though typically such materials yield monolayers unsuitable for vertical deposition.

An important limitation of LBK technology is the amount of time required to build up films of sufficient thickness (>0.5 micrometers, typically, >350 layers) for waveguiding. Three ways that the rate of deposition can be increased without sacrificing film quality are:

1) lower monolayer viscosity by use of higher subphase temperatures, choice of subphase ions, or change of pH, (*10*)
2) utilize monolayer compression schemes which allow continuous processing; such as the spinning roller (*16*), flowing subphase (*17*), or the dynamic thin laminar flow methods (*18*), or
3) use horizontal (Langmuir-Schaefer) deposition of films (*19*).

The objectives of the work reported here are the elimination of the entire electric-field poling step, reduction or elimination of the chromophore dilution and Tg-lowering effects of the hydrophobic alkyl groups, and increase the rate of film growth. With these goals in mind, the pairwise deposition of NLO polymers was conceived.

Description of Pairwise-deposited Films. The polymers in this study are comprised of chromophores linked head-to-head and tail-to-tail (i.e., syndioregically, (*1*)) by bridging groups (Figure 1). Each repeat unit contains two chromophores and two bridging groups. The chromophores are axially electronically asymmetric, with one end being electron-accepting and the other end being electron-donating. Along the mainchain, *every other* bridging group bears an ionic charge so these polymers may be classified as semi-ionomeric syndioregic polymers. In concept, a bilayer of these polymers has each semi-cationomeric polymer chain paired with one complementary semi-anionomeric polymer, thus ionically bound as a polymeric salt bilayer. A stack of the polymeric salt yields a polar film, that is, a film containing chromophores arranged noncentrosymmetrically with respect to the plane of the film (Figure 2).

The polycation (**F**) is essentially *insoluble in water* and the polyanion (**S**) is *water-soluble*. The solubility properties of the two polymers are exploited by spreading an insoluble monolayer of **F** on the surface of a trough filled with an aqueous solution of **S**. **S** ionically bonds with insoluble **F** at the surface forming the polymeric salt bilayer.

The bridging groups are designed to allow the polymer backbone to fold so that the chromophores can easily lie with their long geometric axes at non-zero angles

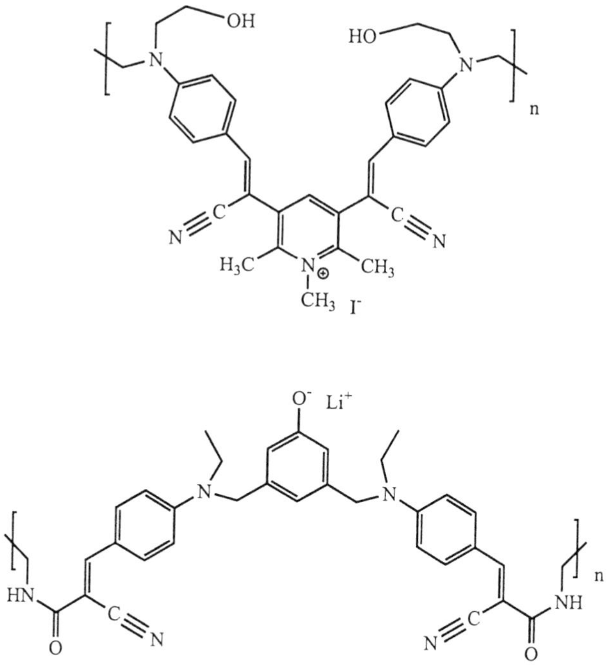

Figure 1. Semi-ionomeric syndioregic nonlinear optical polymers.

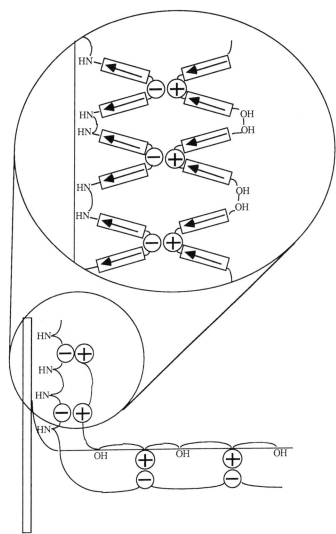

Figure 2. Idealization of pairwise LBK deposition.

with respect to the plane of the film. The ground-state dipole moments of the chromophores in both **F** and **S** are designed to point in opposite directions with respect to their respective ionic bridges, so that they will point in the same direction with respect to the plane of the film. The Z-type pairwise deposition of layers of the above complementary semi-ionomeric syndioregic polymers, forms a polarized film (Figure 3).

Experimental

Synthesis of Polymer S. The following description illustrates the synthesis of polymer **S** (Figure 4).

Synthesis of Monomer A. The following describes the synthesis of monomer **A** which was used in a condensation polymerization with monomer **B** to make polymer **S**.

Step 1 **Dimethyl-5-tosyloxy-phthalate (1).** A mixture of 10 g of 5-hydroxy-dimethylphthalate and 10 g tosyl chloride in 20 ml of pyridine was stirred at room temperature for 5 hours then poured into 200 mL of water. This was extracted with methylene chloride (3 X 100mL) and the organic layer washed with dilute HCl (3 X 150ml) then dried (MgSO4), filtered and evaporated. The solid product was recrystallized from benzene-methanol giving 13.82 g (80 % yield) of a white solid m.p.: 129-30 °C. Analysis: Calcd. for $C_{17}H_{16}SO_7$: C, 56.04; H, 4.40. Found: C, 56.19; H, 4.34.

Step 2 **1-O-Tosyl-3,5-bis (hydroxymethyl)phenol (2).** To a solution of 1.82g of $\underline{1}$ in 30 ml of benzene was slowly added 20 mL of diisobutylaluminum hydride (1.5 M in toluene) and the mixture stirred at room temperature for 4 hours. To this was then added 3-5 mL of methanol followed by 10 mL of benzene and then the mixture was acidified with dilute HCl. The organic layer was separated and the aqueous layer extracted with ether (4 X 75 mL) and the combined organics dried (MgSO4), filtered and then concentrated to about 20 ml. After cooling the solid was filtered, washed with benzene (2 X 20 mL) and air-dried giving 1.30 g (86% yield) of white, crystalline solid m.p.: 107-8 °C. Anal. Calcd. for $C_{15}H_{16}SO_5$: C, 58.44; H, 5.19. Found: C, 58.53; H, 5.01.

Step 3 **1-O-Tosyl-3,5-bis(bromomethyl)phenol (3).** To 6.10 g of diol $\underline{2}$ in 200 mL of benzene was added a solution of 4.4 g phosphorus tribromide in 10ml of benzene followed by 20 drops of dry pyridine. The stirred mixture was heated at 55-60 °C for 5 min the cooled and poured into 400 mL of ice water. To this was added 150 mL of brine. The organic layer was separated and the aqueous layer extracted with 150 mL of benzene. The combined organic layers were dried (MgSO4) and evaporated to a white crystalline solid m.p.: 102-103 °C. Anal. Calcd. for $C_{15}H_{14}SO_3Br_2$: C, 41.76; H, 3.24. Found: C, 4.10; H, 3.16.

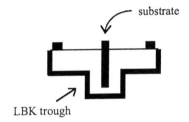

substrate

LBK trough

STEP 1. The LBK trough is filled with aqueous 0.001 M polyanion, the substrate is positioned at the bottom of its stroke, and the barriers are fully open.

STEP 2. A 0.5 M organic solution of water-insoluble polycation is placed dropwise on the liquid surface and the solvent is allowed to evaporate. The polyanion ionically bonds with the polycation at the argon/liquid interface.

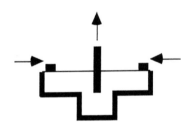

STEP 3. The polycation/polyanion bilayer is compressed laterally. Once the target surface pressure is reached, one bilayer is deposited as the substrate is moved upward to the top of its stroke.

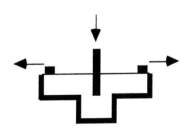

STEP 4. The lateral pressure on the bilayer is released, the substrate is repositioned at the bottom of its stroke, and the dipping cycle is repeated beginning with STEP 3.

Figure 3. Step-by-step diagram of the pairwise-deposition cycle.

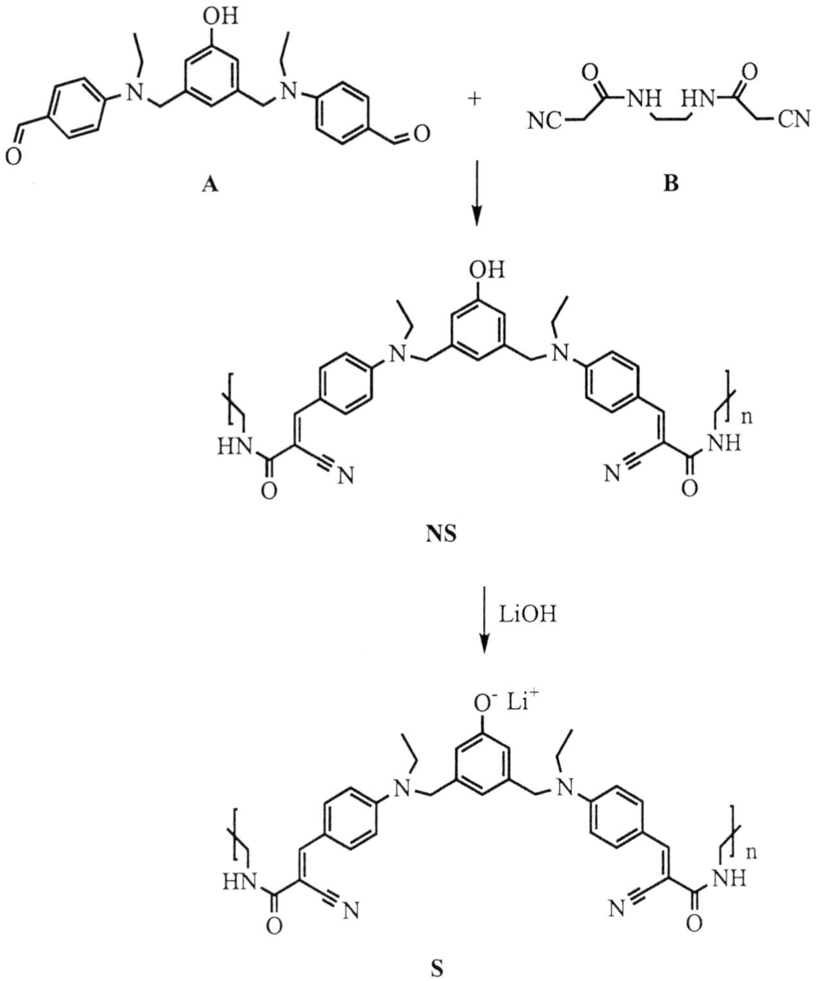

Figure 4. Scheme for synthesis of NLO-active water-soluble semi-anionomeric polymer, **S**.

Step 4   1-0-Tosyl-3,5-bis(N-ethyl-N-phenylaminomethyl)phenol (4). A solution of 4.31 g of dibromide 3 was slowly added to a mixture of 4.81 g N-ethylaniline and 3.36 g of sodium bicarbonate in 100 mL of chloroform heated at reflux. The mixture was heated at reflux for 18 hours, then cooled and poured into 150 mL water plus 50 mL of saturated brine. The aqueous layer was separated and extracted with 100 ml of methylene chloride and the combined organic layers dried (MgSO$_4$) and evaporated at reduced pressure. The resulting oil was triturated with boiling methnol which was decanted after cooling. Repeating this procedure one more time gave 2.316 g of an oil which was pure by NMR. The mother liquor from the first trituration was chromatographed (SiO$_2$/hexanes/CH$_2$Cl$_2$) and the resulting oil triturated with methanol as above giving an additional 0.436 g of product.

Step 5   3,5-Bis(N-ethyl-N-phenylaminomethyl)phenol (5). A solution of 436 mg of tosylate 4 in 10 mL of hot ethanol was slowly added to a solution of 1.0 g of potassium hydroxide in 1.5 mL of water plus 3 mL of ethanol heated at reflux. The resulting mixture was heated at reflux for 40 min, cooled and neutralized with 1.07 g of glacial acetic acid. After concentrating the mixture under vacuum, the resulting thick liquid was extracted with ether (2 X 30 mL). The combined ether extracts were washed with 20 mL of sat'd sodium bicarbonate solution then dried (MgSO$_4$) and evaporated under reduced pressure to 274 mg of a pale yellow oil.

Step 6   3,5Bis(N-ethyl-N-4-formylphenylaminomethyl)phenol (6). A solution of 1.611 g of phenol 5 in ca. 5 mL of 1,2-dichloroethane was added to a mixture of 4.18 g of dimethylformamide and 7.04 g of phosphorus oxychloride then heat to 85°C for 3 hours. After cooling the reaction was poured into a mixture of 5 g of sodium acetate, 100 mL of water and 100 ml of chloroform and then stirred for 3 hours. The layers were separated and the aqueous layer extracted with 50 mL of chloroform. The combined organic layers were washed with water (4 X 75 mL) dried (MgSO$_4$) and evaporated under reduced pressure. The resulting oil was chromatographed (SiO$_2$/CHC1$_3$/1%EtOH) giving 1.354 g of an orange glassy solid.

Synthesis of Monomer B; 1,2-Ethylenediamine *bis*-cyanoacetamide. A solution of 10 mL ethylenediamine (1500 mmoles) and 35 mL ethyl cyanoacetate (329 mmoles) in 200 mL abs ethanol was refluxed for 24 hours. The solution was allowed to cool slowly. The solids were filtered off and washed with abs ethanol then dried to give 22.01 g of slightly pink crystals (76%). ^{1}H NMR (DMSO-d$_6$): 3.58 (4H), 3.13 (4H); ^{13}C NMR (DMSO-d$_6$): 162.36, 116.05, 25.32. IR (KBr): 3419.63 (br), 2257.55 (m), 1667 (s).

Polymerization of A with B to give Polymer NS. Monomer A, 0.4155 g (0.001 mole), 0.1935 g of **B** (0.001 mole) and 0.25 g of dimethyl aminopyridine

(DMAP) (0.0022 mole) were dissolved in 20 mL of pyridine and heated to 120 °C. After 4 days the degree of polymerization by NMR was estimated to be approximately 5, and remained unchanged for 3 more days. At this point 0.01 g (0.00005 moles) more of **B** was added and the solution heated for 8 more days. After the additional 8 days, the degree of polymerization was estimated to be at least 13. The solution was then cooled and precipitated into absolute ethanol and stirred over night, then filtered and dried. The residue was re-dissolved into pyridine and precipitated again into absolute ethanol, filtered and dried *in vacuo* to give 0.33 g (59%) of polymer **NS** (the neutral form of polymer **S**). The glass transition temperature (T_g) of the polymer was found to be 188°C and the average molecular weight was found to be 9,200 g/mole (DP = 16) by NMR end-group analysis.

Conversion of phenol to phenoxide ion. Polymer **NS** (above), 0.15 g, was dissolved in 4 mL of DMSO. Lithium hydroxide hydrate, 0.015 g, was added and the solution stirred. An additional 246 mL of DMSO was added to bring the final concentration of phenoxide ions to 0.001 moles/liter of polymer **S**.

Synthesis of Polymer F. The following illustrates the synthesis of Polymer **F** (Figure 5) including the synthesis of **D**, one of the monomers polymerized to make **F**. The synthesis of monomer **C** has been previously described (*2*).

Synthesis of Monomer D. Monomer **D** was synthesized by the following four step procedure.

    Step 1 2,6-Dimethyl-3,5-pyridinedimethanol. A 100 mL round bottom flask was equipped with a reflux condenser, 25 mL dropping funnel with side arm and a nitrogen inlet and outlet. The reaction flask was cooled to 0 °C in an ice/water bath to which was added 1.0 g (4 mmol) of diethyl-2,6-dimethyl-3,5-pyridine dicarboxylate in 11 mL of toluene. The dropping funnel was charged with 11.0 mL (17 mmol) of diisobutylaluminum hydride (DIBAL-H) and the addition of the DIBAL-H solution to the reaction flask proceeded dropwise over several minutes. The solution was warmed slowly to room temperature , cooled to 0 °C , and an excess of methanol added. The solution was heated to 75 °C , refluxed, cooled to room temperature, the white precipitate filtered. The filtrate rotovapped, remaining yellow oil washed with hot hexanes and dried under vacuum to give an off-white powder (2,6-dimethyl-3,5-pyridinedimethanol) in 0.6 g (91%, mp = 137.5-138.5 oC) 1H NMR (d ppm, MeOD-d3): 2.38 (s, 6H, -CH3), 4.55 (s, 4H, -CH2-) and 7.68 (s, 1H , aromatic).

    Step 2. 2,6-Dimethyl-3,5-pyridinium dimethanol hydrochloride: A 50 mL 3-neck round bottom flask was equipped with a Teflon stopcock for gas inlet and exit and charged with 0.5 g (3 mmol) of 2,6-dimethyl-3,5-pyridinedimethanol dissolved in 10 mL of methanol. Gaseous hydrogen chloride was added slowly to the reaction flask for 20 minutes after which time the solution had become clear. The solvent was removed by rotovapping under reduced pressure and the solid dried under vacuum to give a white powder (2,6-dimethyl-3,5-

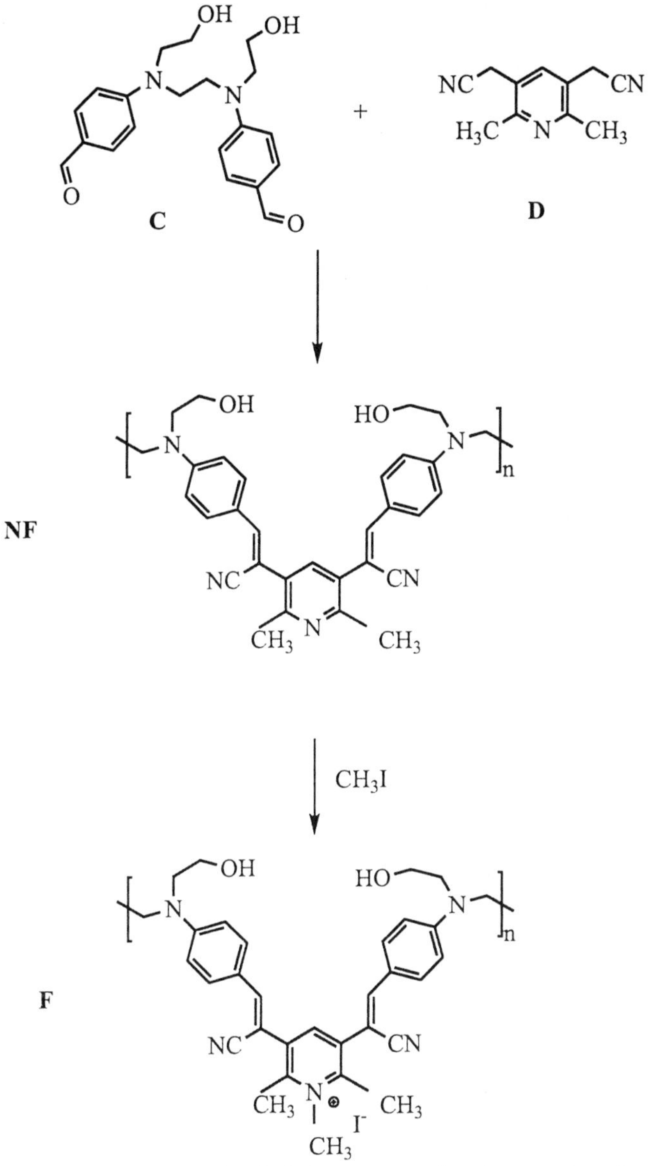

Figure 5. Scheme for synthesis of NLO-active semi-cationomeric polymer, **F**, that is insoluble in water.

pyridinium dimethanol hydrochloride) in 0.6 g (94%, mp = 122-123 °C) 1H NMR (d ppm, MeOD-d3): 2.50 (s, 6H, -CH3), 4.54 (s, 4H, -CH2-) and 8.33 (s, 1H, aromatic).

Step 3 2,6-Dimethyl-3,5-pyridinium dichloromethane hydrochloride. A 50 mL round bottom flask was charged with 0.45 g (2.2 mmol) of 2,6-dimethyl-3,5-pyridinium dimethanol hydrochloride dissolved in 3 mL of thionyl chloride. After reaction complete, solution was stirred for one hour under N_2 and the excess solvent removed, residue dried and an off-white powder was obtained in 0.5 g (2,6-Dimethyl-3,5-pyridinium dichloromethane hydrochloride) (94 %, mp = 175 °C (dec)). 1H NMR (d ppm, MeOD-d3): 2.64 (s, 6H, -CH3), 4.67 (s, 4H, -CH2-) and 8.38 (s, 1H, aromatic).

Step 4 2,6-Dimethyl-3,5-pyridine diacetonitrile. A 50 mL round bottom flask was charged with 0.40 g (1.7 mmol) of 2,6-Dimethyl-3,5-pyridinium dichloromethane hydrochloride, 240 mg of sodium cyanide suspended in 10 mL of dimethylsulfoxide. The solution was heated to 75 °C and refluxed for 4 days under a positive N_2 blanket. The contents were added to a 10 % aqueous solution of Na_2CO_3, washed with $CHCl_3$, DI water, organic phase dried over $MgSO_4$, filtered and the filtrate rotovapped. The residue was dried under vacuum to give an off-white powder in 0.2 g (monomer **D**) (64%, mp = 62-63 °C). GC/MS > 99.9% MS= 185, Elemental Analysis: calc: 71.3 % C; 5.99 % H; 22.7 % N found: 71.25 % C; 5.87 % H; 22.69 % N. 1H NMR (d ppm, CDCl3): 2.61 (s, 6H, -CH3), 3.72 (s, 4H, -CH2-) and 7.69 (s, 1H, aromatic). ^{13}C NMR (d ppm, CDCl3): 20.93, 21.79, 116.38, 122.30, 136 and 155.72.

Polymerization of C with D to give Polymer NF. A 50 ml round bottom flask was charged with 0.15 g (0.81 mmol) of **D** and 0.29 g (0.81 mmol) of **C** (p-xylyl bis-ethanolaminobenzaldehyde)) dissolved in 7 ml of dry pyridine. A catalytic amount (5 drops) of piperidine was added and the mixture was kept at 125 °C for 4.5 days under positive nitrogen pressure. The solution was cooled then added drop-wise into an excess of methanol (about 200 mL) and the precipitated polymer was filtered, dried to yield 0.3 g (80 %) of a yellow powder. Analysis of polymer **NF** (the neutral form of polymer **F**) by ^1H NMR and ^{13}C NMR confirmed the product was obtained. The number average molecular weight was determined to be 4175 g/mole (Dp = 8) by NMR endgroup analysis. Thermal analysis (10 °C/min with N_2 purge) showed the Tg was 173.0 °C and the T_d (2 wt.% loss/N_2) was 330 °C.

Methylation of Polymer NF to give polymer F. A 10 mL round bottom flask was charged with 3.0 ml of dimethylsulfoxide, 90 mg (0.2 mmol) of **NF** and 26.7 mg (0.2 mmol) of iodomethane. The solution was stirred at ambient temperature for 3 days under nitrogen. The solution was added drop-wise to diethyl ether (about 75 ml) and a brown oil precipitated from the ethereal solution. The ethereal solution was decanted and the brown oil dried to yield a brown glassy powder (80 mg, 72 %, polymer **F**). ^1H NMR showed the product was 75 % methylated to the pyridinium accordion ionomer.

Isotherms and Brewster Angle Microscopy. Isotherms for the films were made using a rectangular LB mini-trough (NIMA, Coventry). The mini-trough was kept in the dark in a glove bag continuously purged with nitrogen gas at a temperature of 25 °C. Films were symmetrically compressed at a compression rate of 5 cm^2/min. Isotherms were obtained for **F** on three different aqueous subphases; water (pH=5.5), 0.001 M dihydroxy naphthalene disulfonate, disodium salt (**DHNDS**), and 0.0001 M polymer **S** (pH=9).

Brewster angle microscopy was performed using a Brewster angle microscope built in-house. An argon laser (Lexel Model 95 operated at 0.5 mW) beam (width = 1 mm) plane-polarized by a Glan-Taylor prism (Melles-Griot) incident at the Brewster angle (approx. 53 °) was used to illuminate the nitrogen/water interface. The interface was imaged using a 15 mm focal length lens focused on the video chip of a camera (Sony HVC-2200) with its optics removed. Images were recorded with a beta format video recorder (Zenith VR9800).

Pairwise-deposition of Films. Films were made using a rectangular Langmuir-Blodgett (LB) mini-trough (NIMA, Coventry). The trough was kept in a glove bag (Aldrich, Milwaukee) with argon gas at a temperature of 24 °C in the dark during the bilayer depositions. The substrates were glass slides (Fisher, Catalogue #12-550A) cut to 1.5 cm x 2.5 cm and cleaned with a solution consisting of 30% H$_2$O$_2$ in concentrated H$_2$SO$_4$. For dipping, 2 substrates were clamped back-to-back so that material was deposited on one side of each substrate. Under argon, a 0.026 M solution of polymer **S** in dimethylsulfoxide was diluted to 10^{-4} anionic repeat units per liter of solution with water from a Barnstead Nanopure water purification system (17.9 Mohms resistivity, 0.2 micron filter). The trough was filled with the 10^{-4} M solution of **S** (pH 9). A chloroform/pyridine (3:1) solution of **F** was spread at the argon/aqueous solution interface. The system was allowed to equilibrate for 1 hour before the polymer bilayer was compressed symmetrically at a barrier speed of 5 cm^2/min at a surface pressure of 15 mN/m.

Films were deposited according to the following procedure. The substrates were held at the bottom of the dipstroke during the equilibration and initial compression of the bilayer. Once the system reached the target surface pressure of 15 mN/m, the substrates were moved on the upstroke at 1.5 mm/min. When the substrates reached the top of the stroke, the barriers were moved back at maximum speed until the surface pressure was approximately 0 mN/m. The substrates were then moved on the downstroke at 100 mm/min to minimize the transfer of material. Subsequent bilayers were built up on the substrates by repeating this deposition procedure until a target thickness was achieved.

FTIR ATR of Deposited Films. In order to illustrate that both polymers **S** and **F** are incorporated as a monolayer during each dipping cycle of the multilayer film fabrication method, the following experiment was undertaken.

The attenuated total reflection Fourier transform infrared (ATR FTIR) spectra were obtained with a Nicolet 60SX at 4 cm^{-1} resolution. Si and ZnSe ATR crystals were used as substrates. The crystals were cleaned with DMSO, ethanol, and finally with chloroform before the background spectra were obtained. The end faces of the ATR crystals were wrapped with polytetrafluoroethylene (PTFE) tape before the film deposition.

The LBK film of **F** for the FTIR study was prepared by the following procedure. A chloroform/pyridine (3:1) solution of **F** was spread at an argon/water interface. The polymer monolayer was compressed symmetrically at a barrier speed of 10 cm^2/min to a surface pressure of 15 mN/m. The ZnSe ATR crystal was moved on the upstroke through the polymer film at 1.5 mm/min.

Solution adsorption films of the polyanion for FTIR study were prepared by immersing the ATR crystals in a beaker of 10^{-4} M aqueous solution of the polyanion for 30 minutes under argon gas and protected from room light. Solution adsorption tests showed that **S** will adsorb to the ZnSe but will not adsorb to the Si. Thus, the Si ATR crystal was chosen as the substrate for the bilayer deposition.

The polycation/polyanion bilayer film for FTIR study was prepared by the following procedure. The Si ATR crystal was held at the bottom of the dipstroke in a subphase of 10^{-4} M aqueous solution of **S** during spreading, equilibrating and initial compressing of **F** on the surface of the subphase. Once the system reached the target surface pressure, the substrates were moved on the upstroke at 1.5 mm/min through the argon-aqueous solution interface. The sample was rinsed with ultrapure (17.9 MΩ/cm resistivity, 0.25 μm filtered) water and dried under nitrogen before the FTIR spectrum was obtained.

UV-Visible Spectroscopy and Second Harmonic Generation Characterization of Films. The transmission UV-Vis spectra of the films were obtained with a Cary 5 NIR-Vis-UV spectrophotometer. The films were referenced to air, and the glass background was subtracted to obtain the film spectra .

The SHG signal was generated by transmission of a fundamental beam from a Q-switched Nd:YAG laser (pulse width of 10ns and repetition rate of 10Hz) at an incident angle of 54° from normal. The SHG signal was detected with an intensified Si diode array (Tracor Northern).

Deposition of Polycation/Glucose Phosphate Bilayers. LBK films were made using a circular alternate layer trough (NIMA, Coventry). The trough was kept in a glove box continuously purged with nitrogen gas during the LBK film depositions at 24 °C. The substrates were 1" x 3" glass slides (Fisher, Cat. # 12-550A) cleaned with H_2SO_4/H_2O_2. For dipping, the substrate was clamped to a slide holder so that material was deposited on only one side of the substrate. The trough was filled with the 10^{-3} M solution of glucose phosphate disodium salt (Aldrich) (pH=5.5). A chloroform/pyridine (3:1) solution of **F** was spread at the nitrogen/aqueous solution interface in one compartment of the alternate layer trough (The second compartment contained only the aqueous solution. That is, no polymer was spread in the second compartment; hereinafter referred to as the clean compartment). The system was allowed to equilibrate for 1 hour before the polymer film was compressed asymmetrically at a barrier speed of 10 cm^2/min to a surface pressure of 15 mN/m.

The film deposition was performed using the following procedure. The substrates were held in the nitrogen atmosphere during the equilibration and initial compression of the polymer film. Once the system reached the target surface pressure, the substrates were moved down into the aqueous subphase in the clean compartment. Film deposition occurred as the substrate was moved on the upstroke at 3 mm/min out of the first polymer-containing compartment. Subsequent bilayers were built up on the substrates by repeating this procedure.

Results and Discussion

A comparison of the isotherms of **F** on the different subphases is shown in Figure 6. The isotherm for **F** on water (Figure 6, curve a) is consistent with the cross-sectional area for this type of chromophore in this polymer indicating a monolayer is obtained. Under the Brewster angle microscope (BAM), **F** on ultrapure water appears as irregular shaped domains of random sizes from up to 1 mm in breadth down to the limit of the resolution of the microscope (Figure 7). Lateral compression deforms the domains until a nearly uniform monolayer is seen at high surface pressures.

The isotherm for **F** on **DHNDS** subphase is shifted to higher areas per repeat unit (Figure 6, curve b). In the ideal view of a uniform monolayer of **F** with the dianion ionically bonded to it, the isotherm would be more similar to **F** on pure water. BAM images of **F** on **DHNDS** reveal that the monolayer is not uniform but instead consists of irregular shaped domains which do not deform even under high surface pressures. These irregular shaped rigid domains when packed together leave open gaps which explains the shift in the isotherm to higher areas per repeat unit. The isotherm for **F** on the **S** subphase come closer to that obtained for **F** on pure water with the area per repeat unit at collapse within 1 square Angstrom (Figure 6, curve c). The film appears uniform over the field of view of the BAM and a uniform increase in reflectivity is observed as the film is compressed with no domains appearing. The position of the isotherm of **F** on **S** at higher areas may be due to **S** adsorbed at the interface and mixed (interpenetrated) with **F**.

Figure 8 shows the FTIR ATR spectra for each individual polymer with peak intensities normalized. The polyanion, **S**, exhibits a strong absorbance at 1605 cm^{-1}. The polycation, **F**, has a characteristic band at 1587 cm^{-1}. The bilayer spectra contains the characteristic bands of both polymers. These results show that both polymers are transferred to the Si ATR crystal, but because the FTIR peaks are not calibrated for concentration or thickness, the relative amounts of the two polymers can not be determined quantitatively.

The transferred films of both polymers have broad absorbance peaks due to their chromophores and both polymers have a λ_{max} near 410 nm. For the pairwise deposited film, the UV-visible absorbance at 410 nm is 1.5 times higher than the absorbance for a transferred monolayer of **F** alone. This suggests that more than a monolayer is transferred during each dipping stroke. The qualitative conclusion may be drawn that both of the semi-ionomeric polymers are transferred during the pairwise deposition process.

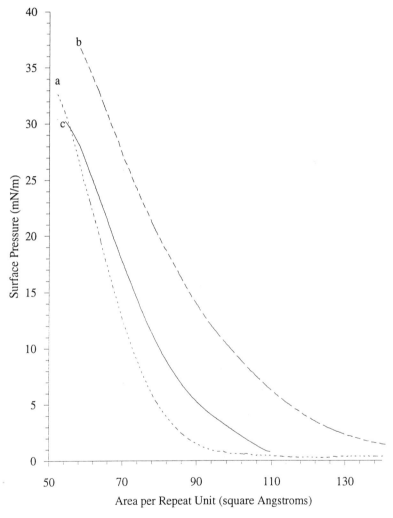

Figure 6. Isotherms obtained at 25 °C. (a) **F** on 18 MΩ-cm water. (b) **F** on 0.001 M dihydroxy naphthalene disulfonate, disodium salt. (c) **F** on 10^{-4} M **S**.

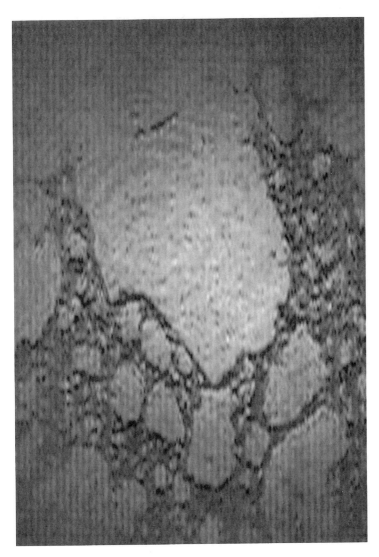

Figure 7. Appearance of films imaged with Brewster angle microscopy. Imaged size 0.80 mm X 0.55 mm.

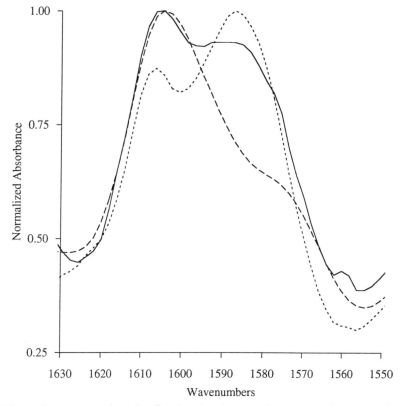

Figure 8. Attenuated total reflection Fourier transform infrared spectra of **F** monolayer on ZnSe (· · ·), **S** adsorbed from aqueous solution on ZnSe (- - -), and pairwise-deposited **F/S** bilayer on Si (———).

The SHG signal is a factor of 12 stronger in the pairwise-deposited films compared to the monolayer deposited films of **F** with glucose phosphate. The twelve-fold enhancement of the SHG signal must include a contribution from **S**. Improved orientational order of the chromophores of **F** may also contribute to the enhanced SHG signal.

A critical factor in the quality of films resulting from a deposition process is the layer to layer uniformity that can be achieved. Three measures indicate the uniformity of LBK deposition; namely, constant transfer ratios, a linear increase in UV-Visible absorbance at λ_{max}, and a quadratic increase in the SHG signal. For the pairwise deposited films, the average transfer ratio was 1.04. The dependence of UV absorbance and the SHG intensity on film thickness is plotted in Figure 9.

Conclusion

These experiments indicate that films with noncentrosymmetric order of chromophores can be produced by a pairwise deposition process which involves an insoluble NLO-active polycation spread on an aqueous solution of a NLO-active polyanion. The polymers ionically bind at the gas/liquid interface to produce a bilayer polymer salt which may subsequently be transferred to a solid substrate in a highly uniform manner with each added bilayer. The main advantages of pairwise deposition of films are: 1) no hydrophobic alkyl groups necessary, 2) ionic bonding between polymer chains contributes to the structural stability of the films, and 3) the rate of film deposition is increased. Future experiments will include ellipsometry, production of thicker films, further characterization of NLO figures of merit, and scanning Kelvin probe microscopy (*20*) of the transferred films.

Acknowledgments

The authors are thankful to R. Y. Yee for useful discussions and M. Seltzer for use of his laser equipment. The Office of Naval Research is acknowledged for funding this work.

Literature Cited

1.   Stenger-Smith, J. D.; Henry, R.; Hoover, J.; Lindsay, G.; Fischer, J.; Wynne, K. J., *U.S. Pat. No. 5,247,055* issued Sep 21, **1993**.
2.   Wynne, K. J.; Lindsay, G. A.; Hoover, J. M.; Stenger-Smith, J.; Henry, R. A.; Chafin, A. P. *U.S. Pat. No. 5,520,968*, issued May 28, **1996**.
3.   *Polymers for Second-Order Nonlinear Optics,* G. A. Lindsay and K. D. Singer, Eds., Am. Chem. Soc. Advances in Chemistry Series 601, Washington, D.C., **1995**.
4.   Butcher, P. N.; Cotter, D. *The Elements of Nonlinear Optics;* Cambridge University Press: Avon, U.K., **1990**.

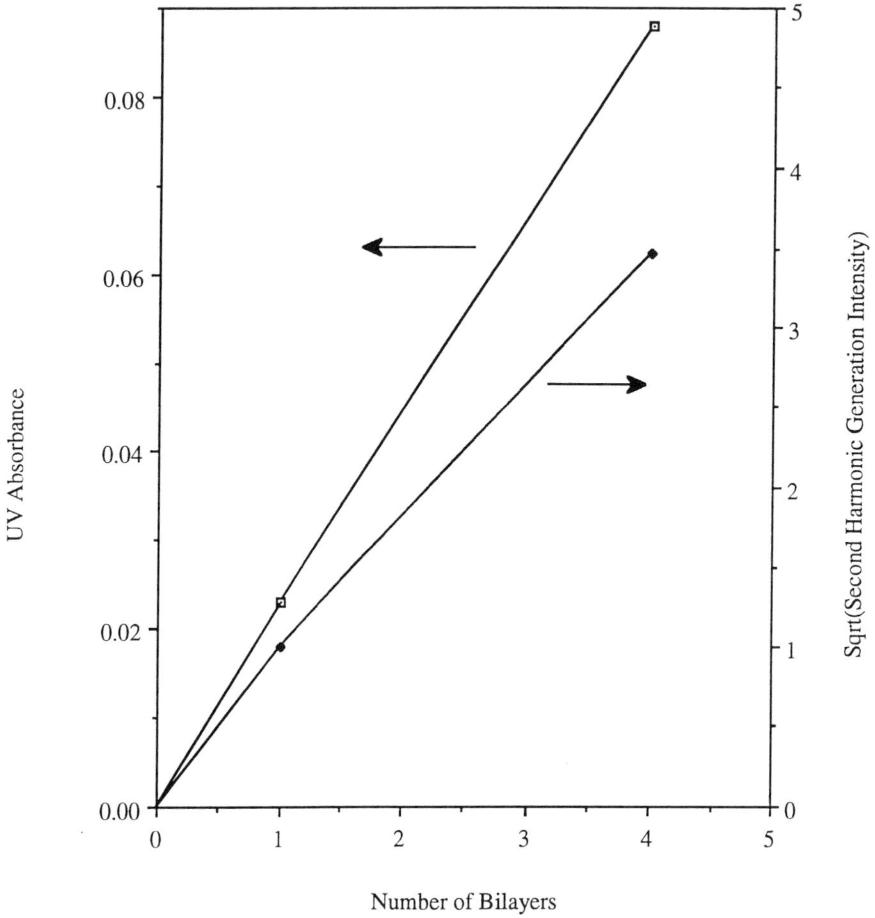

Figure 9. Quadratic enhancement of SHG intensity and linear increase in UV-Visible absorbance as a function of the number of **F/S** bilayers.

5. Lindsay, G. A. In *Polymers for Second-Order Nonlinear Optics,* G. A. Lindsay and K. D. Singer, Eds., Am. Chem. Soc. Advances in Chemistry Series 601, Washington, D.C., **1995**, pp 1-19.

6. Sekkat, Z.; Dumont, M. *App. Phys.* **1992**, B54, 486.

7. Lvov, Y., Yamada, S., Kunitake, T. *Thin Solid Films* **1997**, 300, 107-112.

8. Li, D.; Ratner, M. A.; Marks, T. J. *Journal of the American Chemical Society* **1990**, 112, 7389-7390.

9. *Poled Polymers and their Applications to SHG and EO Devices -(Advances in Nonlinear Optics; v. 4),* Miyata, S.; Sasabe, H. Eds., Gordon & Breach Amsterdam, **1997**.

10. *Insoluble Monolayers at Liquid-Gas Interfaces,* Gaines, G. L., Wiley & Sons, New York, **1966**.

11. Lindsay, G.; Wynne, K.; Herman, W.; Chafin, A.; Hollins, R.; Stenger-Smith, J.; Hoover, J.; Cline, J.; Roberts, J. *Nonlinear Optics* **1996**, 15, 139-146.

12. Hall, R. C.; Lindsay, G. A.; Hoover, J. M. *U.S. Pat. No. 5,162,453* issued Nov. 10, **1992**.

13. Hall, R. C.; Lindsay, G. A.; Hoover, J. M. *U.S. Pat. No. 5,225,285* issued Jul. 6, **1993**.

14. Penner, T. L.; Noonan, J. M.; Ponticello, I. S. *U.S. Pat. No. 4,830,952* issued May 16, **1989**.

15. Ulman, A.; Williams, D. J.; Penner, T. L.; Robello, D. R.; Schildkraut, J. S.; Scozzafava, M.; Willand, C. S. *U.S. Pat. No. 4,792,208* issued Dec. 20, **1988**.

16. Barraud, A.; Vandevyver, M. *Thin Solid Films* **1983**, 99, 221-225.

17. Embs, F.; Funhoff, D.; Laschewsky, A.; Licht, U.; Ohst, H.; Prass, W.; Ringsdorf, H.; Wegner, G.; Wehrmann, R. *Advanced Materials* **1991**, 3(1), 25-31.

18. Picard, G.; Alliata, D., Nevernov, I., Pazdernik,, L. *Langmuir* **1997**, 13, 264.

19. Troitsky, V. I.; Matveeva, N. K. *Proceedings of LB8 - Eighth International Conference on Organized Molecular Films* **1997**.

20. Nonnenmacher, M.; Boyle, M. P.; Wickramasinghe; H. K. *Applied Physics Letters* **1991** 58(25):2921-2923.

Chapter 21

Thermoset Second-Order Nonlinear Optical Materials from a Trifunctionalized Chromophore

Youn Soo RA[1], Shane S. H. Mao[1], Bo Wu[1], Lan Guo[1,2], Larry R. Dalton[1,2,4], Antao Chen[3], and William H. Steier[3]

[1]Loker Hydrocarbon Research Institute, Department of Chemistry, and [2]Department of Materials Science and Engineering, University of Southern California, Los Angeles, CA 90089–1661
[3]Center for Photonic Technology, Department of Electrical Engineering, University of Southern California, Los Angeles, CA 90089–0483

A thermosetting polyurethane has been synthesized from a new trifunctionalized Disperse-Red type chromophore (DRTO). Precuring of this chromophore with excess of tolylene diisocyanate (TDI) at 80°C yielded soluble oligomer, which can be spin-coated onto various substrates to form high quality films. Corona poling of the thin films afforded reproducible electrooptic coefficient, ranging from 10 to 15 pm/V at 1.06 μm depending on the loading density. High temporal thermal stabilities up to 120°C were observed. The advantages of this thermosetting polyurethane are that it produces batch-to-batch reproducible results results such as high electrooptic coefficients and thermal orientational stabilities, and the processing of the material from the precuring step to spin casting of the films is simple.

For second order nonlinear optical (NLO) materials to be used in electrooptic device applications, a large optical nonlinearity, appropriate processibility, long term stability in the working temperature range of 60°C to 120°C and low optical loss are required.(*1*) Organic polymeric second order NLO materials are considered promising for fabrication of electrooptic devices because of their large nonlinearities, high laser damage threshold, low dielectric constants, and excellent processibility. After the non-centrosymmetric order has been poling-induced in a polymer, the NLO chromophores have to be locked into the polymer matrix to prevent them from relaxing back to random orientation.(*2*)

Many approaches have been explored to preserve the poling-induced dipole alignment.(*3-19*) Among them, cross-linking reactions have been proven to be very useful. So far, however, there has not been a polymer prepared that meets all the requirements for electrooptic applications.(*4*)

Recently, a new class of double-ended functionalized NLO chromophores and their crosslinked polyurethanes were prepared by Francis and coworkers at 3M.(*13*) However, the nonlinearity reported was small due to the degradation of the chromophores. Most recently, trihydroxyl functionalized amino-sulfone azobenzene chromophores and a thermosetting polyurethane system was reported.(*14*) In this system both ends of the NLO chromophores were anchored to the network *via*

---

[4]Corresponding author.

urethane reaction between the hydroxyl groups and the isocyanate groups by the crosslinking agent, TDI. The observed low nonlinearity in the polymer was attributed to the use of a chromophore with low nonlinearity. Also, there was a 34% decrease in nonlinearity after 800 h at 70°C due to incomplete crosslinking.

Very few polymer materials possess enough NLO orientational stability, especially at elevated temperature (i.e., 90 - 125°C), and many gain the NLO stability at the sacrifice of optical nonlinearity, film quality, material processibility, chromophore degradation, or optical loss. Therefore, it is necessary to synthesize NLO polymers exhibiting large and stable second order optical nonlinearities as well as excellent film quality, and good material processibility.

In an effort to overcome the shortcomings of the existing polymeric NLO materials, a new trifunctionalized Disperse Red-19 type chromophore was synthesized and crosslinked into a polyurethane network by a one-step processing scheme. The material obtained can be efficiently poled to yield resonable electrooptic coefficients, excellent film quality, high thermal stability, and low optical loss.

Experimental

Anhydrous dioxane, tolylene diisocyanate (TDI) and *N, N*-bis(2-hydroxyethyl)aniline were purchased from Aldrich Co. TDI was distilled and kept in argon atmosphere before use. *m*-Aminobenzyl alcohol was obtained from Pfaltz and Bauer. The synthesis of the trifunctionalized chromophore (disperse-red triol, DRTO) is shown in Figure 1. *m*-Aminobenzyl alcohol was protected by acetylation, followed by nitration with nitronium tetrafluoroborate to give the corresponding nitroaniline derivative. The nitration with nitronium tetrafluoroborate generally gave a higher yield than the conventional acidic nitrating agents. The nitrated compound was deprotected to give the corresponding nitroaniline in 97% yield. Diazotization of the nitroaniline, followed by reaction with *N, N*-bis(2-hydroxyethyl) aniline gave DRTO in 85 % yield after recrystallization from methanol (3x) and drying *in vacuo*. DRTO: m.p.: 162°C. T_d: 270°C. UV-Vis (in dioxane): 470 nm. ^{1}H-NMR (ppm, in acetone-d_6):8.34 (s, 1H), 8.22 (d, 1H), 7.87 (d, 2H), 7.85 (d, 2H), 6.94 (d, 2H), 5.08 (d, 2H), 4.74 (t, 1H), 4.21 (t, 2H), 3.85 (t, 2H), 8.82 (t, 2H), 3.7 (t, 2H). ^{13}C-NMR (ppm, in acetone-d_6): 170.92, 157.00, 130.85, 141.38, 126.92, 126.77, 122.73, 121.08, 114.71, 112.80, 61.84, 60.20, 55.09. Elemental analysis: Found, C 56.70; H 5.60; N 15.47. Theoretical, C 56.66, H 5.59, N, 15.54.

^{1}H-NMR and ^{13}C-NMR were taken using a Bruker-250-FT-NMR spectrometer operating at 250 MHz. Optical spectra were obtained from a Perkin-Elmer Lambda-4C UV/ Vis in a 1 cm^3 quartz cell. The decomposition temperature (T_d) was determined by Shimadzu Thermogravimetric Analyzer, measurements being performed under nitrogen atmosphere and at a heating rate of 5 °C/min. Melting points was determined by Perkin-Elmer DSC-7 under nitrogen atmosphere and at a heating rate of 5 °C/min.

Two types of prepolymers were synthesized, PU-DRTO (as seen in Figure 1) and PU-DRNTO (this prepolymer was designation assigned to PU-DRTO materials processed with triethanolamine (TEA). The PU-DRTO and PU-DRNTO prepolymer solutions were filtered through a 0.2 μm syringe-filter, spin-casted onto ITO-coated glass substrates and dried *in vacuo* for 4 hours or overnight before poling.

The film was optically clear and homogeneous. The film was then poled using a corona-poling setup, with a tip-to-plane distance of ca 1.5 cm. The poling protocols are shown in Figure 2. The poled films were then cooled to room temperature in the presence of the electric field.

The electrooptic coefficient and film thickness were measured from the ATR measurement technique at 1.064 μm. The thermal stability was measured by monitoring the decay of the second harmonic signal as a function of temperature (10°C/min.). Optical loss was measured using the "dipping technique"(20) at 1.3 μm or by imaging the guided light streak at 0.98 μm on a video camera.

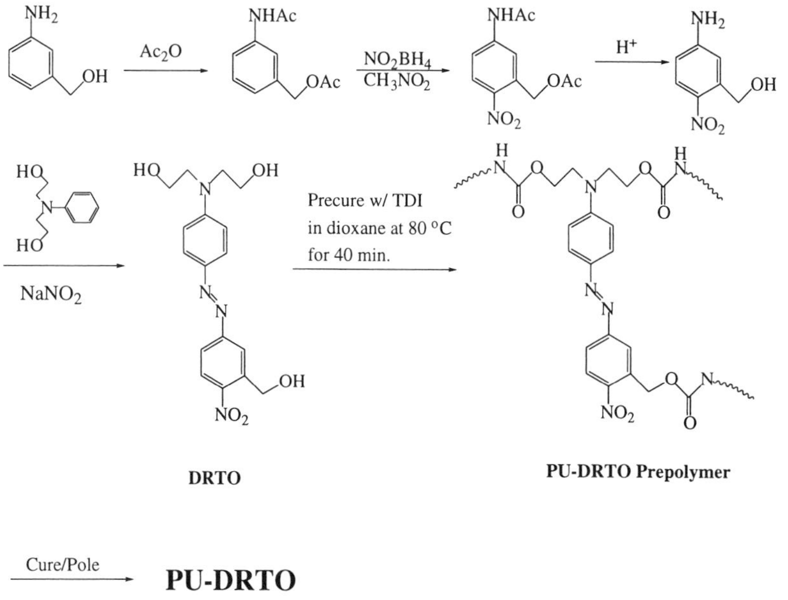

DRTO   **PU-DRTO Prepolymer**

$\xrightarrow{\text{Cure/Pole}}$ **PU-DRTO**

Figure 1. Synthetic scheme for PU-DRTO.

Results/ Discussion

DRTO was prepared by the diazo coupling of *N, N*-phenyldiethanol amine and 3-hydroxy methyl-4-nitroaniline. The resulting solid was recrystallized from methanol to produce shiny red needle like crystals.

Two types of polymers were synthesized , PU-DRTO and PU-DRNTO (Table I), each with different loading densities of DRTO in order to find the optimum loading density to produce the highest obtainable electrooptic coefficient, and to study the polymer behavior in terms of thermal stability and optical loss. No catalyst was used in order to avoid electrical conduction of the finally cured film. PU-DRTO prepolymers were synthesized by precuring two components, DRTO and TDI in anhydrous dioxane at 80°C for 40 minutes. The loading density of the DRTO in the prepolymer solution was controlled by varying the amount of TDI added. The PU-DRNTO prepolymer was synthesized by precuring the PU-DRTO prepolymer solution and triethanolamine (TEA) to eliminate the excess isocyanate functionalities. 35% PU-DRNTO contained stoichiometric ratio of NCO/OH while 30% PU-DRNTO contained excess of TEA. The films spin cast from the prepolymer solutions were soft and soluble in common organic solvents which indicates the oligomeric nature of the polymer. After the films were poled, they became tough and resistant to solvents as crosslinking proceeded.

The films were generally poled at 120°C after a short excursion from 80°C with a poling voltage of 6 kV as it can be seen from Figure 2. Complete curing of the film at 120 °C for 1.5 hours was observed through an FT-IR study.

The measured electrooptic coefficients ranged from 9.7 to 14.5 pm/V depending on the loading density . It was observed that the optimum loading density was 35% by weight. Decrease of chromophore loading density lowered the electrooptic coefficient, which can be explained by the lowering of the chromophore concentration in film. Increase of the chromophore loading to above 35% also lowered the r_{33} to 10 pm/V a feature which is attributed to a decrease of poling efficiency caused by electrostatic interactions between the chromophores in the film.

The dynamic thermal stabilities for both PU-DRTO and PU-DRNTO were in the range of 110 to 120°C except for the case of 30% PU-DRNTO, which was lower 90°C. From these observations it can be seen that the prepolymers containing either an excess of TDI or stoichiometric ratio of all components produced reasonable thermal stability. The prepolymer that contained excess TEA crosslinker exhibited lower thermal stability which can be attributed to faster reaction of the TEA with TDI than DRTO. This leads to fewer reaction sites available for the DRTO molecules to covalently anchor themselves into the hardening polymer lattice, thus leading to lower thermal orientational stability. A 10°C increase in the thermal stability of the films was observed when the films were poled for a long time at high temperatures, as it can be seen from Figure 2. This is attributed to the formation of allophanates. After the formation of urethane linkage, -NHC(O)O-, the nitrogen can further crosslink the remaining isocyanates in TDI at elevated temperatures to crosslink even further by forming an allophanate linkages.(21)

The optical loss for PU-DRTO (35 %) was 6.3 dB/cm, at 1.3 μm, which is too high for device applications. The isocyanate functionalities in excess TDI may have aggregated or reacted with water molecules from air forming numerous scattering centers in the film, which leads to the observed high optical loss. Under careful study of these films under an optical microscope (500 x), white droplets were observed in the PU-DRTO films as well as some needle-like crystals. A slight modification was carried out with PU-DRTO to decrease the optical loss. In order to prevent aggregation of the excess isocyanate groups, one equivalent (based on the remaining isocyanates) of TEA was added to the prepolymer solution to react with the isocyanates during the poling/curing step. After the film was poled by the same method as for PU-DRTO, the optical loss was measured to be 3.4 dB/cm at 0.98 μm. The optical loss was thus lowered by a factor of two even after measured at a wavelength closer to that of the λ_{max} of the film (470 nm). At 1.3 μm optical loss was decreased to 1.0 dB/cm by preheating films before poling.

The PU-DRNTO (35%) films were then preheated in 100°C oven for various times before being poled by one of two different poling protocols. The poling protocol (a) involved raising film the temperature to 120°C from room temperature in 10 minutes, then holding at that temperature for additional 45 minutes while the poling voltage was maintained constant at 6 kV. Poling protocol (b) involved the same film temperature as in (a) but varying the starting poling voltage from 5 kV initially then raising it up to 6 kV within 5 minutes.

When the PU-DRNTO film was first preheated for 5 minutes, and poled using protocol (a), the optical loss was high (10.3 dB/cm at 1.3 μm); after 8 minutes, the optical loss was dropped to 6.7 dB/cm. When the preheating time for PU-DRNTO was 8 minutes and poled under protocol (b), the optical loss was reduced to 3 ~ 6 dB/cm (the latter value is due to poling damage of the film). Preheating the PU-DRNTO film for 10 minutes and employing protocol (b) resulted in an optical loss of 1.0 dB/cm. This low optical loss was observed also on a nonpoled part of the sample.

The refractive indices of the polymer films varied depending on the loading density of the chromophores in the polymer matrix as it can be seen in Table I. The high refractive index of the material could be also due to TDI which is used for high refractive index materials.

Table I.   Linear and nonlinear optical properties of PU-DRTO and PU-DRNTO.

Chromophore (wt % loading)	r_{33} @1.064 μm (pm/ V)	Thermal stability (°C)	Optical loss (dB/cm)	Refractive Index
PU-DRNTO (30 %)	12	90	-	1.675 - 1.676
PU-DRTO (35 %)	14.5	120	6.6 @ 1.3 μm	1.64 - 1.72
PU-DRTO (42 %)	9.7 -10	110 - 120	-	1.72
PU-DRNTO (35 %)	14.8	110	3.6 @ 0.98μm	1.689 - 1.692

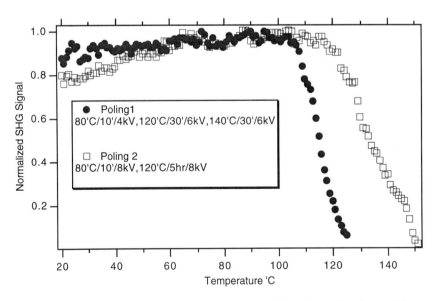

Figure 2.   The thermal stabilities of PU-DRTO oriented using different poling profiles.

Table II. PU-DRNTO optical loss study at 1.3 μm.

Prebaking Time (min.)	Poling Profile[a]	Optical loss @ 1.3 μm
5	(a)	10. 3 dB/ cm
8	(a)	6.7 ± 0.2 dB/ cm
8	(b)	3 ~ 6 dB/ cm
10	(b)	1. 0 dB/ cm

a. poling profiles described in text.

Conclusion

A new trifunctionalized DR type chromophore was synthesized which has been covalently anchored to a three dimensional polyurethane network by a one-step process. The materials thus obtained were efficiently poled to yield consistently r_{33} value of 14.5 pm/V and high thermal orientational stability of 110 - 120°C. The optimum loading level of the system was found to be 35 % loading by weight. A low optical loss of 1.0 dB/cm at 1.3 μm was achieved by hardening of the PU-DRNTO (35%) prepolymer film *via* preheating the film at 100°C for 10 minutes before poling.

Acknowledgment. The authors acknowledge the Air Force Office of Scientific Research (F49620-0201) for financial support. Also, special thanks are given to Deacon Research Inc. for the optical loss measurements at 980 nm.

Literature Cited

1. Lipscomb, G. F.; Lytel, R. S.; Ticknov, A. J.; Eck, T. E. V.; Kwiatkowski, S. L. and Girton, D. G. *Proc. SPIE,* **1990**, 1337, 23.
2. *Nonlinear Optical Properties of Organic Polymeric Materials,* Williams, D. J. Eds.; ACS Symp. Ser. 233; American Chemical Society, Washington, DC, **1983**.
3. Prasad, P.N.; Williams, D. J. *Introduction to Nonlinear Optical Effects in Molecules and Polymers*; John Wiley & Sons: New York, **1991**.
4. Bosshard, C.; Sutter, K.; Prêtre, P.; Hullinger, J.; Flörsheimer, M.; Kaatz, P.; Günter, P. *Organic Nonlinear Optical Materials: Advances in Nonlinear Optics*; Gordon and Breach Publishers: Postfach, Switzerland, 1995, Vol. 1.
5. Dalton, L. R.; Harper, A. W.; Wu, B.; Ghosn, R.; Laquindanum, J.; Liang, Z.; Hubbel, A.; Xu, C. *Adv. Mater.* **1995**, 7, 519.
6. Dalton, L .R.; Harper, A. W.; Ghosn, R.; Steier, W. H.; Ziari, M.; Fetterman, H.; Shi, Y.; Mustacich, R. V.; Jen, A. K.-Y.; Shea, K. J. *Chem. Mater.* **1995**, 7, 1060.
7. Chen, M.; Yu, L. P.; Dalton, L. R.; Shi, Y.; and Steier, W. H. *Macromolecules* **1991**, 69, 8011.
8. Chen, M.; Dalton, L. R.; Yu, L.; Shi, U.; and Steier, W. H., *Macromolecules,* **1992**, 25, 4032.
9. Becker, M. W.; Sapochak, L. S.; Ghosn, R.; Xu, C.; Dalton, L. R.; Shi,, Y.; Steier, W. H. *Chem. Mater.* **1994**, 6, 104.
10. Yu, D.; Gharavi, A.; and Yu, L.; *Macromolecules* **1996**, 29, 6139.
11. Yu, D.; Gharavi, A.; and Yu, L.; *J. Am. Chem. Soc.* **1995**, 117, 11680.
12. Xu, C.; Wu, B.; Dalton, L. R., Shi, Y., Ranon, P. M., and Steier, W., *Macromolecules,* **1992**, 25, 6714.

13. Francis, C. V.; White, K. M.; Boyd, G. T.; Moshrefzadeh, R. S.; Mohapatra, S. K.; Fadcliffe, M. D.; Trend, J. E.; Williams, R. C., *Chem. Mater.* **1993**, 5, 506.

14. Boogers, J. A. F.; Klaase, P. Th. A.; Vlieger, J. J. De.; Tinnemans, A. H. A. *Macromolecules* **1994,** 27, 205.

15. Jen, A. K.-Y; Drost, K. J.; Cai, Y.; Rao, V. P.; Dalton, L. R.*; J. Chem. Soc. Chem. Commun.* **1994**, 965.

16. Jen, A. K.-Y; Liu, Y.-J.; Cai, Y.; Rao, V. P.; Dalton, L. R. *J. Chem. Soc. Chem. Commun.* **1994**, 2711.

17. Hubbard, M. A.; Marks, T. J.; Lin, W.; Wong, G. K.; *Chem. Mater.* **1992**, 4, 965.

18. Burland, D. M.; Miller, R. D.; Walsh. C. A. *Chem. Rev.* **1994,** 94, 31.

19. Lon, J. T.; Hubbard, M. A.; Marks, T. J.; Lin, W.; Wong, G. K. *Chem. Mater.* **1992**, 4, 1148.

20. Teng, C. C., *Applied Optics* **1993,** 32 (7), 1951.

21. Dusek, K., M. Ilansky, and J. Somokovarsky, *Polym. Bull.* **1987**,18, 209.

Chapter 22

Polymer Structural Effects on Sub-Transition Temperature Light-Induced Molecular Movement in High-Temperature Nonlinear Optical Azo-Polyimides

Zouheir Sekkat[1,4], André Knoesen[1], Victor Y. Lee[2], Robert D. Miller[2], Jonathan Wood[3], and Wolfgang Knoll[3,5]

[1]Department of Electrical and Computer Engineering, University of California, Davis, CA 95616
[2]IBM Almaden Research Center, 650 Harry Road, San Jose, CA 95120
[3]Max-Planck-Institute for Polymer Research, 55128 Mainz, Germany

The polymer structural effects on light-induced molecular movement below the glass transition temperature (Tg) of very high Tg (up to 350°C) polyimide polymers containing nonlinear optical azo chromophores are discussed. It is shown to what extent the isomerization reaction itself and the light-induced nonpolar and polar orientation depend on the molecular structure of the unit building blocks of the polymer. We demonstrate photo-induced orientation of azo chromophores in these polyimides at room temperature even though the chromophore is firmly embedded into the rigid polymer backbone, and how Light can alter the optical third-order nonlinearity by changing the chromophore hyperpolarizability through a photo-induced isomer change in shape.

Polymeric nonlinear optical (NLO) films containing azo chromophores with appreciable molecular hyperpolarizabilities have been attracting increasing interest in the last few years in studies of the role of photoisomerization in linear and nonlinear optics (1-3). Light-induced ordering processes of NLO azo chromophores have shown potential as optical data storage and novel optical poling techniques (4-12), and several authors reported studies in Langmuir-Blodgett-Kuhn azo-polymers as multilayer structures and alignment layers for liquid ctystal molecules (13-20), self-assembled monolayers (21-24), dendrimers (25), phospholipids (26), polypeptides (27), peptide oligomers (28), zeolites (29), hybrid polymers (30), amorphous and liquid crystalline polymers (31-40). In the recent years, the effect of the photoisomerization of nonlinear optical azo chromophores in polymer films has been of interest (7-12). Studies of the role of inter-chromophores interactions and molecular addressing have

---

[4]Permanent address: Department of Physics, Faculty of Sciences, University Sidi Mohamed Ben Abdellah, BP 1796, Atlas-Fes, Morocco.
[5]Also with the Frontier Research Program, The Institute of Physical and Chemical Research (RIKEN), Wako, Saitama 351-01, Japan.

been reported *(33,34,40)*, and questions have begun to arise only very recently concerning the relationship of optical ordering processes in amorphous polymers to the Tg and polymer structure including the main chain rigidity, the free volume, and the nature of the connection of the chromophore to a rigid or a flexible main-chain *(41-44)*. Following the pioneering work in the early 1960's of Neoport and Stolbova on photo-induced nonpolar orientation (PIO) in viscous solutions *(45)*, Todorov et al. *(46)* used PIO of methyl orange (an azo chromophore) in a polyvinyl alcohol polymer film as a guest host system to inscribe reversible polarization holograms. Many of the light-induced nonpolar orientation subsequent studies have been performed in liquid crystalline polymers *(31-37)*, and relatively low Tg (around 130 °C) flexible poly(methyl-methacrylate) (PMMA) derivatives polymers containing azo dye *(4-11)*. Recently, optical ordering in higher Tg polymers has been of interest *(42-44, 47,48)*.

In this paper, we review our recent research on photo-induced effects in high Tg NLO polyimides, and we present new insights into the photophysical phenomena which occur upon the photoisomerization of azo chromophores in these particular class of high Tg polymers. We correlate optical ordering (nonpolar and polar) to the polymer structure in a series of very high Tg (up to 350 °C) rigid or semirigid NLO polyimides. We show that light can efficiently induce molecular movement in rigid polyimide polymers well below their Tgs (up to 325 °C). This sub-Tg molecular movement strongly depends on the polymer architecture, a finding which demonstrates a new way of probing sub-Tg polymer dynamics in photochromic NLO polymers. In particular, we show that the molecular movement in polymeric materials, which is generally believed to be governed by the difference between Tg and the operating temperature T *(49)*, depends on the molecular structure of the unit building blocks of the polymer, and that polymers with similar Tgs can exhibit significantly different photo-induced properties. We discuss the influence of the polymer molecular structure on photo-induced linear as well as second and third order nonlinear optical effects. While light does cause isomerization and nonpolar orientation in all of the azo-polyimide derivatives, we find that light does not promote sub-Tg polar orientation if combined with a static electric field when the chromophores are embedded into the polyimide backbone via the donor substituents. We discuss polymer structural effects on the *(i)* cis⇒trans thermal isomerization, *(ii)* light-induced nonpolar orientation, *(iii)* light-assisted polar orientation in poling electric fields, and *(iv)* photo-induced changes in third-order optical nonlinearities, each in a separate section.

Four high Tg NLO polyimide derivatives used in our research, PI-1, PI-2, PI-3a and PI-3b, each with distinct differences in the molecular structure of the unit building blocks, are shown in Fig. 1. The Tg values for PI-1, PI-2, PI-3a and PI-3b were 350 °C, 252 °C, 228 °C, and 210 °C respectively as measured by differential scanning calorimetry. Details of the polymer synthesis and characterization can be found in Refs. 50 and 51. PI-1 and PI-2 are both donor-embedded systems, where the NLO chromophore is incorporated rigidly into the backbone of the polymer without any flexible connector or tether. The difference between PI-1 and PI-2 is the flexibility introduced into the polyimide backbone via the precursor dianhydride that lowers the Tg by almost 100 °C (350 versus 252 °C). PI-3a and PI-3b, on the other hand, are true side chain systems where the NLO azo chromophore is tethered to the main chain via a flexible tether. The size of the diarylene azo chromophore in PI-3a is significantly larger than that of the Disperse Red 1 (DR1) -type chromophore in PI-3b; a feature which requires more free volume for chromophore movement. In comparison to the donor-embedded systems, the flexible side chain systems allow more freedom of movement of the azo chromophore. As a result of the polymer structures and properties, one would anticipate that the intrinsic chromophore mobility should decrease according to the series, PI-3b > PI-3a > PI-2 > PI-1. Samples were prepared by spin-casting onto quartz substrates, and the remaining solvent was removed by baking the films at elevated temperatures. Photoisomerization of the azo units was induced by irradiation with green (530 nm, 30 mW/cm^2; or 544 nm, 8 mW/cm^2) laser light.

When azo chromophores are photoisomerized using polarized light, anisotropic changes in the linear optical properties (e.g. refractive indices, and absorbance) provide some insight on both the isomerization reaction itself as well as on light-induced nonpolar orientation of the chromophores. The components of the linear susceptibility in the directions parrallel and perpendicular to the polarization of the irradiating light are $\chi_{//} = N\alpha(1 + 2aA_2)$ and $\chi_{\perp} = N\alpha(1 - aA_2)$. α is the isotropic molecular polarizability; a is the molecular anisotropy; N is the chromophores density; and A_2 is the expansion parameter of the second order Legendre polynomial which characterizes the orientation of the chromophores. The variations of the refractive index (n_i), and the absorbance (Abs_i) in the principal direction i of the irradiated film are respectively proportional to the real and imaginary parts of the variation of χ_i. The linear polarization per unit volume $(N\alpha)$ and the dichroic ratio, or the order parameter (S), can be respectively computed as:

$$N\alpha = (\chi_{//} + 2\chi_{\perp})/3, \text{ and } S = aA_2 = (\chi_{//} - \chi_{\perp})/(\chi_{//} + 2\chi_{\perp}).$$

UV-vis dichroism and the attenuated total reflection (ATR) waveguide spectroscopy are well suited to investigate both the isomerization reaction itself, and the photo-induced nonpolar orientation through the respective changes in $N\alpha$ and S after irradiation with polarized light. $N\alpha$ and S can be computed from ATR spectroscopy and UV-dichroism measurements respectively as:

$$N\alpha = (n_{//}^2 + 2n_{\perp}^2 - 3)/3, \text{ and } S = (n_{//}^2 - n_{\perp}^2)/(n_{//}^2 + 2n_{\perp}^2 - 3),$$

and

$$N\alpha = (Abs_{//} + 2Abs_{\perp})/3, \text{ and } S = (Abs_{//} - Abs_{\perp})/(Abs_{//} + 2Abs_{\perp}).$$

When the polar NLO azo chromophores are photoisomerized in the presence of a poling dc electric field, their rotational mobility is sometimes enhanced and the light can assist electric field poling below the polymer Tg (photo-assisted poling). Changes in the second order nonlinear optical properties provide information on the polar orientation of the chromophores. Second harmonic generation (SHG) is well suited for probing both photo-assisted polar orientation in the presence of dc electric field as well as induced changes in polymer third-order nonlinearity monitored by the electric field induced second harmonic (EFISH).

Cis⇒Trans Thermal Isomerization

The thermally activated cis⇒trans back isomerization of push-pull substituted azobenzene derivatives is usually complete after only few seconds at room temperature. In particular, this has been observed for DR1 added as a guest to PMMA, attached via flexible tethers to flexible PMMA backbone *(52)*, and for a DR1-type chromophore flexibly tethered to a rigid polyimide backbone, e.g see PI-3b *(42)*. However, we will show that the rate of cis⇒trans thermal isomerization of the push-pull chromophore is reduced in the donor-embedded polyimide systems studied, e.g. PI-1 and PI-2.

ATR and UV-vis dichroism were used to study the isomerization of azo chromophores as a function of polymer molecular structure. Figure 2 shows that guided modes in PI-1 shift their ATR-angular position to lower incidence angles as a result of light-induced changes in the optical properties of the polymer caused by the photoisomerization of the azo chromophores using polarized light. The theoretical Fresnel fits to the experimental reflectivity data allow the extraction of $n_{//}$ and $n_\perp$ of the polymer film. In both PI-1 and PI-2, the quantity $(n_{//}^2 + 2n_\perp^2 - 3)/3$ which is equal to $N\alpha$ at optical frequencies, and which is proportional to chromophore density, is not recovered a few minutes (35 and 20 minutes for PI-1 and PI-2, respectively) after irradiation, and suggests the existence of long-lived cis isomer. At 633 nm probe, the variation of $N\alpha$ was −0.023 for PI-1 and −0.016 for PI-2; the irradiation conditions were: wavelength 532 nm and dose 9J/cm^2; and the minus sign represents a decrease of $N\alpha$ after irradiation.

The dynamics of the photo-induced refractive index changes (birefringence) indicate that there is a fast component on the order of seconds for the cis⇒trans thermal back reaction, but the presence of a persistent cis concentration suggests that there are also slow components to this back isomerization. UV-vis studies further verified that for the donor-embedded polyimide systems there is a longer component of the cis⇒trans back isomerization ranging from minutes to hours which can be fitted by a tri-exponential decay (see Fig. 3 for PI-1). The first point of this figure was taken approximately two minutes after stopping the irradiation, so the fastest component of the cis⇒trans back isomerization is not shown by this figure. The slower components of the thermal back-isomerization of PI-1 and PI-2 result from the stronger coupling between the chromophore and the polymer backbone in comparison to the tethered side-chain system PI-3b where there is only a fast component present. For comparison, for PI-3b, after two minutes the reaction is completed since the value of the mean absorbance $(Abs_{//} + 2Abs_\perp)/3$ prior to irradiation is unchanged 2 minutes and 1 month after stopping the irradiation. In PI-1 and PI-2, thermal isomerization of the azo chromophore requires some correlated motion of a substantial portion of the polymer backbone. This decrease in the rate of the thermal back isomerisation is caused by the additional activation energy required for the backbone rearrangement. This clearly shows to what extent embedding the azo chromophore through the donor substituents into a fairly rigid polymer backbone affects the movement of this chromophore. The observed higher thermal stability of dc field-induced polar order of

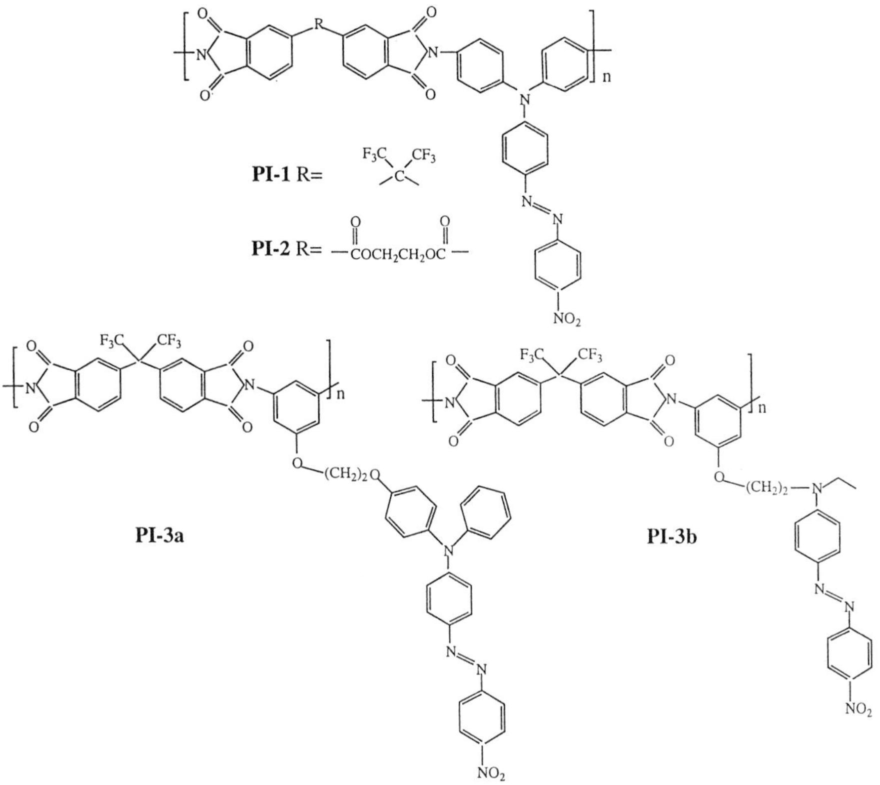

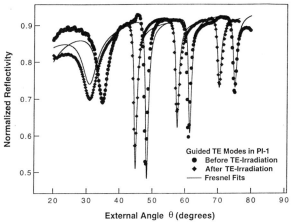

Figure 1. Chemical structure of the azo-polyimide polymers.

Figure 2. Transverse electric (TE) waveguide modes coupled into PI-1 prior to and after TE irradiation with green laser light (530 nm, 9 J/cm²). Note that the modes have shifted their ATR-angular positions to lower incidence angles indicating a decrease of the TE refractive index of the polymer.

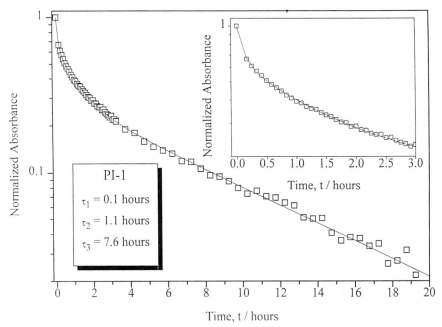

Figure 3. A logarithmic plot of the long time cis ⇒ trans thermal back reaction of PI-1 and with an inset that shows an expanded view of the first points of the figure. The mean absorbance of this figure is normalized and is defined as *(Abs − Abs∞)/(Abs⁰ − Abs∞)*, where *Abs* is the absorbance of the sample and *Abs⁰* and *Abs∞* are the absorbances at the time t = 0 and at the steady state of the back-reaction. Squares indicate the experimental points, and the solid line indicates a tri-exponential theoretical fit with the time constants given on the figure. The time t = 0 corresponds to the state of the system approximately 2 minutes after turning off the irradiation, so the fastest component of the back-reaction is not shown by this figure. (Adapted from Ref. 43.).

the donor-embedded systems *(50)* relative to the flexibly tethered true side chain systems is consistent with the hindered movement of the azo-chromophore mirrored by the "slowed" thermal back isomerization in the donor-embedded systems. Isomerization rates provide a useful information on sub-Tg polymer dynamics in NLO photochromic polymers.

Light-Induced Nonpolar Orientation

When azo chromophores are irradiated with polarized light, they experience successive cycles of trans⇔cis isomerization, and eventually align perpendicular to the irradiating light polarization. Both experimental and theoretical studies of the reorientation mechanism of azo dyes can be found in Refs. 2-11, 45, 46, and 53. Here we focus on the effect of the polymer structure on the photo-induced nonpolar orientation of azo chromophores.

Using the ATR waveguide spectroscopy birefringence measurements discussed previously, we confirmed, for all the azo-polyimide systems studied, that the index of refraction decreases in the direction parallel to the irradiating light polarization and increases in the perpendicular (see Table I) which indicates that the azo molecules become preferentially oriented perpendicular to the polarization direction. In the UV-vis dichroism experiments, prior to irradiation, identical absorbances were recorded for light linearly polarized parallel and perpendicular to the irradiating light polarization, indicating that the sample was optically isotropic in the substrate plane. The dichroic absorbance of all azo-polyimides after polarized irradiation is summarized in Table II. For PI-1, Fig. 4 shows a polar plot of the absorbance of the linearly polarized probe light (at 488 nm) as a function of the angle, Ψ, between the polarization of the probe and the irradiating (530 nm, 9 J/cm^2) light. Nonpolar orientation is clearly generated in PI-1 at room temperature, and similar results were obtained for PI-2. The largest absorption is observed for the irradiated samples when the polarization directions of the probe and irradiation beams are perpendicular, confirming that the azo molecules become preferentially distributed perpendicular to the polarization direction of the irradiating green light.

Table I. Steady-State Birefringence of Irradiated[a] Azo-Polyimides

Azo-Polyimide Polymer	PI-1	PI-2	PI-3b
1. Thickness (μm)	1.99	0.60	0.90
2. Irradiation Dose (J/cm^2)	23.4	12.6	1.35
3. $\Delta n_{//}(\Delta n_y)$	− 3.2	− 3.3	− 4.3
4. $\Delta n_\perp(\Delta n_x)$	+ 0.3	+ 0.9	+ 1.25
5. $\Delta n_\perp(\Delta n_z)$	+ 0.7	+ 0.8	+ 1.25

[a]Irradiation light: Wavelength, 532 nm; Polarization, Linear and parrallel to the y-axis (TE polarized) and perpendicular to the x and z axes. Δn_i represents the variation of the refractive index n_i upon isomerization; and the plus and minus signs respectively represent an increase and a decrease of n_i after irradiation. The films were irradiated until no changes were observed. The data were taken 20 to 35 minutes after the end of the irradiation.

After completion of the cis⇒trans thermal reaction cycle, the quasi-permanent orientation of the trans chromophores in PI-1 induced by irradiation at room temperature, far below the polymer Tg (350 °C), is shown in Fig. 5. This figure shows both the mean absorbance and the dichroic ratio, e.g. $(Abs_{//} - Abs_{\perp})/(Abs_{//} + 2Abs_{\perp})$, of PI-1 prior to and after irradiation. The first point in this figure represents the state of the film before irradiation, the second point was recorded 2 minutes after stopping the irradiation, and the other points were recorded sequentially after the end of the irradiation. From the behavior of the mean absorbance as shown in Fig. 5, it is clear that all of the azo chromophores have reverted to the trans form upon completion of the thermal back reaction (~ 25 hours), while the persistence of the dichroic ratio demonstrates production of a stable orientation upon irradiation at room temperature. The initial increase in the dichroic ratio over the first 25 hours subsequent to irradiation is due to the thermal cis⇒trans isomerization of the azo chromophores. Similar results were obtained for PI-2.

As it has been observed previously for the isomerization reaction itself, the flexible tether which connects the chromophore to the polyimide backbone plays a major role in the light-induced nonpolar orientation process. In contrast to PI-1 and PI-2 where the photo-induced orientation is quite stable, photo-oriented films of PI-3b do show significant relaxation of the light-induced orientation measured by changes in the dichroic ratio in the same manner to PI-1 and PI-2, a feature which is consistent with a weaker coupling of the chromophore movement to the backbone (vide infra). Further evidence of the effect of the flexible tether on photo-induced orientation processes is given by studies of the dynamics of photo-induced birefringence and dichroism in these high Tg polyimides. It was shown that while PI-1 and PI-2 show similar dynamical behaviors of photo-induced anisotropy changes, the flexibly tethered PI-3a and PI-3b polyimides, and DR1-PMMA films do also show a similar dynamical behavior of photo-induced refractive index and dichroism changes (4, 42), and these changes are in contrast to those observed for PI-1 and PI-2. These studies of the dynamics of the photo-induced birefringence changes also indicate that the azo chromophores in the studied true side chain systems can undergo an easy and efficient reorientation under polarized irradiation primarily as a result of the flexible tether which attaches the chromophore to the polymer backbone. These findings are further reinforced by the steady-state values of photo-induced anisotropic absorbance and birefringence in PI-1, PI-2 and PI-3b (see Tables I and II) which show that (i) PI-1 and PI-2 behave similarly as can be clearly seen from Table 1, and (ii) the light-induced alterations in the optical properties observed for PI-1 and PI-2 are smaller than those observed for PI-3b.

Table II. Dichroic Absorbance Properties of Irradiated[a] Azo-Polyimides

Azo-Polyimide Polymer	PI-1	PI-2	PI-3b
1. Initial Optical Density	0.69	1.26	0.28
2. $Abs_{\perp}/Abs_{//}$	1.224	1.333	1.45
3. $(Abs_{//} - Abs_{\perp})/(Abs_{//} + 2Abs_{\perp})$	− 0.065	− 0.091	− 0.115

[a]Irradiation light: wavelength, 532 nm; Irradiation Dose, 9 J/cm^2.

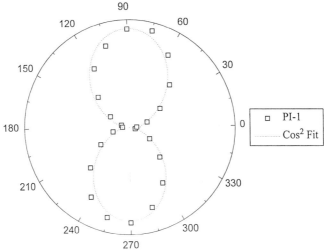

Figure 4. Polar plot depicting the absorbance of PI-1 versus the angle between irradiation (530 nm, 9 J/cm^2) and probe beam polarizations. The markers are experimental data points and the solid curves are second order Legendre polynomial theoretical fits indicating a nonpolar orientation of the chromophores. The optical densities corresponding to the highest and lowest absorbances were approximately 0.65 and 0.53 respectively. At the probe wavelength 488 nm, both cis and trans isomers absorb, and the polar plot which was taken shortly after irradiation (starting at 2 minutes after irradiation) shows both the cis and trans form.

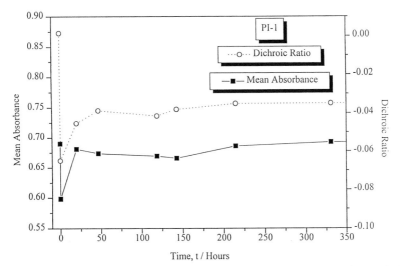

Figure 5. Mean absorbance and dichroic ratio of PI-1 versus time after end of irradiation. (Adapted from Ref. 43.).

The light-induced orientation in PI-1 and PI-2 is not erased even after heating at 170 °C for one hour. This orientation can be completely randomized by heating the samples above their respective Tgs for 10 minutes. The sign of the photo-induced dichroism in PI-1 and PI-2 is inverted from negative to positive when the irradiation light polarization is rotated through 90° and the horizontal and vertical absorbances were exactly interchanged by this procedure. This inversion in the sign of the dichroism shows that photoisomerization can easily reorient the chromophores at a temperature far from Tg (325°C and 225°C below the Tg of PI-1 and PI-2, respectively), while heating to temperatures near Tg is necessary to efficiently thermally induce such significant polymer molecular movement. This strongly suggests that the photoisomerization process, at least to some extent, is capable of moving the polyimide backbone via the coupling to the photo-induced isomerization and orientation of the azo chromophores.

Additional proof that photoisomerization is capable of inducing significant molecular movement in high Tg rigid polyimides even at room temperature, is provided by the rapid photo-induced depoling of the donor-embedded systems which had been oriented by classical corona poling at a temperature near Tg. While the thermal stability of the polar order in such systems was found to range from decades to centuries at room temperature, we find that photoisomerization can significantly depole such oriented samples only in few minutes. This light-driven relaxation phenomenon is a polarization-independent photo-induced depoling process, resulting from photoisomerization of the azo chromophores (42).

Light-Assisted, Electric Field Induced Polar Orientation

In the previous two sections, we have demonstrated that the thermal azo isomerization reaction itself and light-induced nonpolar orientation depend on the structure of the polymer. Next we will show that sub-Tg photo-assisted electric field induced polar orientation processes also strongly depend on the molecular structure of the polymer.

Photo-assisted poling (PAP) occurs when azo chromophores are photoisomerized in the presence of a poling dc electric field. Light enhances the mobility of the chromophores, and poling occurs at temperatures well below the polymer Tg (7-12,54). The initial report of PAP (7,8) has stimulated additional studies in different NLO molecular systems, e.g. Langmuir-Blodgett-Kuhn layers (55), liquid crystalline polymers (56), and a variety of amorphous polymers (9,42,47,57-62). In amorphous polymers, PAP processes have been studied primarily in relatively low Tg (around 130 °C) PMMA derivatives containing NLO azo chromophores. The first example of PAP at room temperature of a high Tg rigid polyimide polymer was reported by our group for PI-3b (42), and subsequent studies on other polyimides containing azo dyes followed (47). To study the effect of the molecular structure on the PAP process, we have extended our PAP studies to the two donor-embedded polyimides PI-1 and PI-2, and the flexibly tethered side-chain polyimide PI-3a. PI-3a represents a true side-chain polymer with a high volume chromophore. The bulky chromophore in PI-3a results in an increase in Tg of ~ 20 °C relative to that in PI-3b. In-situ PAP experiments were performed by exposing the film samples to a (5.5 kV) corona discharge (63) while irradiating with a green (544 nm, 8 mW/cm^2) laser light. The thicknesses of the films of

PI-1, PI-2, and PI-3a were approximately 0.72 μm, 0.14, μm, and 0.88 μm. The respective optical densities were approximately 0.79, 0.3, and 1.34 at 544 nm. Fig. 6 shows the evolution of the second harmonic (SH) intensity for PI-1 (Fig. 6a), and PI-3a (Fig. 6b) during a typical PAP cycle at room temperature. These measurements are for a TM-polarized (transverse magnetic-polarized) probe with TM-polarized irradiation, but similar dynamic behaviors were observed for TE-polarized irradiation. The initial SH signal present in the corona field prior to irradiation is due to the EFISH signal *(-2ω, ω, ω, 0)* which is proportional to the third order susceptibility of the polymers. This signal is observed immediately when the corona voltage is applied to the sample and remains at a constant level prior to irradiation, which shows that no poling occurs in PI-1, PI-2 and PI-3a simply by the application of a corona voltage alone at room temperature. Furthermore, no residual SH signal can be observed for these polymers within few minutes of turning off the corona voltage. At the moment when the sample is irradiated, the EFISH signal shows a sudden decrease. This decrease of the EFISH signal is not due to a decrease of the electric field across the sample, since we could not detect any photoconductivity and the same voltage was maintained on the sample during all the phases of the experiment *(vide infra)*. The origin of this drop in the SH signal is due to light-induced change in the third order nonlinearity of the polymers.

PI-1 and PI-2 behave similarly in the PAP experiments. For each, the SH signal remains stable during irradiation for the time of irradiation (up to one hour and 20 minutes in some cases) at temperatures ranging from room temperature to 150 °C. On the other hand, for PI-3a there is a noticeable growth in the SH signal during irradiation, both at room temperature and at higher temperatures (see Fig. 6b for the data at 25 °C). These experiments show that, in contrast to PI-3a, PI-1 and PI-2 do not undergo PAP during irradiation. It is noteworthy that in spite of the relatively small difference in Tg between PI-2 and PI-3a (252 versus 228 °C), PAP of PI-2 even at 150 °C does not produce any polar order. This is a clear example to what extent polymer molecular structure influences sub-Tg polymer molecular movement. This is in contrast to the predictions of the Williams-Landel-Ferry (WLF) theory which models sub-Tg behavior primarily by the difference *Tg – T (49)*. After the cis⇒trans thermal back reaction for PI-1 and PI-2, the initial level of EFISH signal (see Fig. 6a for PI-1) is restored, while PI-3a achieves appreciable polar orientation with the SH signal exceeding that of the initial EFISH signal observed prior to irradiation (Fig. 6b). PI-3b also undergoes appreciable polar orientation by PAP at room temperature *(42)*. This further demonstrates that embedding the azo chromophore into a fairly rigid polymer backbone, substantially affects the movement of the chromophore. This is further confirmation of the findings of the photo-induced nonpolar orientation experiments, the azo chromophore is more strongly coupled to the polymer backbone in PI-1 and PI-2 than in PI-3a, and reorientation of the azo chromophore in the embedded system requires a correlated motion of a portion of the polymer backbone.

We found that the sub-Tg PAP efficiency of PI-3a was improved at elevated temperatures up to 100 °C below Tg. This effect is due to the thermal enhancement of photo-induced mobility of the chromophore. With the operating temperature ranging from room temperature to 100 °C below Tg, we also found that the difference between the polar order obtained by PAP and pure corona poling without irradiation increases

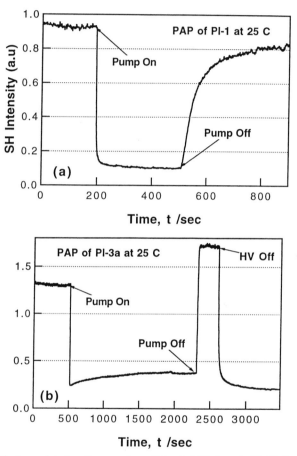

Figure 6. Photo-assisted poling cycles of the **(a)** PI-1 and **(b)** PI-3a polymers when both corona voltage (5.5 kV), and the green irradiating light (544 nm, 8 mW/cm^2) were applied to the samples. The operating temperature corresponds to room temperature and is indicated in the figures. HV Off refers to the moment where the corona voltage was turned off. (Fig. 6 (a) is adapted from Ref. 66.).

with temperature. This confirms that the polar order achieved in this particular polymer PI-3a is mainly due to β relaxation processes (side-chain movement), and that PAP is initiated by first moving the azo side chains. A sharp increase in the SH signal was observed near Tg confirming our conclusions about side-chain movement, and suggesting that efficient sample poling either by PAP or thermal poling, clearly is facilitated by significant movement of the polymer main chain (α relaxation process).

The molecular structure of the polymer determines the extent that main chain movement is coupled to photo-induced side chain movement. With DR1-type chromophores flexibly tethered in true side chain nonlinear optical polymers such as PMMA or polyimides with Tgs in the 100-265 °C range, PAP induces polar orders which are comparable to those produced in thermally poled films *(8,42,47,57)*. This is not the case for the PI-3a polymer studied here. The molecular size of the diarylene azo chromophore of PI-3a is substantially larger than that of the DR1-type molecules in the polymers studied previously (e.g. see PI-3b); a feature which requires more free volume for chromophore movement thereby decreasing mobility. The size of the azo chromophore adds an additional complexity to the polymer structural requirements for efficient orientation of the chromophores by PAP. For the DR1-co-polyimide studied in Ref. 47, the efficiency of the polar order obtained by PAP is comparable to that obtained by us for PI-3b, and is apparently due to the large free volume available to the chromophore, since the structure of this NLO DR1-co-polyimide implies a chromophore for every other repeating unit.

The fact that PAP does not induce any polar order in donor-embedded polyimides PI-1 and PI-2 even at elevated temperatures is somewhat surprising considering how easily both light-induced nonpolar orientation and photo-induced depoling occur in these polymers. This suggests that the processes of light-induced nonpolar orientation and photo-assisted poling are seemingly different; each requiring some correlated motion of main chain and side chain but to a different extent.

Optical Control of the Third Order Molecular Polarizability

Recent studies performed by absorption saturation and the z-scan technique or degenerate four waves mixing, indicated that photoisomerization of azo chromophores in polymer hosts can contribute to the observed third order nonlinearities of such polymer films through the nonlinear refractive index *(64,65)*. Here we review our experiments that show that photoisomerization of NLO azo chromophores actually manipulates the third order nonlinearity of azo-polyimide polymers *(66)*. The observation is made directly through resonant EFISH which results from electronic nonlinearity contributions dominated by the azo chromophore at second harmonic wavelength within the UV-vis spectral region *(67)*.

If polar order is present, the EFISH signal of an NLO azo-polymer has contributions from the second- and third- order nonlinearities. For centrosymmetric systems only third order nonlinearities are present. In the donor-embedded systems, neither electric field poling nor photo-assisted poling occurs at room temperature. This permits us to study in isolation the effect of photoisomerization on the third order nonlinearities. We will demonstrate that the third order molecular polarizability (γ) of the azo chromophore decreases significantly with the molecular shape change from the trans to the cis form. EFISH was used as an *in-situ* probe to monitor the photochemical induced change in the third order susceptibility of the polyimide films. The effect of irradiation on the EFISH signal observed in PI-1 at room temperature is shown in Fig. 6a, and similar results were obtained for PI-2. Again in PI-1 and PI-2, noncentrosymmetry does not contribute to the observed EFISH signal since application of a strong corona field to either polymer even at 150 °C does not result in any polar order. The photo-induced change of the molecular geometry of the NLO-azo chromophore in going from the trans to the cis form results in a drastic change in the third order susceptibility of the PI-1 and PI-2 films.

The drop of the nonlinearity at the moment when the sample is irradiated by the pump light is not due to photo-induced change in electric field. We do not observe a change in either the corona voltage or the current during irradiation. The observed decrease in EFISH upon pump irradiation is too large to be due to a slight change in voltage or current which would be undetectable in our experiments. The EFISH signal can be nearly completely erased under high pump irradiation intensity, an effect which would require the voltage to drop to zero in presence of photoconductivity. The fast response rules out charge injection and charge migration. Such processes are much slower than the fast decrease in nonlinearity that we observed. In photorefrative polymers which must contain photoconductive units such as carbazole, photoconductivity occurs in the minutes time scale *(68)*. Further evidence of how fast these changes in nonlinearity can be induced is given in Ref. 9 where it is shown that, in marked contrast to photoconductive materials, the observed photo-induced decrease in nonlinearity can be induced by nanoseconds pulses, inducing photoisomerization that is known to occur within the picosecond time scale *(69)*. It was also shown that right after the nanoseconds pulse excitation, the second harmonic signal is enhanced to a level that exceeds the initial EFISH signal, and this cannot be explained by a decrease in the electric field across the polymer. In addition, while all the NLO azo-polymers studied so far show a fast photo-induced decrease in nonlinearity by EFISH, these polymers do not show the photo-induced decrease in nonlinearity when they are studied by electro-optic measurements through electric field induced Pockels effect which is the equivalent of EFISH *(8,41,42)*. In such experiments, the NLO polymer sample is sandwiched between two electrodes, and if the polymer is photoconductive, the drop in nonlinearity should be observed also in this type of experiments when the sample is irradiated by the pump beam.

Figure 6a demonstrates that when the irradiation light is turned off, the EFISH signal increases due to the recovery in the third order macroscopic susceptibility caused by the cis⇒trans thermal back reaction. It can also be seen from Fig. 6a that the EFISH signal is not completely recovered because of the existence of slow components of the cis⇒trans spontaneous recovery *(vide infra)*. A much more clear presence of the long components of the cis⇒trans recovery during EFISH can be found in Ref. 66. This reversible erasure of the hyperpolarizability γ can be conducted many times leading to all-optical modulation of the SH light (see Fig. 7 for PI-2). The slower component of the cis⇒trans recovery is not shown in the experiments described in Fig. 7 because the initial EFISH level in this figure does not correspond to a fully relaxed sample. Biexponential fits to the observed EFISH recovery during the cis⇒trans thermal back isomerization for up to 500 sec over several recovery cycles,

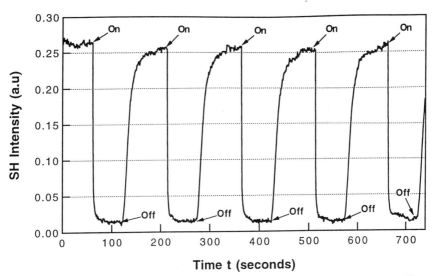

Figure 7. All-optical light modulation of the SH signal of the PI-2 polymer. The moments of turning the irradiating light on and off are indicated by arrows. (Adapted from Ref. 66.).

yielded mean values of 28 and 14 sec, respectively, for the fastest component in isomerized PI-1 and PI-2. It is noteworthy that this fastest component which has also been observed by dynamical refractive index change by the ATR technique, is not shown in Fig. 3 since the first point in this figure corresponds to the PI-1 sample 2 minutes after the end of irradiation *(vide infra)*.

The contribution of photo-induced anisotropy to the effects observed in the EFISH decrease is negligible. The same decrease is observed for a TE or TM polarized probe regardless of the pump polarization. The photochemically induced molecular shape change of the NLO dye blows out the strong optical field driven anharmonic movement of the electronic cloud of the NLO dye. The effect is reversed upon back-isomerization of the dye to the trans form. This change in nonlinearity may be rationalized by the conformation of the cis isomer of the NLO dye which is more globular and less conjugated (twisted) than the trans form. It has been shown experimentally that twisted organic compounds exhibit smaller γ *(70)*, and theoretical calculations indicate that both the ground state dipole moment and the second order molecular polarizability of DR1, a chromophore structurally related to the chromophores studied here, are appreciably greater for the trans form *(52)*. Furthermore, in the case of conjugated organic compounds, the third order polarizability γ can be expressed as the product of the linear polarizability, α, and a nonlinear term, f, corresponding to an anharmonicity factor *(71)*. So, the decrease upon photoisomerization of the absorption of the azo chromophore at the SH wavelength, ca. 526.5 nm, which is due to a smaller α for the cis form, demonstrates that γ should also decrease upon trans⇒cis isomerization as much as the anharmonicity factor f decreases with the twisted cis form of the chromophore. A further proof of the photochemical reduction of the third order molecular polarizability is provided by a theoretical model which assumes that the third order polarizability of the chromophore decreases upon trans to cis molecular shape change. The model predicts an EFISH intensity at the photostationary state which varies hyperbolically relative to the irradiating light intensity *(66)*. Indeed, hyperbolic functions were adjusted to the experimental data showing the variation of the SH intensity at the steady state of the irradiation versus the irradiation intensity for both PI-1 and PI-2. This shows that the azo chromophores in PI-1 and PI-2 behave consistently with the model, and validates the concept of the reversible rapid photochemical erase of the third order molecular polarizability of these photochromic NLO dyes.

Summary

We have shown that light-induced molecular movement in high Tg nonlinear optical polymers provides insight into sub-Tg polymer dynamics in nonlinear optical azo-containing polymers. All of the azo-polyimide systems studied isomerize upon irradiation, and the isomerization reaction itself, as well as the light-induced polar and nonpolar orientation depend on the molecular structure of the unit building blocks of the polymer. In particular, we have shown that the presence of the flexible tether which connects the chromophore to the polyimide backbone facilitates the orientation of the chromophore either by an optical or a dc electric field, and the direct embedding of the chromophore without flexible connectors or tethers into the polyimide backbone strongly hinders the chromophore's movement. More importantly, we have shown that the generally believed WLF-law describing sub-Tg molecular movement as being governed by Tg-T is not always valid, instead, the polymer molecular structure strongly influences the sub-Tg molecular movement. We also have demonstrated that the third order molecular polarizability of azo chromophores can be rapidly reduced with the attending change in the shape of the chromophores from the trans to the cis form leading to all-optical light manipulation utilizing third order nonlinear optical effects. We have also shown that the combination of photo and thermal isomerization is capable of inducing molecular orientation in high Tg azo-containing polymers, in one case as much as 325 °C below Tg of a nonlinear optical polyimide containing no flexible connector or tether. While thermal processes at temperatures near Tg (350 °C) are required to completely erase this orientation, light can however control it at room temperature. In this regard, the light-induced orientation that induces a stable anisotropy in this class of exceptionally thermally stable polyimides, could lead to applications in optical data storage.

Acknowledgments

Z. Sekkat thanks the Office of Naval Research for research support during his stay at the University of California at Davis, and Z. Sekkat and J. Wood acknowledge support from the Max-Planck Society during their stay in the MPI-Polymerforschung in Mainz. Z. Sekkat, R. D. Miller, W. Knoll, and A. Knoesen who are also with the Center on Polymer Interfaces and Macromolecular Assemblies acknowledge the support from the National Science Foundation under Award No. DMR-9400354. The authors would like to thank P. Prêtre, L. M. Wu, and D. Yankelevich for helpful discussions.

Literature Cited

1. Sekkat, Z.; Knoll, W. in Advances in Photochemistry, Wiley & Sons, D. Neckers, D. Volmann, G. von Bunau, Eds. **1997**, Vol. 22, pp.117-195.
2. Sekkat, Z.; Knoll, W. SPIE Proceeding, Andrews, M. P. Ed. **1997**, 2998, 164.
3. Sekkat, Z.; Knoesen, A.; Knoll, W; Miller, R. D. Critical Reviews SPIE Proceeding, Najafi, I.; Andrews, M. P. Eds. **1997**, Vol. CR68, pp. 374-398.
4. Sekkat, Z.; Dumont, M. SPIE Proceeding, Lessard, R. Ed. **1992**, 1774, 188.
5. Natansohn, A.; Rochon, P.: Gosselin, J.; Xie, S. Macromolecules **1992**, 25, 2268.
6. Shi, Y.; H. Steier, W.; Yu, L.; Shen, M.; R. Dalton, L. Appl. Phys. Lett. **1991**, 58, 1131.
7. Sekkat, Z.; Dumont, M. Appl. Phys. B. **1992**, 54, 486.
8. Sekkat, Z.; Dumont, M. Mol. Cryst. Liq. Cryst. Sci. Technol. - SecB: Nonlinear Optics, **1992**, 2, 359.
9. Hill, R. A.; Dreher, S.; Knoesen, A.; Yankelevich, D. Appl. Phys. Lett. **1995**, 66, 2156.
10. Charra, F.; Kajzar, F.; Nunzi, J. M.; Raimond, P.; Idiart, E. Optics Letters, **1993**, 12, 941.

11. Fiorini, C.; Charra, F.; Nunzi, J. M.; Raimon, P. J. Opt. Soc. Am. B. **1997**, 14, 1984.
12. Nishida, F.; Tomita, Y. J. Appl. Phys. **1997**, 81, 3348.
13. Sawodny, M. ; Schmidt, A.; Stamm, M.; Knoll, W.; Urban, C.; Ringsdorf, H. Polym. Adv. Technol. **1991**, 2, 127.
14. Büchel, M.; Sekkat, Z.; Paul, S. Weichart, B.; Menzel, H.; Knoll, W. Langmuir **1995**, 11, 4460.
15. Stumpe, J. Fischer, Th.; Menzel, H. Macromolecules **1996**, 29, 2831.
16. Wang., R.; Iyoda, T.; Hashimoto, K.; Fujishima, A. J. Phys. Chem. **1995**, 99, 3352
17. Schönhoff, M.; Mertesdorf, M.; lösche, M. J. Phys. Chem. **1996**, 100, 7558.
18. Seki, T.; Skuragi, M.; Kawanishi, Y.; Suzuki, Y.; Tamaki, T.; Fukuda, R.; Ichimura, K. Langmuir **1993**, 9, 211.
19. Seki, T.; Ichimura, K.; Rukuda, R.; Tanigaki, T.; Takashi, T. Macromolecules **1996**, 29, 892.
20. Sekkat, Z.; Büchel, M.; Orendi, H.; Knobloch, H.; Seki, T.; Ito, S.; Koberstein, J.; Knoll, W. Opt. Commun. **1994**, 111, 324.
21. Ichimura, K.; Hayashi, Y.; Akiyama, H.; Langmuir **1993**, 9, 3298.
22. Wolf, M. O.; Fox, M. A. Langmuir **1996**, 12, 955.
23. Sekkat, Z.; Wood, J.; Geerts, Y.; Knoll, W. Langmuir **1995**, 11, 2856.
24. Sekkat, Z.; Wood, J.; Geerts, Y.; Knoll, W. Langmuir **1996**, 12, 2976.
25. Junge, M.; McGrath, D. V. Chem. Commun. **1997**, 9, 857.
26. Song, X.; Geiger, C.; Vaday, S.; Perlstein, J.; Whitten, D. G. J. Photochem. Photobiol. A: Chem. **1996**, 102, 39.
27. Fissi, A. Pieroni, O.; Balestreri, E.; Amato, C. Macromolecules **1996**, 29, 4680.
28. Berg, R. H.; Hvilsed, S.; Ramanujamm, P. S. Nature **1996**, 383, 505.
29. Hoffmann, K.; Marlow, F.; Caro, J. Adv. Mater. **1997**, 9, 567.
30. Böhm, N.; Materny, A.; Kiefer, W.; Steins, H.; Müller, M. M.; Schottner, G. Macromolecules **1996**, 29, 2599.
31. Eich, M. Wendorff, J. H. J. Opt. Soc. Am. B. **1990**, 7, 1428.
32. Anderle, K.; Bach, H.; Fuhrmann, Th.; Wendorff, J. H. Macromol. Symp. **1996**, 101, 549.
33. Anderle, K.; Birenheide, R.; Werner, M. J. A.; Wendorff, J. H. Liq. Cryst. **1991**, 9, 691.
34. Wiesner, U.; Reynolds, N.; Boeffel, C.; Spiess, H. W. Makromol. Chem. Rapid Commun. **1991**, 12, 457.
35. Wiesner, U.; Reynolds, N.; Boeffel, C.; Spiess, H. W. Liq. Cryst. **1992**, 11, 251.
36. Fisher, Th.; Lasker, L.; Stumpe, J.; Kostromin, S. G. J. Photochem. Photobiol. A: Chem. **1994**, 80, 453.
37. Hvilsed, S.; Andruzzi, F.; Ramanujam, R.; Opt. Lett. **1992**, 17, 1234.
38. Lagugné Labarthet, F.; Sourisseau, C. J. Ram. Spectro. **1996**, 27, 491
39. Buffeteau, T.; Pézolet, M. Appl. Spectro. **1996**, 50, 948.
40. Buffeteau, T.; Natansohn, A.; Rochon, P.; Pézolet, M. Macromolecules **1996**, 29, 8783.

41. Sekkat, Z.; Kang, C.-S.; Aust, E. F.; Wegner, G.; Knoll, W. Chem. Mater. **1995**, 7, 142.

42. Sekkat, Z.; Wood, J.; Aust, E. F.; Knoll, W.; Volksen, W.; Miller, R. D. J. Opt. Soc. Am. B. **1996**, 13, 1713.

43. Sekkat, Z.; Wood, J.; Knoll, W.; Volksen, W.; Miller, R. D.; Knoesen, A. J. Opt. Soc. Am. B. **1997**, 14, 829.

44. Sekkat, Z.; Wood, J.; Knoll, W.; Volksen, W.; Lee, V. Y; Miller, R. D.; Knoesen, A. Polymer Preprints, Am. Chem. Soc. Div. Polym. Chem **1997**, 38, 977.

45. Neoport, B. S.; Stolbova, O. V. Opt. Spectrosc. **1961**, 10, 146.

46. Todorov, T.; Nicolova, L.; Tomova, T. Appl. Opt. **1984**, 23, 4309.

47. Chauvin, J.; Nakatani, K.; Delaire, J. A. SPIE Proceeding, Andrews, M. P. Ed. **1997**, 2998, 205.

48. Lee, T. S.; Kim, D.-Y.; Jiang, X. L.; Li, L.; Kumar, J.; Tripathy, S. Macromol. Chem. Phys. **1997**, 198, 2279.

49. Williams, M. L.; Landel, R. F.; Ferry, J. D. J. Am. Chem. Soc. **1955**, 77, 3701.

50. Verbiest, T.; Burland, D. M.; Jurich, M. C.; Lee, V. Y.; Miller, R. D.; Volksen, W. Science **1995**, 268, 1604.

51. Miller, R. D.; Burland, D. M.; Jurich, M. C.; Lee, V. Y.; Moylan, C. R.; Twieg, R. J.; Thackara, J.; Verbiest, T.; Volksen, W. Macromolecules **1995**, 28, 4974.

52. Loucif-Saibi, R.; Nakatani, K.; Delaire, J. A.; Dumont, M.; Sekkat, Z. Chem. Mater. **1993**, 5, 229.

53. Sekkat, Z.; Wood, J.; Knoll, W. J. Phys. Chem. **1995**, 99, 17226.

54. Sekkat, Z.; Knoll, W. J. Opt. Soc. Am. B. **1995**, 12, 1855.

55. Palto, S. P.; Blinov, L. M.; Yudin, S. G.; Grewer, G.; Schönhoff, M.; Lösche, M. Chem. Phys. Lett. **1993**, 202, 308.

56. Anneser, H.; Feiner, F.; Petri, A.; Bräuchel, C.; Leigeber, H.; Weitzel, H. P.; Kreuzer, F. H.; Haak, O.; Boldt, P. Adv. Mater. **1993**, 5, 556.

57. Blanchard, P. M.; Mitchell, G. R. Appl. Phys. Lett. **1993**, 63, 2038.

58. Bauer-Gogonea., S.; Bauer, S.; Wirges, W.; Gerhard-Multhaupt, R. J. Appl. Phys. **1994**, 76, 2627.

59. Haase, W.; Grossmann, S.; Saal, S.; Weyrauch, T.; Blinov, L. M.; OSA Technical Digest Series, **1995**, 21, PD5-1.

60. Dalton, L. R.; Harper, A. W.; Zhu, J.; Steier, R.; Salovey, R.; Wu, J.; Efron, U. SPIE Proceeding, Yang, S. C.; Chandrasekhar, P. Eds. **1995**, 2528, 106.

61. Jiang, X. L.; Li, L.; Kumar, J.; Tripathy, S. K. Appl. Phys. Lett. 1996, 69, 3689.

62. Yilmaz, S.; Wirges, W.; Bauer-Gogonea, S.; Bauer, S.; Gerhard-Multhaupt, R.; Michelotti, F.; Toussaere, E.; Levenson, R.; Liang, J.; Zyss, J. Appl. Phys. Lett. **1997**, 70, 568.

63. Mortazavi, M. A.; Knoesen, A.; Kowel, S. T.; Higgins, B.; Dienes, A. *J. Opt. Soc. Am.* **1989**, *B6*, 733.

64. Egami, C.; Suzuki, Y., Sugihara, O.; Okamoto, N.; Fujirama, H.; Nakagawa, K.; Fujiwara, H. Appl. Phys. B. **1997**, 64, 471.

65. Dong, F.; Koudoumas, E.; Courtis, S.; Shen, Y.; Qiu, L.; Fu, X. J. Appl. Phys. **1997**, 81, 7073.

66. Sekkat, Z.; Knoesen, A.; Lee, V. Y.; Miller, R. D.; J. Phys. Chem. B. **1997**, 101, 4733.
67. Levine, B. F.; Beteha, C. G.; J. Chem. Phys. **1975**, 63, 2666.
68. Moerner, W. E.; Grunnet-Jepsen, A.; Thompson, C. L. Annual Review of Materials Science **1997**, 27, 585.
69. Lednev, I. K.; Ye, T. Q.; Hester, R. E.; Moore, J. J. Phys. Chem. **1996**, 100, 13338.
70. Heflin, J. R.; Cai., Y. M.; Garito, A. F. J. Opt. Soc. Am. B. **1991**, 8, 2132.
71. Brédas, J. L.; Adant, C.; Tackx, P.; Persoons, A. Chem. Rev. **1994**, 94, 243.

Chapter 23

The Influence of the Molecular Structure on the Second-Order Nonlinear Optical Properties of Pyroelectric Liquid Crystalline Polymers

M. Trollsås[1,3], F. Sahlén[1,4], P. Busson[1], J. Örtegren[1], U. W. Gedde[1], A. Hult[1,5], M. Lindgren[2], D. Hermann[3], P. Rudquist[3], L. Komitov[3], B. Stebler[3], and S. T. Lagerwall[3]

[1]Department of Polymer Technology, Royal Institute of Technology, S-100 44 Stockholm, Sweden
[2]National Defence Research Establishment, S-581 83 Linköping, Sweden
[3]Physics Department, Chalmers University of Technology, S-412 96 Göteborg, Sweden

Pyroelectric Liquid Crystal Polymers (PLCP), a novel class of material with intrinsic polar order developed in our laboratory, are discussed. Thin films (2-4 µm) of PLCP were prepared by photopolymerization of ferroelectric liquid crystalline monomers in the chiral smectic C (SmC*) phase. The poly(acrylate) materials have been structurally modified in order to improve their nonlinear optical and thermal properties. Further are their longterm nonlinear optical properties discussed based on results from second harmonic generation (SHG), spontaneous polarization (P_S), and dielectric spectroscopy.

Ferroelectric liquid crystals (FLC) is a group of organic materials that possesses spontaneous polar order (*1*). This unique property has increased the interest to use liquid crystals in the field of second-order non-linear optics (NLO). The intrinsic thermodynamically stable polar order separate this group of materials from other polar organic materials which usually are poled by external electrical fields which is subsequently locked in by either lowering the temperature below the glass transition temperature of the material or by a chemical cross-linking reaction (*2*).

[3]Current address: IBM Research Division, Almaden Research Center, 650 Harry Road, San Jose, CA 95120–6033.
[4]Current address: Institut für Polymere, ETH-Zentrum CNBE 94, Universitätsstrasse 6, CH-8092 Zürich, Switzerland.
[5]Corresponding authors.

FLCs aimed for second-order nonlinear optics were first synthesized by Walba et al (3) who synthesized a low molar mass FLC **I** (Figure 1) which generated a second harmonic signal (d_{22}) of 0.6 ± 0.3 pm/V in the chiral smectic C (SmC*) phase. The NLO-units (NLO-chromophore) which generate the SHG-signal, and are aligned perpendicular to the long axis of the molecules in the direction of the polarization, were later successfully modified by Schmitt et al (4) in an attempt to increase the second harmonic signal, **II**.

The thermal stability of the polar order in these materials is limited, since the polar order only is present in the chiral smectic C phase. In addition to this, the mechanical stability of these low molar mass compounds is a limiting factor. Subsequently to Walba, Zentel et al (5) synthesized a ferroelectric liquid crystalline side-chain polymer and recently Keller et al (6) presented the first ferroelectric liquid crystalline main-chain polymer.

The thermal and mechanical stability and the long-term properties of NLO-materials are known to be improved by cross-linking (2). A pyroelectric liquid crystalline polymer (PLCP) was therefore synthesized. This was realized by the synthesis of a crosslinkable monomer mixture that possessed chiral smectic C (SmC*) mesomorphism over a wide temperature range. This ferroelectric mixture was polar aligned and subsequently photo-cross-linked into the pyroelectric liquid crystal polymer **III**, (Figure 2) (7). The cross-linked material, which did not exhibit a ferroelectric behavior displayed a clear Pockels effect and a small but clear second harmonic signal (d-coefficient = 0.02 pm/V) (7). To increase the NLO-activity in this new material the molecular structure had to be modified. The possibilities for modification of the monomer system are, however, limited due to the demands of ferroelectricity and cross-linking. However, new monomer systems have recently been developed (8). The spontaneous polarization of these new materials have been measured and the results have been compared with the results from second harmonic generation. This has generated an understanding of the influence of the molecular structure on the properties of these complex materials. The cross-linked materials have also been investigated by dielectric spectroscopy (9) and the long-term properties of the polar order is under investigation by second harmonic generation (SHG) (10). In this paper these results are combined and discussed in order to generate an understanding of how the molecular structure ought to be altered in order to improve the over all properties of these materials.

Experimental

The synthesis of 4''-[(11-acryloyloxy) undecyloxy]-4'-biphenyl 4-[(R)-(+)-2-octyloxy]-3-nitrobenzoate **A1**, 4-(11-acryloyloxyundecyloxy) phenyl 4-(4-(11-acryloyloxyundecyloxy) phenyl) benzoate **A2**, 4''-{(R)-(-)-2-octyloxy}-3''-nitro phenyl 4-{4'-[11-acryloyloxyundecyloxy] phenyl} benzoate **A1b**, 4''-{11-acryloyloxyundecyloxy}-3''-nitro phenyl 4-{4'-[11-acryloyloxyundecyloxy] phenyl} benzoate **A2b**, and 4''-{(R)-(-)-2-[(10-Acryloyloxy)decyl]oxy}-3-nitro phenyl 4-{4'-[(11-acryloyloxy)-undecyloxy] phenyl} benzoate **A2c** have been described elsewhere (7-

*The SH-signal for a 2μm sample of a 35/65 mol-% A1/A2-mixture was reported to be about 1000 times smaller than the signal from a quartz reference (7). In order to compare the signals, that of the sample cell should be extrapolated to the signal that would be obtained from a sample where the interaction length is half the coherence length of the interaction. The d-coefficient based on quartz should therefore be $d_{A1/A2}<0.017$ pm/V(10). The reported higher of about 0.4 pm/V in reference 7 was due to an overestimated coherence length.

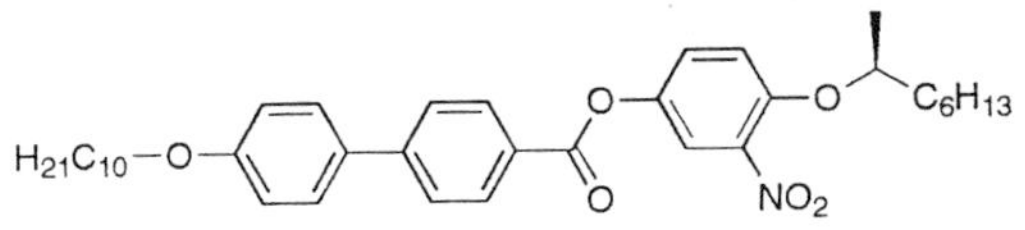

Figure 1 Structures of low-molar mass liquid crystals specially devoted to nonlinear optics which can possess macroscopic polar order in the chiral smectic C (SmC*) phase.

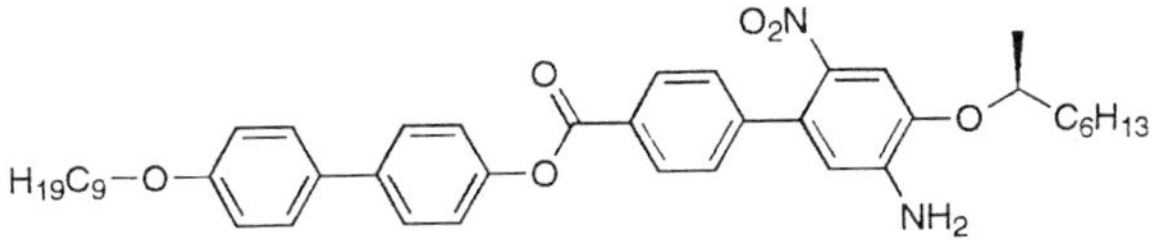

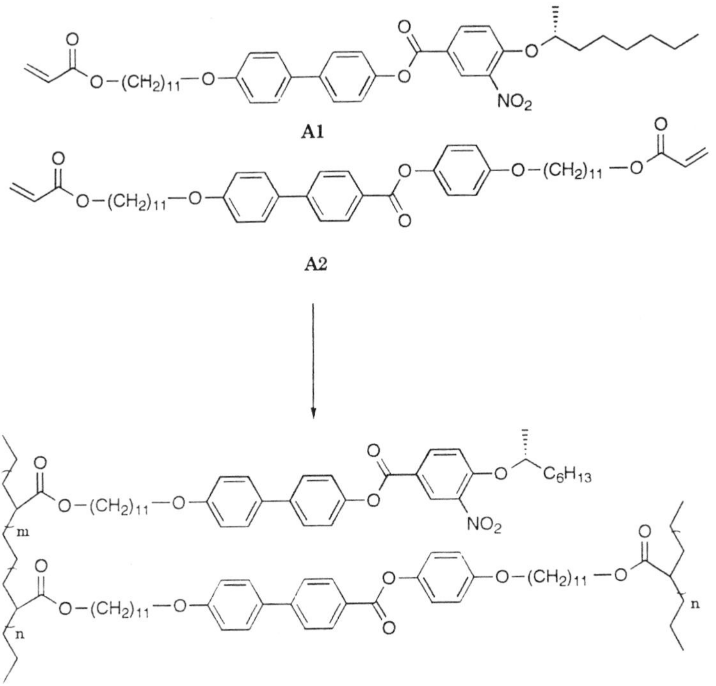

Figure 2 Structures of acrylate monomers (**A1** and **A2**) in the crosslinkable ferroelectric monomer mixture and structure of the crosslinked acrylate polymer.

8, 11) and the synthesis of 4″-{[(R)-(-)-1-methyl-8-acryloyloxynonyloxy] carbonyl} phenyl 4-{4′-[(11-acryloyloxyundecyloxy) phenyl] benzoate **A2e** and 4″-{[(R)-(-)-1-methylheptyloxy] carbonyl}-3″-nitro-6″-dimethylamino phenyl 4-{4′-[11-acryloyloxyundecyloxy] phenyl} benzoate **A1e** will be published elsewhere (*12*).

Polar Orientation in the Surface Stabilized Ferroelectric Liquid Crystal (SSFLC) Cell. Cells of a conventional sandwich type, consisting of two parallel glass substrates kept 2 μm apart by evaporated SiO_x spacers, were used for the ferroelectric, poling, polymerization and non-linear optical experiments. The substrates were prepared from ITO-coated glass sheets (Balzers Baltracon) on which an electrode pattern was formed. An insulating layer of SiO_x about 1000 Å thick was deposited onto the electrodes. The uniform bookshelf alignment of the liquid crystal material in the cell (smectic layers being essentially perpendicular to the plates) was achieved by using a thin unidirectionally rubbed polyimide aligning layer deposited on top of the insulating layer.

The liquid crystalline substance was introduced into the cell in the SmA* or in the isotropic phase by capillary forces. The cell was inserted into a Mettler FP 52 hot stage with the temperature controlled to an accuracy of ± 0.1 °C and the liquid crystalline substances were examined in a polarizing microscope (crossed polarizers). At the temperatures at which the ferroelectric electro-optic response indicated that the SmC* phase was fully developed, a dc electric field (E ≈100 V/mm) was applied in order to orient the spontaneous polarization in the whole cell in one direction. This gave a unique direction to the optical axis, tilted with respect to the smectic layer normal and thus resulting in the formation of a ferroelectric mono-domain between the electrodes. The uniformity of the orientation was examined in the polarizing microscope and the applied dc field was sufficient to cause full extinction of the transmitted light when the optical axis of the cell was either parallel to or perpendicular to the transmission direction of the polarizer.

Thermal Stabilization by in-situ Photo-Polymerization. After obtaining a ferroelectric mono-domain structure for the monomer/photo initiator mixture at the temperature at which the ferroelectric electro-optic response in the SmC* phase was fully developed, the cell was irradiated with UV-light, while keeping the dc-field on, for about ten minutes, during which the liquid crystalline mixture polymerized. The ratio of photo-initiator to monomer was 1:200 (0.5 mol%). The low concentration of photo-initiator did not destabilize the SmC* phase. In order to avoid undesired photopolymerization of the liquid crystalline material in the cell during the preparation and investigative procedures, the work was performed under yellow light.

Measurement of the SHG. Measurements of the second-harmonic generation were carried out on two different set-ups with pulsed Nd:YAG lasers. The first set-up used a Spectron laser operating at 1064 nm wavelength with a pulse width of 10 ns at a repetition rate of 10 Hz. The other set-up used a Continuum Sunlite OPO laser system with a pulse width of 7-8 ns at a repetition rate of 10 Hz, which in the present experiment was set to operate at 1100 nm wavelength. In both cases one or several visible-cut filters were placed in front of the sample allowing only fundamental light to reach the sample, and an IR-cut filter was placed behind the sample eliminating any remaining fundamental light after the sample. In the Sunlite OPO set-up, one or several narrow band pass filters

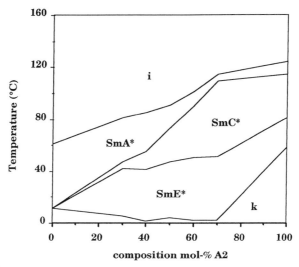

Figure 3   Phase diagram obtained on first cooling for mixtures of **A1** and **A2**.

were also placed in front of the detector (Dl=5-10 nm) to assure that only the second-harmonic light was detected. The signal from the photo multiplier could thus be attributed to the visible second-harmonic light generated in the sample. The intensity of the fundamental light was simultaneously measured by a second photo multiplier or photo diode.

Dielectric Measurements: The dielectric measurements were made on 50 μm or 10 μm thick samples with the lateral dimensions 10 mm x 10 mm held between two glass plates coated with conductive indium-tin-oxide and rubbed poly(imide) films of submicron thickness placed between the conductive indium-tin-oxide layers and the polymer. The poly(imide) film showed no measurable dielectric loss in the temperature and frequency range used in this study. The dielectric apparatus was an IMASS TDS time domain spectrometer equipped with a Hewlett Packard Series 300 computer. All measurements were carried out by first cooling the sample to 90 K and then heating it while making measurements at progressively higher temperatures. Temperature equilibrium was established prior to each measurements.

Results and Discussion

To incorporate the demands that are required to obtain a pyroelectric liquid crystal polymer in one monomer is possible but synthetically challenging. Therefore a two-monomer-system was used in the first PLCP developed, **III** (7). The different mixtures of the two acrylate monomers (**A1** and **A2**) displayed enantiotropic liquid crystalline behaviour. Depending on the temperature and the proportions of the two monomers, different phases were revealed including isotropic, SmA*, SmC*, SmE* and crystalline phases. The desired enantiotropic SmC* phase, indicated by the appearance of the typical Schlieren texture appeared only in mixtures with a maximum of 35% of the NLO-active monomer **A1**, as seen in the phase diagram, Figure 3.

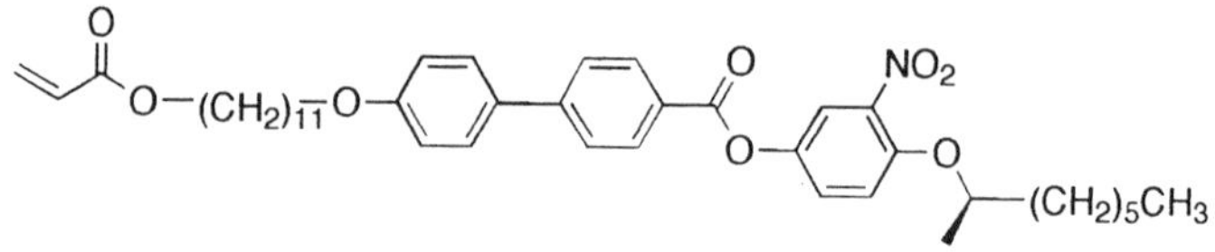

A1b

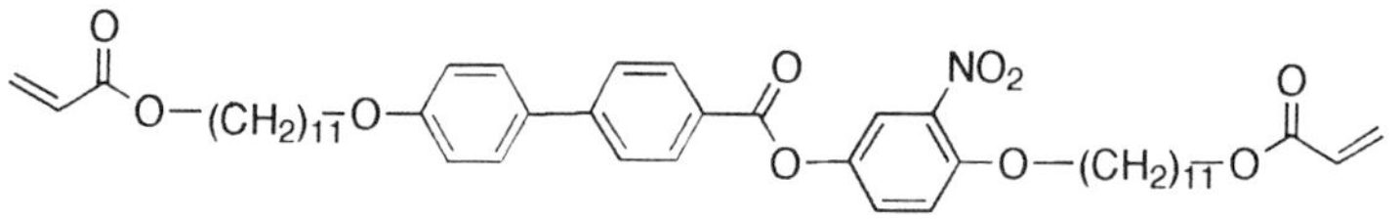

A2b

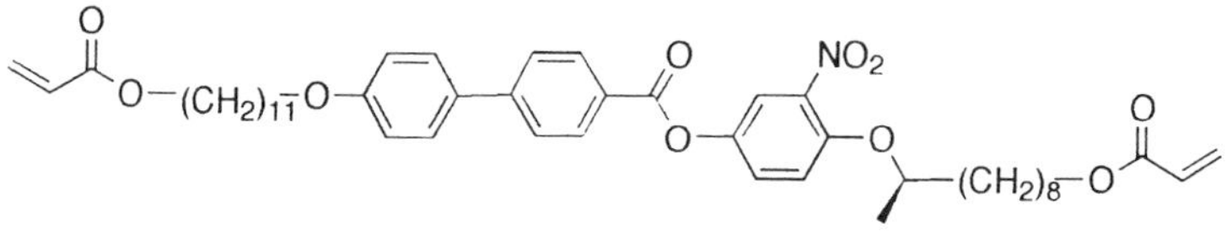

A2c

Figure 4   Structures of acrylate monomers **A1b**, **A2b**, and **A2c**.

The molecular structure had therefore to be modified in order to generate a SmC* phase with the widest possible temperature range and the highest possible concentration of lateral NLO groups. This led to the development of monomer **A1b** (Figure 4), which possessed SmA* and SmC* mesomorphism (8). When **A1b** and **A2** were mixed, all concentrations consisted the SmC* phase, Figure 5.

Monomer **A1b** was synthesized to increase the concentration of lateral NLO groups in the monomer system. Since the monomer **A1b** only contains one acrylate group the cross-linking density will be remarkably reduced with the increased concentration of this monomer. This means that the thermal stability of the orientational order will be lost at higher temperatures (13). This knowledge led to the development of monomer **A2b**. This monomer was made to replace monomer **A2**. All mixtures of **A1b** and **A2b** gave wide SmC* phases and in addition the concentration of lateral NLO groups was maximized.

Spontaneous Polarization. The spontaneous polarization, P_S, was measured in mixtures of **A1**, **A2**, **A1b**, **A2b** and **A2c** to observe the changes in P_S in the new systems. The spontaneous polarization is a measure of the net polarization in the material, when it is not influenced by an external electric field. The P_S was measured since it was anticipated that a higher value might be a sign of a higher $\chi^{(2)}$-value (3, 14). Table 1 lists the results of the spontaneous polarization in different mixtures of **A1**, **A2**, **A1b** and **A2b**. The listed

Table 1 The spontaneous polarization (Ps) obtained for different mixtures of chiral and non-chiral monomers and their concentrations of NLO-groups and NLO groups adjacent to chiral centers.

Mixture	NLO, Chiral	P_S^* (nC/cm^2) (mol%)	NLO (mol%)
A2/A1 (70/30)	30	22	30
A2/A1b (70/30)	30	21	30
A2b/A1 (70/30)	30	19	100
A2b/A1b (70/30)	30	18	100
A2/A1 (65/35)	35	28	35
A2b/A1b (50/50)	50	57	100
A1b	100	181	100

*All spontaneous polarization measurements were performed at the same reduced temperature (t_r, in this case t_r= 0.96).

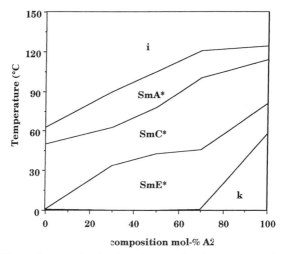

Figure 5 Phase diagram obtained on first cooling for mixtures of **A1b** and **A2**.

values were measured at the same reduced temperature, t_r=0.96 (*10*). Table 1 also lists the concentration of lateral NLO groups and lateral NLO groups adjacent to the chiral center in the different mixtures.

The four first mixtures in Table 1 (**A1/A2**, **A1b/A2**, **A1/A2b**, and **A1b/A2b**) contain 30 mol% of lateral NLO groups adjacent to a chiral center but a different total amount of lateral NLO groups. In mixtures **A1/A2** and **A1b/A2** 30 mol% of the monomers contain lateral NLO groups, whereas in mixtures **A1/A2b** and **A1b/A2b** all monomers carry a lateral NLO group. The P_S-value however, was of the same magnitude for the mixtures, with an average value of 20 nC/cm^2. This means that the contribution to the P_S-value originates only from the chiral monomers and not from the monomers with only lateral NLO groups. A comparison between the P_S-values in the four bottom mixtures in Table 1, where the concentration of chiral monomer **A1** or **A1b** is raised from 30 mol% to 100 mol%, shows that a significant increase in the spontaneous polarization was detected. The P_S-value increased from 20 nC/cm^2 to 181 nC/cm^2. The tilt angles in the samples were around 20° (at the particular reduced temperature) except in pure monomer **A1b**, where the tilt angle was 31°. The values of the P_S in Table 1 was plotted versus the concentration of chiral monomers in the different mixtures, Figure 6 (*10*). The spontaneous polarization showed to have a non-linear dependence on the concentration of chiral monomers.

The results obtained generated an understanding suggesting that a optimized pyroelectric liquid crystal polymer (PLCP) should have all required properties built in to a single monomer. Monomer **A2c** was therefore designed and synthesized. Monomer **A2c** comprises a lateral NLO group adjacent to a chiral center and has two acrylate groups that allow the monomer to be cross-linked. **A2c** was found to have two liquid crystalline phases between the solid (crystal) and the isotropic liquid states. The low-temperature phase is a SmC* phase, where a clear ferroelectric switching could be observed. The high temperature phase, however, has not yet been possible to identify clearly. The transition temperature between the SmC* phase and the high-temperature phase was found to be about 33°C on heating, and about one degree lower on cooling. Both phases appear to be highly twisted phases. The spontaneous polarization of **A2c**, measured at the same reduced temperature as used in Table 1, was calculated to be 164 nC/cm^2 (*8,10*). The result is thus in accordance with the other results plotted in Figure 6.

Second Harmonic Generation. A correlation between the temperature dependence of the spontaneous polarization and that of the NLO-*d*-coefficients has been reported several times in the literature for ferroelectric liquid crystals (*3,14*). The following relationship has been derived (*10*),

$$\frac{P_s(x_2)}{P_s(x_1)} = \frac{\chi^{(2)}(x_2)}{\chi^{(2)}(x_1)} \tag{1}$$

which implies that if the spontaneous polarizations (P_s) and the second-order nonlinear optical susceptibilities ($\chi^{(2)}$) at two different mole fractions $x_A = x_1, x_2$ are compared, the respective relationship should be equal. From the measurements of the spontaneous polarization in Figure 6, the susceptibility for **A2c** is expected to be about 8 times greater than for the mixture **A1/A2** 30/70 and about 6.4 times greater than for the mixture **A1/A2** 35/65.

The second harmonic signal for a cross-linked sample of monomer **A2c** were measured as a function of the angle of incidence. The results of this sample showed a symmetric profile and were compared with the SH-signal from a LiNbO$_3$-crystal. Through a detailed analysis and using $d_{33}=40$ pm/V for LiNbO$_3$, the result $d_{22}=0.08$ pm/V was obtained. The second harmonic signal for another cross-linked sample of **A2c,** detected with another setup, was compared to a poled polymer of PMMA-DR1 of a similar thickness to that of the **A2c**-sample. For PMMA-DR1 $d_{33}=2.5$ pm/V (*15*), giving the estimation of $d_{A2c} \approx 0.2$ pm/V for poly(**A2c**), Figure 7, which is of the same order of magnitude as $d_{22} \approx 0.1$ pm/V as determined by the detailed analysis (*10*).

Second-harmonic generation was previously demonstrated for both cross-linked and monomeric samples of mixtures of **A1** and **A2** in the molar proportions 30/70 and 35/65, respectively (*7*). The SH-signal of this sample was much weaker and more difficult to detect than that from the pure **A2c**-sample, in accordance with the expectations from Equation 1. The upper limit for the *d*-coefficient of the **A1/A2** 35/65 mixture is 0.017

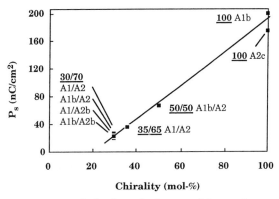

Figure 6   The spontaneous polarization of mixtures of the acrylate monomers shown in Table 1, as a function of the content of chiral monomers.

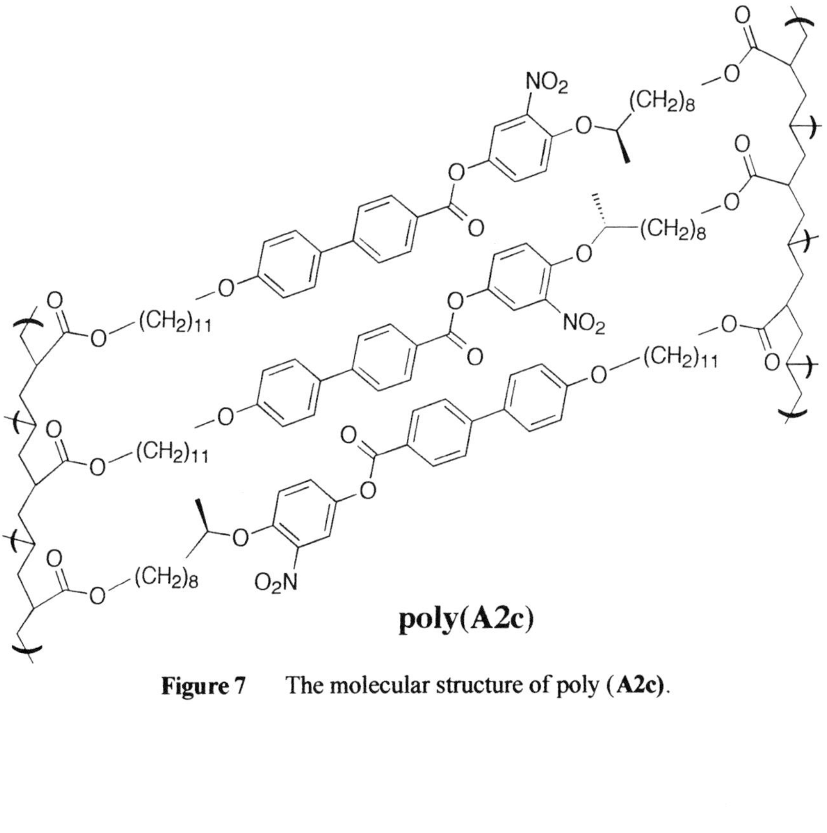

poly(A2c)

Figure 7   The molecular structure of poly (**A2c**).

pm/V*, which gives $d_{22, A2c}/d_{A1/A2} > 4.9$. This is consistent with the spontaneous polarization data, giving $Ps_{A2c}/Ps_{A1/A2} = 160/25 = 6.4$ for the **A1/A2** 35/65 mixture.

The long-term stability of the polar order may decay with time by co-ordinated torsions about bonds near or in the mesogens causing reorientation of the lateral dipoles of the mesogens. This was investigated by remeasuring the SH-signal in cross-linked **A2c** one month after the first measurement. A clear SH-signal was still obtained, albeit being about half as strong as the first signal. In contrast, no SH-signal whatsoever could be detected for the PMMA-DR1 poled polymer one month after the first measurement. In comparison to conventional poled polymers, the Pyroelectric Liquid Crystal Polymer has a good thermal stability. It should also be pointed out that no changes of the orientational order of the mesogenic groups could be detected in the PLCP by polarized light microscopy.

Dielectric Spectroscopy. Since no changes in the orientational order could be observed the long-term relaxation decay of the SH-signal should originate from some kind of intra-molecular motion i.e. a torsion of the lateral NLO-group around the molecular long axis (β-process), Figure 8. If this motion is present it should lead to the randomization of the polar orientation of the lateral dipoles in the mesogenic groups. Dielectric spectroscopy data of cross-linked mixtures of **A1**, **A2**, **A1b**, **A2b**, and **A2c** indicate that the mesogenic ester groups in the units containing no lateral nitro groups undergo conventional reorientation according to the β mechanism and that the β mechanism also is present but considerable suppressed in polymers with lateral nitro groups in the mesogenic structure (9). These data indicate that this β process, indeed weak in the nitro-group containing polymers, may be responsible to the observed decay in the SH-signal.

Conclusions

A long-term decay of the SH-signal has been observed. Experimental data indicate that the origin of this decay fmay be the torsion of the lateral NLO-group around the molecular long axis (β-process), Figure 8. This β motion should therefore in some way be reduced or completely hindered. A new monomer **A1f** (Figure 9) has therefore been synthesized (12). Monomer **A1f**, which has not yet been fully characterized, has only one acrylate group but also a bulky dimethylamino group in the para position to the nitro group of the lateral NLO chromophore. This strongly electron donating group should improve the second-order nonlinear optical susceptibility of the material and even more important this bulky group should lower the possibility for the lateral NLO group to rotate around the molecular long axis. Another monomer **A2e** which is cross-linkable and antiferroelectric has also been synthesized (12). The maximum spontaneous polarization of the new and crosslikable anti-ferroelectric monomer **A2e** is 138 nC/cm^2 and the tilt angle at 35°C is 32°. The idea with this antiferroelectric monomer is to see whether the large tilt angle may reduce the β motion or if it will be increased due to the absensce of large pendant side groups. Both of the two new monomers (**A1f**, **A2e**), shown in Figure 9, should give information about structural changes needed in order to improve the over all nonlinear optical properties of the pyroelectric liquid crystal polymers (PLCP).

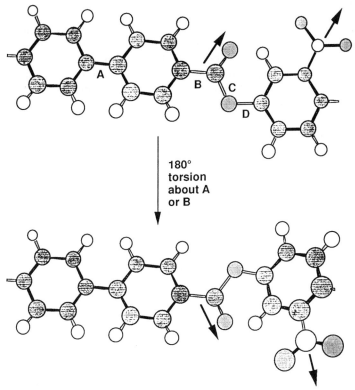

Figure 8 Simple model of the b mechanism in the lateral NLO group containing monomers. Atoms: hydrogen-white; carbon-shaded; oxygen- large dots; nitrogen- small dots. The arrows indicate major dipol moments.

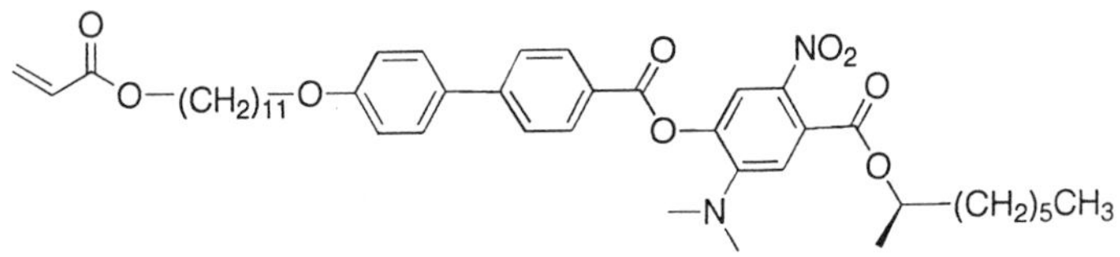

A1f

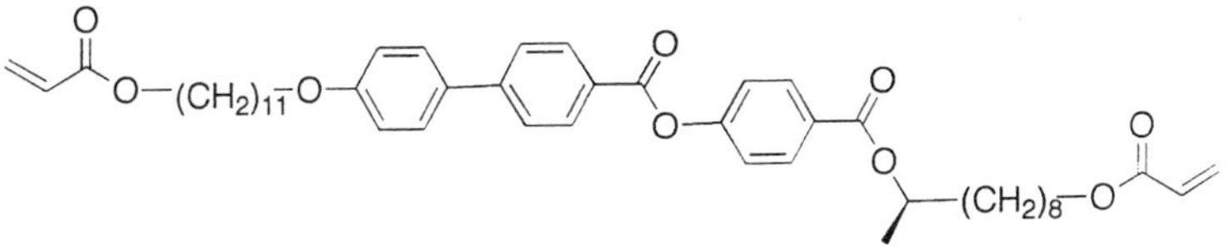

A2e

Figure 9 Molecular structure of the two novel monomers **A1f**, and **A2e**.

326

Acknowledgments

We are grateful to Dr Christian Orrenius and Didier Rotticci at the Department of Organic Chemistry, Royal Institute of Technology for help with the synthesis of chiral alcohols. We also gratefully acknowledge Professor K. Yoshino and his collaborators at the University of Osaka, Japan for giving us the opportunity to do SHG measurements in their lab and Professor R. Boyd University of Utah, USA for giving us the opportunity to do dielectric measurements in his lab.

Financial support from the Wenner-Gren Foundation (scholarship, MT) and from the Swedish Board for Technical and Industrial Development (NUTEK; grant 86-03476P) is gratefully acknowledged as well as grants from the Swedish Research Council for Engineering Sciences (TFR).

Literature Cited

1. Lagerwall; S. T.; Dahl, I. *Mol. Cryst. Liq. Cryst,* **1991**, *114*, 151.
2. Burland, D. M.; Miller, R. D.; Walsh, C. A. *Chem. Rev.* **1994**,*94*, 31.
3. (a) Walba, D. M.; Ros, M. B.; Clark, N. A.; Shao, R.; Johnson, K. M.; Robinson, M. G; Liu, J. Y.; Doroski, D. *J. Am. Chem. Soc.* **1991**,*113*, 5471. (b) Walba, D. M.; Ros, M. B.; Sierra, T.; Rego, J. A.; Clark, N. A.; Shao, R.; Wand, M. D.; Vahra, R. T.; Arnett, K. E.; Velsco, S. P. *Ferroelectrics* **1991**,*121*, 247. (c) Walba, D. M.; Ros, M. B.; Clark, N. A.; Shao, R.; Johnson, K. M.; Robinson, M. G; Liu, J. Y.; Doroski, D. *Mol. Cryst. Liq. Cryst.* **1991**,*198*, 51. (d) Liu, J.Y.; Robinson, M. G.; Johnson, K. M.; Walba, D. M.; Ros, M. B.; Clark, N. A.; Shao, R.; Doroski, D.*J. Appl. Phys.,* **1991**, *70*, 3426.
4. Schmitt, K.; Herr, R.-P.; Schadt, M.; Fünfschilling, J.; Buchecker, R.; Chen, X. H.; Benecke, C. *Liq. Cryst.* **1993**,*14*, 1735.
5. Kapitza, H.; Zentel, R.; Twieg, R. J.; Nguyen, C.; Vallerien, S. U.; Kremer, L. F.; Wilson, C. G., *Adv. Mater.* **1990**, *11*, 539.
6. Keller, P.; Shao, R.; Walba, D. M.; Brunet, M.; *Liq. Cryst.* **1995**, *18*, 915.
7. (a) Trollsås, M.; Sahlén, F.; Gedde, U. W.; Hult, A.; Hermann, D.;Rudquist, P.; Komitov, L.; Lagerwall, S. T.; Stebler, B.; Lindström, J.; Rydlund, O. *Macromolecules,* **1996**, *29*, 2590. (b) Hult, A.; Sahlén, F.; Trollsås, M.; Lagerwall, S. T.; Hermann, D.; Komitov, L.; Rudquist, P.; Stebler, B. results presented at the European Liquid Crystal Winter Conference, Bovec, Slovenia, March 1995, *Liq. Cryst.* **1996**, *20*, 23.
8. (a) Trollsås, M.; Orrenius, C.; Sahlén, F.; Gedde, U. W.; Norin, T.; Hult, A.; Hermann, D.; Rudquist, P.; Komitov, L.; Lagerwall, S. T.; Lindström, J.; *J. Am. Chem. Soc.* **1996**, *118*, 8542. (b) Sahlén, F.; Trollsås, M.; Hult, A.; Gedde, U. W.; Hermann, D.; Komitov, L.; Rudquist, P.; Lagerwall, S. T. *Liq. Cryst.* submitted 1996.
9. Krupicka, A; Åberg, J.; Trollsås, M.; Sahlén, F.; Hult, A.; Gedde, U. W. *Polymer* **1997**, *14*, 3463..
10. Hermann, D.; Rudquist, P.; Lagerwall, S. T; Komitov, L.: Stebler, B.; Lindgren, M.; Trollsås, M.; Sahlén, F.; Hult, A.; Gedde, U. W.; Orrenius, C.; Norin, T., *Liq. Cryst.* submitted 1997.
11. Sahlén, F.; Trollsås, M.; Hult, A.; Gedde, U. W. *Chem. Mater.* **1996**, *8*, 382.

12. Busson, P.; Hult, A., manuscript in preparation.
13. Sahlén, F.; Andersson, H.; Ania, F.; Hult, A.; Gedde, U. W. *Polymer* **1996**.
14 (a) Shtykov, N. M.; Barnik, M. I.; Beresnev, L. A., Blinov, L. M., *Mol. Cryst.Liq. Cryst.* **1985**, *124*, 379, (b) Stanley, M.; Day, S. E.; Dunmur, D. A.; Grayson, M., *Ferroelctrics* **1996**, *179*, 249.
15 Prasad, P. N.; Williams, D. J., *Introduction to Nonlinear Optical Effects in Molecules & Polymers*, (Wiley, 1991).

Chapter 24

Optical Dispersion Properties of Tricyanovinylaniline Polymer Thin Films for Ultrashort Optical Pulse Diagnostics

Ph. Prêtre, E. Sidick[1], L.-M. Wu[1], A. Knoesen[1], D. J. Dyer[2,3], and R. J. Twieg[2,4]

[1]NSF Center on Polymer Interfaces and Macromolecular Assemblies, University of California, Davis, CA 95616
[2]IBM Research Division, Almaden Research Center, San Jose, CA 95120

We have investigated a series of tricyanovinylaniline (TCV) polymer thin films for their use in ultrashort optical pulse (USP) diagnostics of femtosecond Ti:Sapphire lasers. These thin films are ideally suited for USP diagnostics since they eliminate the angle tuning associated with birefringent phase-matched crystals, minimize pulse distortion introduced by group velocity dispersion, and exhibit excellent photochemical stability. The linear optical dispersion of these polymers can be tailored over a wide range for efficient and distortionless frequency conversion. Coherence lengths between 420 nm and 54 microns at a wavelength $\lambda = 800$ nm have been found for the two extreme cases of dispersion in these materials. Film thicknesses of at least two microns are tolerable without introducing any significant pulse distortion at the same wavelength ($\lambda = 800$ nm).

The development of lasers producing ultrashort pulses (USP) of 10 fs or less duration has created a need for a simple and complete diagnosis of these pulses. Second Harmonic Generation Frequency Resolved Optical Gating (SHG FROG) allows for a direct determination of phase and amplitude of femtosecond pulses *(1)*. Since all the information of the laser pulse is converted onto the second harmonic signal, any distortion by the NLO material will reduce the accuracy of the diagnosis. Pulse distortion occurs due to dispersion of the linear optical properties over the large frequency bandwidth of the fundamental and second harmonic pulses.

In phase-matched frequency doubling crystals the group velocity mismatch may cause severe distortion of the generated pulses over long interaction lengths. Even for the thinnest crystals (~ 50 microns), obtaining the correct alignment is iterative and time consuming and presents a significant complication, especially when an unknown pulse shape must be measured. In contrast, because poled nonlinear optical polymer (NLOP) films are highly nonlinear, the interaction length is on the order of a few microns, and they do not suffer as much from pulse broadening and the need of precise angle tuning. In addition, films consisting of electric field oriented organic chromo-

---

[3]Current address: Division of Chemistry and Chemical Engineering, California Institute of Technology, Pasadena, CA 91125.
[4]Current address: Division of Chemistry and the Liquid Crystal Institute, Kent State University, Kent, OH 44242.

phores are easily prepared by spin coating and corona poling *(2)*. The critical align-ment is eliminated since the angle of incidence onto the NLOP only needs to be set within a few degrees of the Brewster angle. Recently, we reported on SHG FROG measurements of 13 fs pulses from a Ti:Sapphire oscillator using a TCV NLOP *(3)*.

We have presented a detailed analysis of USP SHG including effects of group velocity mismatch (GVM) and intrapulse group velocity dispersion (IGVD) else-where *(4, 5)*. Here, we demonstrate that NLOPs are uniquely suited for USP applica-tions with advantages not matched by any other material. We examine relevant dispersion properties and compare high conversion with low distortion in NLOP based USP SHG.

Synthesis of NLO Polymers for USP SHG

A number of NLO chromophores possess large first hyperpolarizabilities and large ground state dipole moments for high conversion efficiencies and electric field poling, respectively. USP SHG applications require specific properties not found in most common NLOPs. In particular, the material must be sufficiently transparent at both fundamental and second harmonic wavelengths, and photochemically stable as it will be subjected to large peak intensities.

We found that the tricyanovinylanilines (TCV) *(6, 7)* exhibited the requisite characteristics for Ti:Sapphire USP SHG pulse diagnostics. In TCV, the charge transfer (CT)-band is red shifted more than 100 nm relative to the nitroanilines. This shift of the CT-band gives rise to a transparency "window" in the 400 nm region.

We investigated spin cast films of the TCV side chain polymers and 5, 10 and 15% by weight guest-host polymers of N,N-diphenyl-4-tricyanovinylaniline (PhTCV) in polymethylmethacrylate (PMMA). PhTCV was prepared from triphenylamine and tetracyanoethylene in dimethylformamide at 80 °C, see also Figure 1. The synthesis of the TCV side chain polymers is shown in Table 1 and described as follows.

Table I.   Nomenclature, Substitution Patterns, Molecular Weights and Glass Transition Temperatures of TCV Side Chain Polymers.

Polymer	m	n	R	Mw	dye conc. [wt. %]	T_g [°C]
dd4-11	1	2	CH_3	38000	44	135
dd4-14	1	0	-	-	70	132
dd4-15	1	1	CH_3	38300	54	137
dd4-20	1	1	$CH_3(CF_2)_3CHF_2$	63700	37	106
dd4-21	1	1	(image)	53.7	40	159

N-ethyl-N-phenyl-3-[(2-methyl-1-oxo-2-propenyl)oxy]ethylamine (dd4-4).
Dicyclohexylcarbodiimide (15.63 g, 75.7 mmol) was added to a stirred solution of the
N-ethyl-N-phenyl-ethanolamine (10.0 g, 60.6 mmol), methacrylic acid (5.74 g, 66.6
mmol), and N,N-dimethylaminopyridine (50 mg, 0.41 mmol) in 500 ml
dichloromethane. The reaction mixture was stirred for 48 hours and was then filtered.
Silca gel ~30ml was added to the solution and the crude was evaporated to dryness.
The crude was purified via flash chromatography over silica gel with gradual elutions
from 95:5 to 80:20 (hexanes:ethyl acetate). Evaporation of solvent yielded 9.843 g
(70%) of a colorless oil: ^{1}H NMR (250 MHz, CDCl3) δ 1.17 (t, J=7.03 Hz, 3H), 1.93
(s, 3H), 3.41 (q, J=7.05 Hz, 2H), 3.59 (t, J=6.37 Hz, 2H), 4.30 (t, J=6.40 Hz, 2H),
5.55 (m, 1H), 6.09 (s, 1H), 6.70 (m, 3H), 7.21 (m, 2H).

Polymer dd4-11. Monomer **dd4-4** (4.87 g, 20.9 mmol) and methylmethacrylate
(4.19 g, 41.8 mmol) in chlorobenzene (25 ml) were degassed under nitrogen for 2
hours at 70 °C. 2,2'-azobisisobutyronitrile (AIBN, ~30 mg) as radical initiator for
polymerizations was added and the reaction mixture was stirred 12 hours under N2 at
70 °C. More AIBN was added and the reaction mixture was stirred for an additional
12 hours. The reaction mixture was then dissolved in 20 ml dichloromethane and this
solution was added dropwise to a vigorously stirred solution of cold methanol (400
ml). The resulting white solid was dissolved in 40 ml dichloromethane and
reprecipitated from cold methanol (400 ml) to yield 5.68 g (63%) of a white powder.
The resulting polymer (2.0 g, 4.6 mmol) and tetracyanoethylene **TCNE** (769 mg, 6.0
mmol) were stirred in N,N-dimethylformamide (90 ml) at 70 °C for 20 hours. The
reaction mixture was then cooled and added dropwise to a vigorously stirred solution
of cold methanol (500 ml). The resulting red solid was dissolved in 50 ml
dichloromethane and reprecipitated twice from cold methanol (500 ml) to yield 2.03 g
(82%) of a red powder (**dd4-11**): MW 38,000; Tg 135 °C (DSC); ^{1}H NMR (250
MHz, CDCl3) δ 0.5-2.0 (m), 3.2-4.2 (m), 6.5-7.0 (m), 8.03 (s).

Polymer dd4-14. The homopolymer was synthesized according to the procedure for
copolymer dd4-11. The crude polymer was dissolved in dimethylformamide (DMF)
and reprecipitated once from cold methanol to yield (99%) of a red powder. The
polymer was not soluble enough in common organic solvents for GPC or NMR
analysis: Tg 132 °C (DSC).

Polymer dd4-15. Synthesized according to the procedure for polymer **dd4-11** to
yield (89%) of a red powder: MW 38,300 (GPC); Tg 137 °C (DSC); ^{1}H NMR (250
MHz, CD2Cl2) δ δ 0.64 (s), 0.81 (s), 1.20 (s), 1.55 (s), 3.47 (s), 3.69 (s), 4.04 (s),
5.24 (s), 6.83, (s), 7.97 (s).

Polymer dd4-20. Synthesized according to the procedure for polymer **dd4-11** to
yield (89%) of a red powder: MW 63,700 (GPC); Tg 106 °C (DSC); ^{1}H NMR (250
MHz, acetone-d6) δ δ 0.90 (s), 1.29 (s), 3.75 (s), 3.90 (s), 4.25 (s), 4.60 (s), 6.70 (t),
7.10, (s), 8.10 (s).

Polymer dd4-21. Synthesized according to the procedure for polymer **dd4-11** to
yield (93%) of a red powder: MW 53,700 (GPC); Tg 159 °C (DSC); ^{1}H NMR (250
MHz, CDCl3) δ 0.5-2.0 (m), 3.2-4.0 (m), 4.0-4.5 (m), 6.9 (s), 8.0 (s).

Linear Optical Dispersion and USP SHG

It is well known that high SHG conversion efficiencies over long interaction lengths
can be achieved for negligible phase mismatch Δk between the center frequencies of
fundamental and SH pulses. NLOPs are, in general, not phasematchable and the

intensity of the generated SH signal will oscillate along the material with a period given by the coherence length

$$l_c = \frac{\pi}{\Delta k} = \frac{\lambda}{4\left|\tilde{n}_{2\omega} - \tilde{n}_{\omega}\right|}, \tag{1}$$

where $\tilde{n}_{\omega}$ and $\tilde{n}_{2\omega}$ are the incident angle-dependent refractive indices at the fundamental and harmonic frequencies and lambda the fundamental wavelength. Hence, thicknesses in excess of one coherence length will not improve the efficiency. Note that l_c is entirely determined by the linear optical dispersion in a given material.

Consider now a 10 fs fundamental pulse with large spectral bandwidth Δf_{B1} of about 70 nm at a wavelength of 800 nm. Over this large bandwidth, group velocity mismatch (GVM) due to phase mismatch between corresponding spectral parts of the fundamental and SH pulse away from the pulse center will act as a band-limiting spectral filter superimposed on the generated SH signal; this unavoidably lengthens the SH pulse. Earlier, we introduced the pulse-width-preservation length, L_τ, as a new reference length to quantify the pulse broadening that occurs during harmonic conversion of USP(4). It is defined as the distance at which the SH pulse width τ_{p2} (intensity FWHM) becomes equal to the pulse width τ_{p1} of the fundamental, i.e. $\tau_{p2} = \tau_{p1}$. In case of a transform-limited hyperbolic secant fundamental pulse shape we have the relation

$$L_\tau \propto \left(\Delta f_{B1} \cdot \left| \frac{\partial \tilde{k}_2}{\partial \omega_2} - \frac{\partial \tilde{k}_1}{\partial \omega_1} \right| \right)^{-1} \tag{2}$$

where $\partial \tilde{k}_i / \partial \omega_i$ is the incident angle-dependent group velocity at fundamental and harmonic frequency (1, 2, respectively). Again, linear optical dispersion determines the maximum thickness of the NLOP so that the fundamental pulse is not distorted.

For pulsewidths ~ 10 fs or shorter, group velocity dispersion within the bandwidth of each of the fundamental and the SH pulse (IGVD) introduces additional pulse distortion. The IGVD is a function of the derivative of the group velocity, therefore of the second derivative of the index of refraction. However, in this case a closed form expression for L_τ is not available but can be found in numerical simulations (4) if linear optical properties are known.

As seen from the expression for the coherence length and the pulse width preservation length, knowledge of the dispersion of the refractive index as well as the first derivative with respect to the light frequency is needed.

Determination of the Linear Optical Dispersion Properties of TCV Polymers

Most easily, the dispersion properties of thin films are measured directly by means of spectroscopic ellipsometry and derivatives are calculated numerically. However, one would prefer to predict the dispersion and therefore the usefulness of a new NLOP for USP SHG without involved chemistry, film formation and various methods of characterization. We have introduced earlier a method of calculation of optical properties of NLOPs based on simple absorption measurements of the chromophore in solution (8). We demonstrate here the use of the conjugate Fourier series method (CFSM) (9) in order to obtain not only linear optical dispersion properties but also all the needed derivatives.

The CFSM is closely related to the Kramers-Kronig approach to obtain refractive indices from measured extinction coefficients. It makes use of the fact that, through the causality condition, the complex dielectric constant and therefore also the complex refractive index, $\tilde{n} = \sqrt{\varepsilon}$, are analytic functions in the upper half complex frequency plane. It can readily be shown that $\tilde{n}$ can be expressed as *(9)*

$$\tilde{n}(\omega) - 1 \quad \rightarrow \quad \tilde{n}\left(-\cot\frac{\theta}{2}\right) - 1 \equiv n(\theta) - 1 - ik(\theta) \tag{3}$$

with

$$n(\theta) - 1 = \frac{a_0}{2} + \sum_{m=1}^{\infty} a_m \cos m\theta \tag{4}$$

and

$$k(\theta) = - \sum_{m=1}^{\infty} a_m \sin m\theta. \tag{5}$$

Eq. (4) and (5) can now be recognized as conjugate Fourier series for functions with appropriate symmetry properties. In practice, the Fourier series coefficients are most conveniently determined by a fast Fourier transformation of a finite number of points, N *(10)*. Knowing the spectrum of the imaginary part determines all coefficients a_m except a_0. However, from the asymptotic behavior of the real part it follows that

$$\lim_{\substack{\theta \to 0 \\ \omega \to \infty}} n(\theta) - 1 = \frac{a_0}{2} + \sum_{m=1}^{N} a_m = 0 \tag{6}$$

and hence

$$a_0 = -2 \sum_{m=1}^{N} a_m. \tag{7}$$

With all coefficients known, Eq. (4) allows then the calculation of the dispersion mode. Note that there are no integral transformations involved in this procedure. Since fast Fourier algorithms operate based on equal spacing for the value of θ in Eq. (4), the spacing on the real (frequency) wavelength axis will be of a (co)tangential functional form according to Eq. (3). In practice, 1024 points equally spaced from $0 \le \theta < \pi$ representing a wavelength range $0 \le \lambda \le 500 \ \mu m$, will ensure sufficient wavelength resolution in the visible. k-values outside the measured range are most conveniently set equal to zero. Derivatives of the refractive index are easily obtained from Eq. (4): even/odd j^{th}-order derivatives are given by the cos/sin series with Fourier coefficients

$$a_m^{(j)} = (-1)^l m^j \cdot a_m, \text{ where } l = \frac{1}{2}\left[j + \frac{1}{2}\left(1 - (-1)^j\right)\right]. \tag{8}$$

PhTCV/PMMA Guest-Host Polymers. In order to demonstrate the predictive power of the CFSM, we calculate here all relevant linear optical dispersion properties for the PhTCV/PMMA guest-host systems. For comparison, thin films of 5, 10 and 15 % by wt. chromophore were prepared by standard spin cast techniques and their linear optical dispersion was measured with a variable angle spectroscopic ellipso-meter (Woollam VASE, 300 nm $\le \lambda \le$ 1700 nm) using silicon substrates. Extinction coefficients of these samples were linearly downscaled to a (fictive) 1 % by weight sample and averaged in order to provide a "reference" for the calculation of any arbitrary dye concentration. Note that an absorption measurement of PhTCV in some low polarity solvent can also provide sufficient information for the CFSM calculation

(8). The shape of this dye's main absorption band in the visible spectrum is very much Gaussian. We therefore used a Gaussian fit to the reference data as input for the calculation of the refractive index following Eq. (3) - (7).

For the cases of interest discussed in this paper, $n(\omega) - 1$ will have to be replaced by $\Delta n \equiv n(\omega) - n_b$. The introduction of the background index n_b becomes necessary since in most cases no information about the dissipative mode at ultraviolet (UV) wavelengths is available. n_b arises mainly from these absorption bands of the polymer backbone and the chromophore in the deep UV. To first order, we therefore will have

$$n_b = n_{PMMA} + const \cdot f, \qquad (9)$$

where f is the dye weight fraction in the polymer. Δn, the contribution from the dye's main absorption band, is also expected to be proportional to the weight fraction f. The dye dependent constant follows directly if the index is known at a specific wavelength. However, 0.083 as determined in the case of PhTCV/PMMA is a good estimate for other systems too, since this factor introduces only a small correction, especially for low dye concentrations. Refractive indices for common host materials like PMMA, polystyrene or some polyimides may be found in the literature *(11)*. The dispersion in these polymer hosts at visible wavelengths is usually well described by a Sellmeier-type formula,

$$n^2 - 1 = B + \frac{q}{1/\lambda_{max}^2 - 1/\lambda^2}. \qquad (10)$$

B, q, and λ_{max}, the wavelength of maximum absorption, are Sellmeier parameters. For PMMA, we determined these parameters by ellipsometry: $B = 0.6958$, $q = 2.202 \cdot 10^{-6}$ nm^{-2} and $\lambda_{max} = 150$ nm. Here, we point out that ellipsometric measurements are not necessary in order to obtain Δn of the NLO polymer or n of the host material. The former can also be calculated from an absorption measurement of the chromophore in solution *(8)* and the latter by transmission spectroscopy *(12, 13)* and Eq. (10).

Figure 1 shows extinction coefficients as measured by ellipsometry (data points). The (fictive) 1 % sample is the concentration scaled average of the 5, 10 and 15 % PhTCV/PMMA sample. Refractive indices as measured and calculated according to the CFSM scheme are drawn in Figure 2. Linear scaling of absorption and index dispersion with dye concentration up to at least 15 % by weight (and possibly to a reasonable maximum of 20 % by wt.) is observed. Notice the "blue transparency window" at $\lambda = 390$ nm in these systems, with a concentration normalized extinction coefficient $k/f = 1.7 \cdot 10^{-4}$ %$^{-1}$.

TCV Side Chain Polymers. The TCV side chain polymers from Table 1 have been investigated in a similar way as the PhTCV/PMMA guest-host systems. Figures 3 and 4 show extinction coefficients and refractive indices as measured by ellipsometry. Again, a "blue transparency window" at $\lambda = 370$ nm is observed with a residual extinction coefficient $k = 4.7 \cdot 10^{-3}$ for dd4-14. The refractive index dispersion in these side chain systems is much larger than for the guest-host polymers, a result of the higher chromophore concentration (cf. Table 1). Extinction coefficients scale well with chromophore concentration except for dd4-14. We attribute this difference to dye-aggregations at this high doping level. The introduction of fluorinated side groups in dd4-20 results in a overall lower refractive index, especially at short wavelengths.

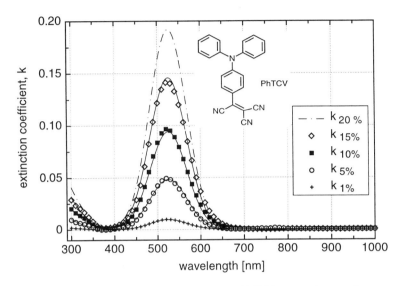

Figure 1. Extinction coefficient for PhTCV/PMMA guest-host polymers at different dye concentrations as measured by ellipsometry. Solid lines: concentration scaled Gaussian curves based on the 1 % sample, see text. Dash dotted line: linear extrapolation for 20 % PhTCV/PMMA.

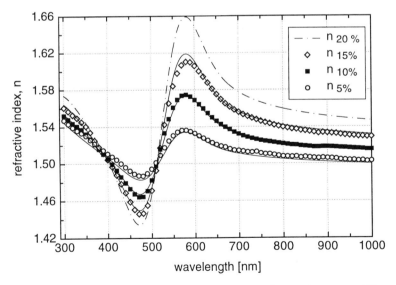

Figure 2. Refractive index for PhTCV/PMMA guest-host polymers at different dye concentrations as measured by ellipsometry. Solid lines: CFSM calculations based on the 1 % sample, see text. Dash dotted line: extrapolation for 20 % PhTCV/PMMA.

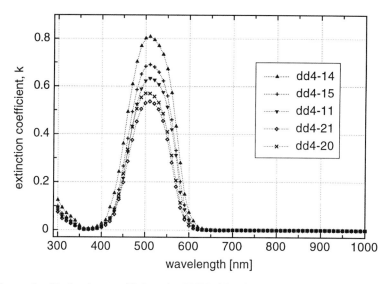

Figure 3. Extinction coefficient for TCV side chain polymers as measured by ellipsometry. Notice the different ordinate range compared to Figure 1 due to the much higher dye concentration in these side chain systems.

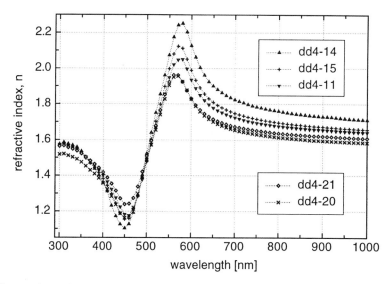

Figure 4. Refractive index for TCV side chain polymers as measured by ellipsometry. Notice the large dispersion in these highly loaded NLO polymers compared to Figure 2.

Second Harmonic Generation and NLO Polymers

In thin NLO polymer films orientational order necessary for SHG can be induced by an electric field poling process. The applied electric field couples to the dipole moment of the NLO dyes and orients them in the polymer matrix breaking the centrosymmetry of an unpoled sample. Most commonly, the poling process is performed using a corona discharge at an elevated temperature where the material softens in the vicinity of the glass transition temperature T_g (2). For the details the reader is referred to the literature (14, 15). Due to this poling-induced ∞mm-point group symmetry, SHG is only observed at oblique angles of incidence. Highest signal levels are achieved for transversal electromagnetic (TM) fundamental fields close to Brewster's angle of incidence. All the relevant properties in the following sections are calculated at the angle of optimized signal output.

By comparing different NLO polymers, we also have to take into account the dependency of the nonlinear coupling coefficient ρ on the chromophore concentration in the polymer. The SHG conversion efficiency is a quadratic function of ρ, which in turn is proportional to the more widely used effective nonlinearity d_{eff}.

Nonlinear Optical Susceptibilities in TCV Polymers. NLO susceptibilities d_{33} in all TCV polymers have been determined as follows: corona poled films in the thickness range $0.15 < L < 6.2$ µm were exposed near Brewster angle to a 100 fs fundamental beam at $\lambda \approx 800$ nm from a mode-locked Ti:Sapphire laser cavity. The SHG signal was measured with a photomultiplier tube and compared to the output of a type I phase matched reference BBO crystal ($d_{11} = 1.6$ pm/V (16)). During the measurement care was taken that the fundamental wavelength and intensity didn't change by on-line monitoring the fundamental spectrum and the SHG output of a second reference KTP crystal. For the evaluation of the polymers' d_{33}, we assumed Kleinman symmetry and $d_{33} = 3 \cdot d_{31}$. Both conditions may not be fulfilled a priori in this wavelength range. As shown below, values for the different polymers are consistent and justify these assumption a posteriori for comparison purposes. Figure 5 shows the result for the PhTCV/PMMA guest-host polymers and Table II summarizes the results for the TCV side chain polymers.

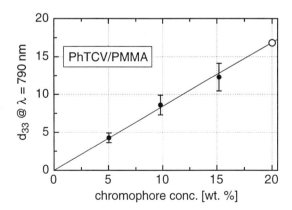

Figure 5. NLO susceptibility in PhTCV/PMMA guest-host polymers as function of dye concentration. Solid line: least squares fit to measured data: $d_{33}/f = 0.84$ pm/V%$^{-1}$. Open circle: expected value for a fictive 20 % by wt. sample.

Table II. NLO Susceptibilities of TCV Side Chain Polymers at $\lambda = 807$ nm

Polymer	dd4-11	dd4-14	dd4-15	dd4-20	dd4-21
d_{33} [pm/V]	37 ± 6	46 ± 9	49 ± 9	31 ± 6	34 ± 6

Linear scaling of d_{33} with chromophore concentration is found in the guest-host systems. Proportionality is also found between the nonlinearities in Table II and the chromophore concentration of the side chain polymers in Table I, except for dd4-14 where most probably dye aggregation as already in the case of linear absorption prevents linear upscaling. Assuming similar mass densities in the two systems, the chromophore density ratio (dye/volume) is about 1.5 times higher for the side chain polymer at the same doping level. On the other hand, the microscopic hyperpolarizability of PhTCV exceeds the value of the diethyl analog by 50 % (17). Therefore, a direct comparison of nonlinearities on the base of weight fraction is possible. $d_{33} = 4.3 \pm 0.7$ pm/V at $\lambda = 790$ nm for the 5 % by wt. PhTCV/PMMA sample would result in $d_{33} = 46$ pm/V for a 54 % by wt. sample, remarkably close to the value for dd4-15.

Coherence Length vs. Pulse Width Preservation Length in TCV Polymers

As pointed out earlier, the coherence length l_c of a NLO polymer is a measure of film thickness at which SHG conversion efficiency is highest. Film thicknesses in USP SHG are additionally limited by the pulse width preservation length L_τ. We will compare the two limiting length scales in the guest-host as well as the side chain polymers and show how the polymers can be tailored to the needs of a specific application by varying dye concentration or degree of copolymerization. In the following, L_τ is calculated for a 10 fs transform limited hyperbolic secant (sech) pulse (see Eq. (2)).

l_c vs. L_τ in PhTCV/PMMA Guest-Host Polymers. Using the CFSM with Fourier coefficients from Eq. (8) we calculated all necessary derivatives of refractive indices from the extinction coefficients of the 1 % reference sample and the Sellmeier equation for the index of PMMA. Figure 6 shows the coherence length and pulse width preservation length for the wavelength range of USP Ti:Sapphire lasers. Depending on dye concentration and wavelength, film thicknesses are either limited by the coherence or pulse width preservation length. Both length scales are decreasing with increasing doping level but ρ, the nonlinear coupling coefficient is accordingly higher. Table III lists the product $\rho \cdot l_c$, on which the overall efficiency depends quadratically, for the fundamental wavelength $\lambda = 790$ nm.

Table III. Normalized Coupling Coefficient $\rho \cdot l_c$ for PhTCV/PMMA Guest-Host Polymers at Fundamental Wavelength $\lambda = 790$ nm

dye conc. [% by wt.]	1	5	10	15	20
$\rho \cdot l_c$ [pm/V]	14	194*	159	104	90

*$\rho \cdot L_\tau$ is given here, since L_τ is the thickness limiting factor in this case

A chromophore concentration of about 5 % by wt. is ideal at this wavelength with a $\rho \cdot L_\tau$ value comparable to that of inorganic BBO (4) . However, the NLOP has not to be phase-matched and therefore no specific sample preparation is necessary.

338

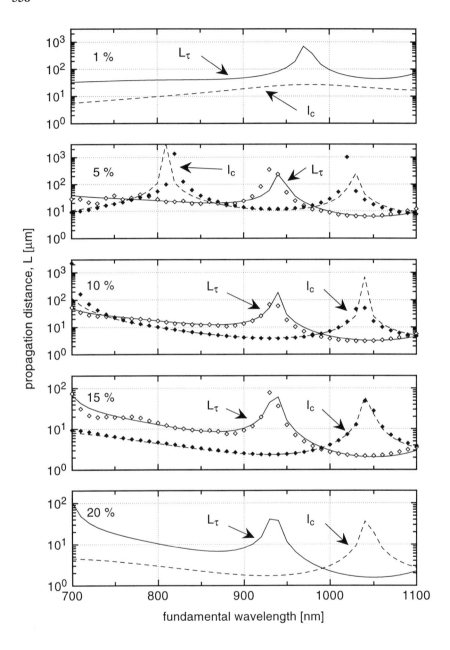

Figure 6. Coherence length l_c, pulse width preservation length L_τ of PhTCV/ PMMA guest-host polymers as function of dye concentration. Symbols: data from linear optical dispersion properties as measured with ellipsometry. Lines: as calculated with the CFSM. L_τ is calculated for a 10 fs transform limited hyperbolic secant pulse.

l_c vs. L_τ **in TCV side chain Polymers.** Figure 7 shows the coherence length and pulse width preservation length for the wavelength range of USP Ti:Sapphire lasers in the case of the TCV side chain polymers.

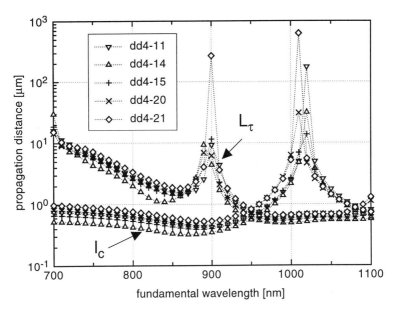

Figure 7. Coherence length l_c, pulse width preservation length L_τ of TCV side chain polymers. Points connected: data from linear optical dispersion properties as measured with ellipsometry. L_τ is calculated for a 10 fs transform limited hyperbolic secant pulse.

Film thicknesses are limited by the coherence length which is relatively small for fundamental wavelengths $\lambda < 950$ nm due to the large linear optical dispersion in these polymers. Almost phase-matched around $\lambda \approx 1$ μm, the pulse width preservation length becomes the thickness limiting factor. Notice the ordinate scale difference compared to the guest-host systems. Again, a higher nonlinear coupling coefficient will compensate for the smaller film thicknesses as seen in Table IV, where $\rho \cdot l_c$ is given for the fundamental wavelength $\lambda = 810$ nm.

Table IV. Normalized Coupling Coefficient $\rho \cdot l_c$ for TCV Side Chain Polymers at Fundamental Wavelength $\lambda = 810$ nm

Polymer	dd4-11	dd4-14	dd4-15	dd4-20	dd4-21
$\rho \cdot l_c$ [pm/V]	40	36	48	39	46

No large difference within the side chain polymers is found but their performance is inferior compared to the guest-host systems. Notice that normalized coupling coefficients in Table IV are accomplished with films of only a few hundred nanometer thickness. Optical quality and poling efficiency are in general superior to films of several micron thickness as required by the guest-host polymers for optimal conversion efficiencies.

Absorption and USP SHG

The considerations of the foregoing sections have to be revised if considerable absorption either at fundamental or second harmonic wavelength is present in the NLOP. In general, absorption at the fundamental wavelength ($\lambda \approx 800$ nm) is negligible but it is not at the second harmonic. Figure 8 compares the extinction coefficient of a widely used NLOP, disperse red 1 (DR1) attached to PMMA, with a PhTCV/PMMA guest-host polymer scaled in concentration for equal peak absorption.

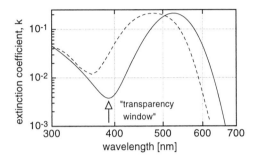

Figure 8. Extinction coefficient in a typical NLO polymer, disperse red 1 attached to PMMA (dashed line), and 22 % by wt. PhTCV in PMMA (solid line).

The "transparency window" at the second harmonic wavelength is essential for conversion efficiency. The one order of magnitude larger extinction coefficient of the DR1-PMMA side chain polymer at around 400 nm reduces the second harmonic signal drastically for a film of only a micron thickness. The residual absorption in the TCV polymers will affect the conversion process to a certain extent too. However, for a 10 % by wt. PhTCV/PMMA guest-host polymer the film thickness for optimal second harmonic conversion in the fundamental wavelength range 750 nm $\leq \lambda \leq 900$ nm has to be reduced by less than 20 % compared to the coherence length .

Photochemical Stability and USP SHG

We have evaluated a number of NLO chromophores for Ti:Sapphire applications (with an appropriate "blue transparency window"), but several were found to be unsuitable for USP SHG because of photochemical instabilities either at ambient conditions or during USP SHG. In contrast, photodegradation effects in poled PhTCV / PMMA films were not observed following several hours exposure times. The second harmonic signal level remained constant at a fundamental wavelength peak intensity of $I_p = 8$ GW/cm^2 and at an average power density of $I_{avg} = 9$ kW/cm^2. Thus, the photochemical stability of the PhTCV chromophores is quite remarkable and technologically important.

Orientational Order Stability and USP SHG

One concern of NLO polymers is the thermal relaxation of the poling-induced order at room temperature. Side chain polymers feature higher orientational stability than guest-host systems even if their glass transition temperature is comparable *(13)*. In the context of USP SHG, orientational relaxation is not an important issue since relaxation times in the polymers described in this paper at room temperature are of the order of months to years, while a typical USP measurement is accomplished within minutes. High temperature excursion during processing and packaging procedures of integrated devices do not occur in this application. Even if a film has been in use for several months it can easily be repoled to regain its maximal performance.

Conclusions

We have investigated a series of tricyanovinylaniline (TCV) chromophores either as guest-host type polymer in PMMA or covalently attached to PMMA backbones for their use in ultrashort optical pulse diagnostics of femtosecond Ti:Sapphire lasers. We demonstrated their photochemical stability under intense laser irradiation. The linear optical dispersion of these polymers has been determined either by ellipsometry or by a computational method based on a conjugate Fourier series approach and it can be tailored over a wide range for efficient and distortionless frequency conversion. This allows for ultrashort optical pulse diagnostics using frequency resolved optical gating and the elimination of time consuming angle tuning associated with birefringent phase-matched crystals. Since such thin NLO films are not limited by pulse distortion, they are ideally suited for extremely short pulse diagnostics.

Acknowledgment

The authors acknowledge support for this research from the National Science Foundation MRSEC program under award DMR-9400354.

Literature Cited

(1) DeLong, K. W.; Trebino, R.; Hunter, J.; White, W. E., *J. Opt. Soc. Am. B* **1994**, *11*, 2206 - 2215.
(2) Mortazavi, M. A.; Knoesen, A.; Kowel, S. T.; Higgins, B. G.; Dienes, A., *J. Opt. Soc. Am. B* **1989**, *6*, 733 -741.
(3) Yankelevich, D. R.; Prêtre, P.; Knoesen, A.; Taft, G.; Murnane, M. M.; Kapteyn, H. C.; R.J. Twieg, *Opt. Lett.* **1996**, *21*, 1487 - 1489.
(4) Sidick, E.; Knoesen, A.; Dienes, A., *J. Opt. Soc. Am. B* **1995**, *12*, 1704 - 1712.
(5) Sidick, E.; Dienes, A.; Knoesen, A., *J. Opt. Soc. Am. B* **1995**, *12*, 1713 - 1722.
(6) Stamatoff, J. B.; Buckley, A.; Calundann, G.; Choe, E. W.; DeMartino, R.; Khanarian, G.; Leslie, T.; Nelson, G.; Stuetz, D.; Teng, C. C.; Yoon, H. N., In *Molecular and Polymeric Optoelectronic Materials: Fundamentals and Applications*; Proc. SPIE, **1987**, *682*, 85.
(7) Cheng, L. T.; Tam, W.; Stevenson, S. H.; Meredith, G. R.; Rikken, G.; Marder, S., *J. Phys. Chem.* **1991**, *95*, 10631 - 10643.
(8) Prêtre, P.; Wu, L.-M.; Knoesen, A.; Swalen, J. D., *J. Opt. Soc. Am. B* **1998**, *101*
(9) King, F. W., *J. Opt. Soc. Am.* **1978**, *68*, 994 -997.
(10) Press, W. H.; Flannery, B. P.; Teukolsky, S. A.; Vetterling, W. T., *Numerical Recipes - the Art of Scientific Computing*, Ed., University Press: Cambridge, 1992.
(11) Ishigure, T.; Nihei, E.; Koike, Y., *Appl. Opt.* **1996**, *35*, 2048.
(12) Manificier, J. C.; Gasiot, J.; Fillard, J. P., *J. Phys. E: Sci. Instrum.* **1976**, *9*, 1002-1004.
(13) Prêtre, P.; Kaatz, P.; Bohren, A.; Günter, P.; Zysset, B.; Ahlheim, M.; Stähelin, M.; Lehr, F., *Macromolecules* **1994**, *27*, 5476-5486.
(14) Bosshard, C.; Sutter, K.; Prêtre, P.; Hulliger, J.; Flörsheimer, M.; Kaatz, P.; Günter, P., *Organic Nonlinear Optical Materials*, Garito, A. F., Kajar, F. Eds., Gordon and Breach Science Publishers: Amsterdam, 1995.
(15) Singer, K. D.; Kuzyk, M. G.; Sohn, J. E., *J. Opt. Soc. Am. B* **1987**, *4*, 968 - 976.
(16) Eimerl, D., *Ferroelectrics* **1987**, *72*, 95-139.
(17) Jen, A. K.-J.; Chen, Y.-A.; Rao, V. P.; Cai, Y.-M.; Liu, Y.-J.; Drost, K. J.; Mininni, R. M.; Dalton, L. R.; Bedworth, P.; Marder, S. R., In *Thin Films for Integrated Optics Applications*; Mat. Res. Soc. Symp. Proc., **1995**, *392*, 33 - 41.

MICROLITHOGRAPHY AND PACKAGING

Chapter 25

Mechanistic Studies of Chemically Amplified Resists

W. Hinsberg[1], G. Wallraff[1], F. Houle[1], M. Morrison[1], J. Frommer[1], R. Beyers[1], and J. Hutchinson[2,3]

[1]IBM Research Division, Almaden Research Center, San Jose, CA 95120
[2]Electronics Research Laboratory, University of California, Berkeley, CA 94720

Chemically amplified resists are now finding widespread use for the fabrication of advanced integrated circuits. A fundamental understanding of the chemistry and physics of chemically amplified resist processing enables the rational design of improved materials and process protocols. We describe here studies aimed at improving our understanding of the chemical and physical processes that influence the performance of advanced resist materials based on acid-catalyzed imaging chemistry.

The rate of progress in the microelectronics industry in large part is gated by advances in microlithography, the collection of technologies that is used to form miniaturized patterns of conductors, insulators and other electronic elements on wafers of silicon. The ability to fabricate integrated circuits with increasingly smaller features leads to improvements in performance and function, and to reductions in cost. In concert with advances in lithographic exposure tooling and optics, photoresists, the polymeric imaging materials used in the microlithographic process, have undergone extensive evolution and refinement(1).

The mainstay photoresist chemistry that has been in use for many years by the microelectronic industry is based on the photochemistry of a diazonaphthoquinone (DNQ) photoactive compound dissolved in a phenolic polymer matrix. This combination is known generically as a DNQ-novolac photoresist(2) (Figure 1(a)). Though they are insoluble in water, novolac films readily dissolve in aqueous alkaline solutions, a consequence of the weak acidity of the phenol group which undergoes deprotonation to form a soluble ionic phenolate species. Addition of a DNQ to novolac reduces this solubility in base by orders of magnitude. Irradiation of the resist converts the DNQ to a carboxylic acid and destroys its inhibitory effect in the exposed areas.

[3]Current address: Components Research, Intel Corporation, Santa Clara, CA 95052.

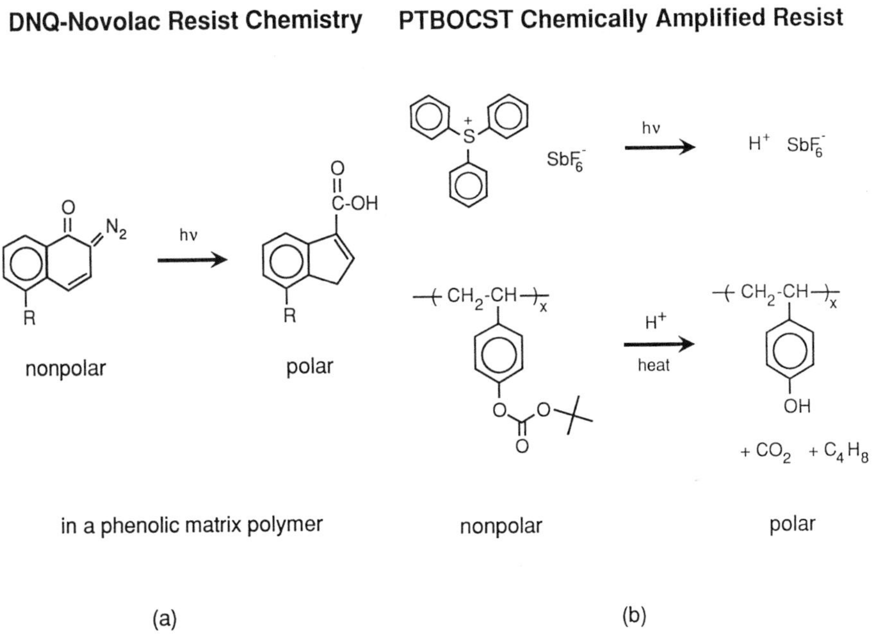

DNQ-Novolac Resist Chemistry   **PTBOCST Chemically Amplified Resist**

Figure 1. The chemistry of DNQ-novolac photoresists, and poly(TBOC-styrene) resist, a chemically amplified resist system. Both systems have been used in microelectronics fabrication.

The exposed regions are then soluble in aqueous alkaline solutions while the unexposed, inhibited areas remain essentially insoluble. Through careful structure-activity studies and materials optimizations, DNQ-novolac photoresists have been refined to the point that they are capable of forming images at dimensions of ca. 0.3 µm using projection exposure tools operating at a wavelength of 365 nm.

The fabrication of features at dimensions below this has required the microelectronics industry to shift to improved exposure tools using short-wavelength ultraviolet light ($\lambda = 248$ nm, and more recently $\lambda = 193$ nm). These new tools provide much weaker illumination of the resist pattern than earlier tool designs. DNQ-novolac resists do not perform adequately with these new exposure tools, a consequence of their intense optical absorbance at the exposing wavelength, combined with their requirement for an intense light source to produce adequate chemical conversion in the photoresist film in a practical exposure time.

A new class of lithographic materials, termed chemically amplified (CA) resists, has been developed specifically to address these shortfalls(1,3). Figure 1(b) shows the chemistry of an early example of a chemically amplified photoresist system(4). In this case the photoactive species is a triarylsulfonium salt, dissolved in a matrix of poly(t-butoxycarbonyloxystyrene) (PTBOCST). Upon exposure to UV light, the sulfonium cation undergoes a multistep reaction, leading ultimately to the formation of a strong Brönsted acid. In a subsequent heating step, the TBOC groups pendant to the polymer chain undergo facile acid-catalyzed fragmentation, producing poly(hydroxystyrene) (PHOST) accompanied by evolution of the gaseous products CO_2 and isobutylene from the film. The initial PTBOCST polymer and the product PHOST polymer differ sharply in solubility properties, allowing either the exposed or unexposed regions to be selectively dissolved during the final development step. Since the acid catalyst is regenerated upon fragmentation, a large number of TBOCST groups undergo reaction for each acid molecule generated. Catalytic chain lengths of up to 1000 have been measured in this resist system(5).

With DNQ-novolac resists, the shape of the final developable latent image is determined by the conditions of the initial exposure. Its detailed form is controlled by the characteristics of the projected optical image, by the optical and photochemical properties of the resist film, and by the optical properties of the substrate. With CA resists, however, though the exposure conditions still play an analogous role, the structure of the developable latent image also is strongly influenced by the chemistry and physics that take place during post-exposure thermal processing. The thermal deprotection step transforms the initial latent image of photogenerated acid into a developable chemical latent image. There is evidence that the kinetics of chemistry and acid diffusion, the presence of airborne substances that inhibit the catalytic reaction, and volatilization of acid catalyst from the film all play a role in this transformation(6-10). Ultimately, these factors influence resist functional properties including radiation sensitivity, spatial resolution, image line shapes and process latitude. We have undertaken a series of studies to characterize at a fundamental level the processes active during CA resist thermal processing.

Chemical Deprotection Kinetics of Chemically Amplified Resists

Our understanding of the chemistry occurring during post-exposure bake (PEB) is based in large part on knowledge of reactant and product structure rather than detailed mechanistic analysis. With a goal of better understanding mechanistic details of the PEB step, we have performed in-situ, high data rate, accurate measurements of the chemical kinetics that occur in prototype CA resists during PEB. In this work we have examined the effects of acid strength and concentration, the structure of the acid-labile protecting group, the impact of absorbed airborne inhibitors, and the role of the uncatalyzed thermal deprotection path.

Methods. We have devised procedures for the measurement and analysis of the kinetics of chemical processes in thin resist films. These are described in detail elsewhere(*11*) and are briefly summarized here. The experimental methodology employs IR or UV absorption spectroscopy under carefully controlled isothermal conditions to monitor resist film composition as a function of time. In a typical experiment, a 1.0 μm-thick resist film was cast from cyclohexanone solution and post-apply baked at 100 °C for 90 seconds. Following a blanket exposure with 254 nm light to a total dose of 100 mJ/cm^2, the substrate was immediately placed in a pre-heated high-temperature spectroscopic cell. After a short delay to allow the wafer to reach thermal equilibrium, the IR or UV absorption spectrum of the film was recorded at regular intervals until all reaction ceased. This procedure has general applicability to a variety of different resist chemistries. Deprotection kinetics were measured at various selected temperatures in the range of 40-200 °C, and also by programmed temperature thermogravimetric analysis. Experimental data were analyzed and interpreted using a stochastic chemical kinetics simulator(*12*) that facilitates the simulation of systems that change volume during reaction, and those where the reaction temperature varies in a predetermined programmed manner with time.

Uncatalyzed Thermal Decomposition. A priori, the uncatalyzed thermal decomposition of PTBOCST was expected to be a simple first-order reaction, paralleling the gas-phase behavior of analogous compounds such as t-butyl acetate(*13*) (Figure 2, Scheme (a)). The dashed line in the graph of Figure 2 shows the predicted time behavior for such a process. Experimentally, this behavior is not observed. Instead, the rate is found to increase sharply as the reaction proceeds, indicating the existence of an auto-accelerating pathway(*14*) in which the reaction products act to increase the overall reaction rate. The simplest possible mechanism incorporating auto-acceleration is depicted in Figure 2, Scheme (b). We find that our experimental results are consistent with Scheme (b) over the experimentally-accessible range of temperatures. The solid line in the graph of Figure 2 represents the results of a simulation of Scheme (b) where the stoichiometric coefficient n=2. The auto-acceleration step, which is third-order overall, likely represents the deprotection of a complex of one reactant group and two product groups formed during a series of sequential bimolecular steps rather than a discrete termolecular event. The uncatalyzed thermal decomposition of PTBMA shows the same functional behavior.

Acid-catalyzed Decomposition. Figure 3, Scheme (a) depicts the simplest possible representation for the acid-catalyzed deprotection of PTBOCST. Here, the protonated TBOCSTH$^+$ intermediate undergoes unimolecular decomposition, followed by rapid reprotonation of one of the remaining TBOCST moieties until all TBOCST groups are reacted. This simple mechanistic description predicts an essentially constant rate of

Scheme (a)

TBOCST $\longrightarrow$ HOST + CO_2 + C_4H_8

Scheme (b)

TBOCST $\longrightarrow$ HOST + CO_2 + C_4H_8

TBOCST + n HOST $\longrightarrow$ (n+1) HOST + CO_2 + C_4H_8

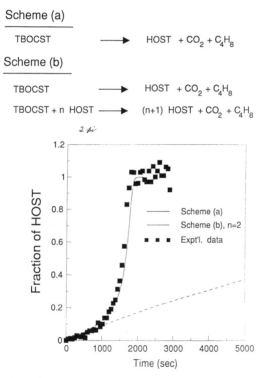

Figure 2. Experimental and simulation data for uncatalyzed decomposition of PTBOCST at 150 °C.

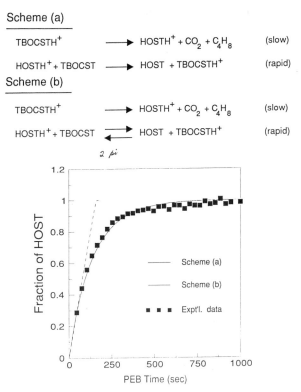

Figure 3. Experimental and simulation data for acid-catalyzed decomposition of PTBOCST at 60 °C.

increase in HOST concentration with time. The dashed line in the graph of Figure 3 represents the results of a simulation of Scheme (a). Experimentally, the acid-catalyzed thermolysis of PTBOCST does not display this behavior, exhibiting instead a strong inhibition of the fragmentation reaction as the reaction proceeds.

For Scheme (a) in Figure 3 to apply, the unreacted TBOCST must have a much stronger tendency than HOST to bind the proton; that is, TBOCST must be more basic than HOST. On the basis of reported basicities of phenols and esters(15), however, HOST and TBOCST are expected to be roughly equal in basicity. Hence, these species will compete for the limited acid available in the film. Figure 3, Scheme (a) can be modified to account for this competition by considering the rapid proton transfer step to be reversible and in equilibrium. Figure 3, Scheme (b) shows the modified mechanism. The graph in Figure 2 displays the fit of a simulation of Scheme (b) (solid line) to experimental data, demonstrating that the proposed mechanism is consistent with experimental observation. Scheme (b) is an example of specific acid catalysis(16).

Temperature Dependence of Reaction Rates. To assess the activation parameters and equilibrium constants for the proposed reactions, kinetic data were acquired over a range of temperatures, and then fit to the mechanisms described above using stochastic simulation. For both uncatalyzed and acid-catalyzed thermolysis of PTBOCST and PTBMA, excellent agreements between the model and experimental data were obtained. Table I summarizes the temperature ranges studied and the activation and thermodynamic parameters derived via the fitting procedure. For a condensed phase system where the properties of the medium change during reaction, a detailed molecular interpretation of the activation parameters is difficult. Several general observations can be made in the absence of such analysis. The 40.9 kcal/mole activation energy for k_1 for PTBMA agrees well with the value reported for the gas-phase decomposition of t-butyl acetate(40.5 kcal/mole)(17). The E_a for PTBMA is higher than the value for TBOC, as expected based on the relative reactivities of esters and carbonates(13). the PTBOCST E_a of 32 kcal/mole is similar to the value reported for thermal degradation of PTBOCST by Long et al.(18). The entropy of activation, calculable from the measured A factors, provides information on the geometry and structure of the transition state. In condensed phases, the structure of the activated complex generally includes molecules in the surrounding medium which contribute to the energetics; for many systems this makes a detailed analysis impractical(19). In the cases at hand, the values shown in Table I are in the range expected for reactions of this type but any further interpretation is unwarranted.

Table 1.   **Rate constants and thermochemical data for CA resist polymers**

Polymer	Rate constant	T range °C	A l/mol-s	E_a kcal/mol	ΔH kcal/mol	ΔS cal/mol-K	ΔG kcal/mol
Thermal							
TBOC	k_1	120-150	$7.3{\times}10^{12}$	32.7	31.9	-2.3	--
	k_2	120-150	$6.3{\times}10^{11}$	29.9	29.1	-7.2	--
TBMA	k_1	180-220	$7.9{\times}10^{14}$	40.9	40	+6.8	--
	k_2	180-220	$2.9{\times}10^{13}$	39.8	38.9	+0.2	--
Acid							
TBOC	k_3	40-60	$2.1{\times}10^{18}$	28.7	28	+23.3	--
	K_4	40-60	1.3	--	--	--	-0.2
TBMA	k_3	60-80	$2.3{\times}10^{21}$	35	34.3	+37.0	--
	K_4	60-80	2	--	--	--	-0.5

Effects of Photo-acid Strength on Deprotection Kinetics . The proton-donating ability of the photogenerated acid is an important factor determining the overall rate of deprotection. In the mechanism shown in Figure 3 Scheme (b), the overall deprotection rate is proportional to the concentration of the protonated repeat unit TBOCSTH$^+$. That concentration in turn is determined by the concentration of acid and

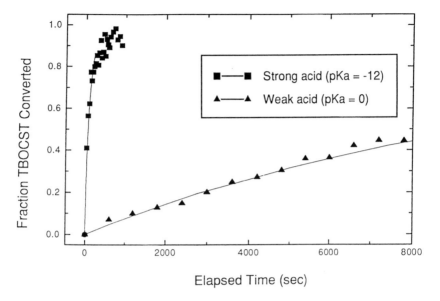

Figure 4. Experimental and simulation data for acid-catalyzed decomposition of PTBOCST at 60 °C. Rapid decomposition occurs when a strong acid (trifluoromethanesulfonic acid, pK_a ~ -12) is present; decomposition is slow when a weak acid (methanesulfonic acid, pKa ~ 0) is used.

the relative basicities of the TBOCST repeat unit and all other species in the film, including the counterion of the acid. If the counterion is sufficiently basic so that it competes with TBOCST for the limited amount of acid, then the concentration of protonated repeat unit will be decreased by this competition and the rate of deprotection will be slowed(20). Figure 4 shows experimental results and simulations for the cases of acid-catalyzed thermolysis of PTBOCST by a very strong and a very weak photogenerated acid (CF_3SO_3H and CH_3SO_3H, respectively). This illustrates the impact of varying acid strength on the rate of deprotection in films of PTBOCST. While in these limiting cases experiment and simulation appear to be in agreement, in general the precise pK_a value of an acid in the polymeric environment must be known to predict with precision its influence on the deprotection kinetics.

Chemical Contamination of Chemically Amplified Resists. While CA resists have been shown to exhibit very high photosensitivities due to their inherent gain, it has been observed that very low levels of certain organic vapors can severely degrade the performance of such systems(21, 22). For example, exposure of a film of PTBOCST/ Ph_3S-SbF_6 resist (imaged in negative tone) to an airstream containing N,N-dimethylaniline at a concentration of 30 ppb causes a significant deterioration of the resist imaging properties within fifteen minutes. Figure 5 illustrates this. Shown are scanning electron micrographs of resist images for exposed and developed wafers which were stored in different environments prior to processing. Those wafers

352

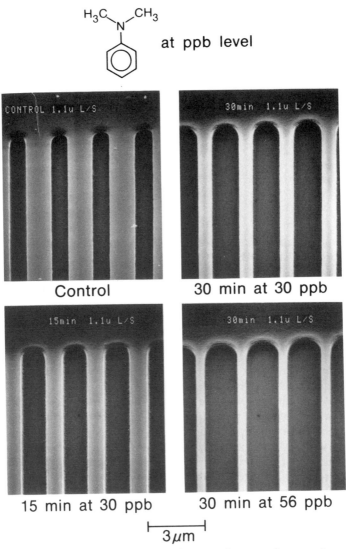

Figure 5. SEM photographs of negative tone images after a wafer coated with PTBOCST/Ph$_3$SSbF$_6$ stood in N,N-dimethylaniline-contaminated air prior to lithographic processing.

exposed to airborne base show significant shifts in linewidth, the magnitude of which increases with storage time and with concentration of base. In positive-acting CA resists, the degradation often is manifested by the formation of a poorly soluble lip or skin on the relief image at the resist-air interface (21,22).

The resist chemistry suggests a reason for the observed degradation in performance. A central feature of any chemically amplified resist is the use of a small number of photogenerated acid molecules to initiate a catalytic reaction. Because of the long catalytic chain length, any basic impurity which reduces the activity of the photogenerated acid will have a correspondingly large effect that degrades lithographic performance. N-methylpyrrolidone (NMP), a liquid in common use as a stripper and coating solvent in semiconductor manufacturing, induces a pronounced and rapid deterioration in the imaging properties of many CA resists. Using an airstream containing radiolabeled NMP, we have shown that the molar amount of airborne base absorbed by a CA resist at the onset of degradation is comparable to the amount of photogenerated acid, consistent with an acid-base interaction(23).The impact of an acid-base neutralization reaction between a basic contaminant and the photogenerated acid can be considerable. To quantify this, small, known quantities of dimethyloctadecylamine were introduced into films of $PTBOCST/Ph_3S-SbF_6$ resist during the coating process. The films were then exposed to DUV light, and the deprotection kinetics of the adulterated films compared to the standard resist(20). Figure 6 shows the impact of the added base contaminant on the resist's deprotection kinetics. The data points represent the behavior of the films determined by experiment. The solid lines are the results of simulations in which acid-base equilibria involving a base contaminant were added to the set of reactions in Scheme (b) of Figure 2. The impact of the base on the deprotection kinetics is large and can be predicted quantitatively.

Overall Deprotection Mechanism in Chemically Amplified Resists. Our kinetic studies thus far indicate that the deprotection kinetics in CA resists can be accurately described by considering the chemical processes listed in Figure 7. In systems examined thus far, the chemical deprotection mechanism is complex. Kinetic measurements by IR and UV absorption spectroscopy, coupled with computer simulation, provide evidence that acidolysis in these systems occurs via specific acid catalysis. In addition, the basicity of the deprotection product, the acidity of the photogenerated acid relative to the other resist components, and presence of basic airborne environmental contaminants all have been shown to be factors influencing the overall rate of deprotection.

While these studies yield an increased understanding of the details of CA resist thermal chemistry, a complete description of the lithographic process must include other factors. Thus far we have studied blanket-exposed films with homogeneous composition, but lithographic imaging by its nature entails a highly nonuniform spatial distribution of reactants and products. The kinetics of transport of various species, in particular the photogenerated acid catalyst, are important elements that affect the appearance of the final relief image and determine the ultimate resolution that can be attained. The study of transport kinetics in these systems is a challenge. Outlined below is one analytical method now under development that is intended to facilitate the characterization of diffusive transport of photogenerated acid in CA resists.

354

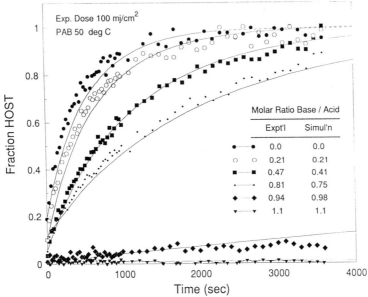

Figure 6. Experimental and simulated data showing the extent of polymer deprotection at 50 °C as a function of added base $(H_3CNH(CH_2)_{17})CH_3$ for the PTBOCST/Ph$_3$S-SbF$_6$ resist system. Experimental and simulated data showing the extent of polymer deprotection at 50 °C as a function of added base $(H_3CNH(CH_2)_{17})CH_3$ for the PTBOCST/Ph$_3$S-SbF$_6$ resist system.

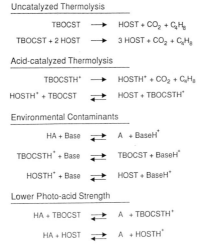

Figure 7. Proposed reaction scheme describing the chemistry occurring in CA resist films

A Method for Visualizing Acid Latent Images in Polymer Films

Current understanding of photo-acid diffusion in CA resist films is imprecise. Since the matrix changes chemically and physically during thermal processing, acid diffusion in this instance is a complex process that in general is not characterizable by a single value of the diffusion coefficient. Traditional methods for characterizing diffusion(24) are not easily applied in this instance, prompting the development of alternate methods for establishing the acid's diffusion properties.

In general, its experimental study is hampered by the lack of techniques suitable for the detection of subtle compositional variations in organic matrices with a suitable spatial resolution. In the case of latent images in a CA resist, a typical resist composition consists of 1-5 wt% of the photoacid-generating sensitizer; a typical imaging dose photolyzes 10-50% of this sensitizer to produce a very low concentration of acidic photoproduct.The acid is confined to regions of a few hundred nanometers scale, and is expected to diffuse over distances of a few to a few hundred nanometers.

We are examining several avenues for characterization of the latent image of acid in CA resist films. One promising approach is derived from a technique termed photoinitiated interfacial cationic polymerization (PICP)(25) (Figure 8(a)). In this method, a thin polymer film containing a photo-acid generator (PAG) is exposed so that a latent image of acid is created at the surface of the film. The film is then treated with a reactive monomer that undergoes acid-catalyzed (i.e. cationic) polymerization. The end result is the formation of polymeric relief images on the surface of the original film. Our goal is to use these relief images as a means for visualizing, and thereby quantifying, the form of the acid latent image in CA resists (Figure 8(b)).

To be useful for that purpose, this technique must satisfy several requirements: (a) the thickness of the deposited film must be related in a simple and unambiguous way to the acid concentration at the interface of the underlying film; (b) the method must be sufficiently sensitive that acid levels comparable to those found in typical CA resist formulations can be quantified; and (c) the spatial resolution must be such that the detailed study of acid transport over small length scales is possible.

In the original work(25), the rate of polymer growth in PICP was found to be complex and nonlinear with time, showing first an induction period, then a period of rapid film growth, and finally a plateau region. By appropriate selection of conditions, we have found that the thickness of the deposited polymer has a simple linear relation to the local acid concentration. To demonstrate this, films of poly(styrene) containing 5 wt% di(4-t-butyliodonium) trifluoromethanesulfonate (TBIT) were coated onto silicon wafers, and large pads ca. 10 mm square were exposed to varying doses of light at 248 nm wavelength. The films were then treated with a solution of glycidoxypropyltrimethoxysilane (GPTS) is hexane solution for 60 seconds. The thickness of the deposited film was measured using a Nanospec optical reflectance film thickness analyzer. In Figure 9 is plotted the measured film thickness as a function of exposure dose. The accuracy of the film thickness measurement was verified by profilometry. In the dose regime shown here, the amount of photogenerated acid formed is essentially linear with exposure dose. These data demonstrate that, under these conditions, an essentially linear relation between the thickness of deposited polymer and the concentration of acid at the film interface exists. It is clear from this plot that very low quantities of acid may be detected. The

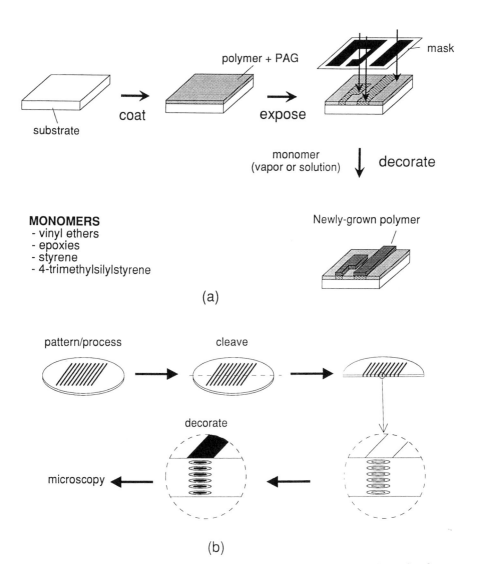

Figure 8. Photoinitiated interfacial cationic polymerization. (a) schematic of process flow; (b) proposed application to the visualization of acid latent images in CA resist films.

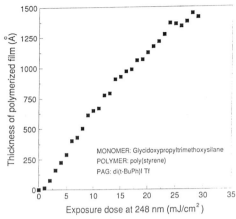

Figure 9. Film growth as a function of exposure dose using a PICP process.

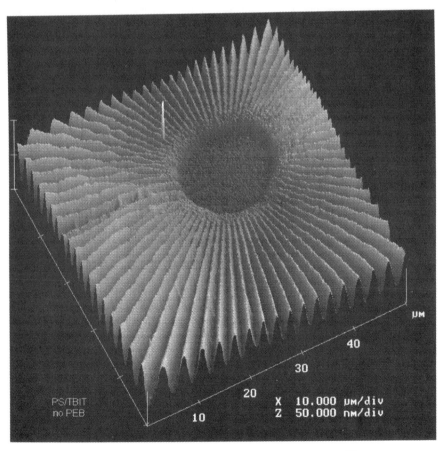

Figure 10. Surface profile of deposited polymer using a PICP process.

exposure doses over which linear response is observed cover the range of interest for CA resist films. While the surface polymers are expected to grow at the interface in an isotropic manner, by restricting the thickness of the deposited film to a fraction of the dimensions of the exposed features, it should prove possible to grow and characterize features at high spatial resolution. Early experiments provide an indication of the capability for high resolution. Using the TBIT/poly(styrene) model system, a series of resolution test patterns were exposed with a Nikon step-and-repeat exposure tool using 248 nm light. The exposed films were treated with GPTS, depositing polymer on the acidic regions of the film surface. The surface topography of the deposited pattern was then examined using atomic force microscopy. Figure 10 shows one feature from such a wafer. This test structure is comprised of a series of lines and spaces of equal width arranged in a radial format, where the linewidth shrinks continuously as one moves toward the center of the circular pattern. The smallest features that can be resolved in the test patterns are ca 0.3 μm, roughly the practical resolution limit of this exposure tool. As anticipated, the heights of the peaks and the depths of the valleys decrease as the line space dimension shrinks, a consequence of increasing diffraction as the feature dimensions approach the wavelength of exposure. Further characterization of the ultimate resolution limits of the method and its applicability to the characterization of acid transport in polymer films is now underway.

Acknowledgments

This work was been supported in part by National Science Foundation Grant DMR-9400254 (CPIMA).

Literature Cited

1. C. G. Willson, *Introduction to Microlithography*, Second Ed., C.G. Willson, L. Thompson and M. Bowden, eds., American Chemical Society, Washington, DC, 1994; pp 139-267.

2. R. Dammel, *Diazonaphthoquinone-based Resists,* SPIE Optical Engineering Press, Bellingham, WA, 1993.

3. E. Reichmanis, F. Houlihan, O. Nalamasu and T. Neenan, *Chem. Matls.*, **1991**, *3*, 394.

4. H. Ito and G. Willson, *Polym. Eng. Sci.*, **1982**, *23*, 1012; J. Frechet, H. Ito, C. G. Willson, *Proc. Microcircuit Engineering*, **1982**, *82*, 260.

5. D. McKean, U. Schaedili and S. MacDonald, *Polymers in Microlithography*, ACS Symposium Series 412, E. Reichmanis, S. MacDonald and T. Iwayanagi, eds., American Chemical Society, Washington, DC, 1989; pp 27-38.

6. S. MacDonald, N. Clecak, R. Wendt, G. Willson, C. Snyder, C. Knors, N. Deyoe, J. Maltabes, J. Morrow, A. McGuire and S. Holmes, *Proc. Soc. Photo-Opt. Instr. Eng.*, **1991**, *1466*, 2.

7. J. Sturtevant, S. Holmes and P. Rabidoux, *Proc. Soc. Photo-Opt. Instr. Eng.*, **1992**, *1672*, 114.

8. H. Roschert, C.Eckes, H. Endo, Y. Kinoshita, T. Kudo, S. Masuda, H. Okazaki, M. Padmanaban, K. Pryzbilla, W. Spiess, N. Suehiro, H. Wengenroth and G. Pawlowski, *Proc. Soc. Photo-Opt. Instr. Eng.,* **1993**, *1925*, 14.

9. T. Fedynyshyn, C. Szmanda, R. Blacksmith and W. Houck, *Proc. Soc. Photo-Opt. Instr. Eng.,* **1993**, *1925*, 2.

10. M. Zuniga, G. Wallraff and A. Neureuther, *Proc. Soc. Photo-Opt. Instr. Eng.,* **1995**, *2438*, 113.

11. G. Wallraff, J. Hutchinson, W. Hinsberg, F. Houle, P.Seidel, R. Johnson and W. Oldham, *J. Vac. Sci. Tech. (B),* **1994**, *12*, 3857.

12. Chemical Kinetics Simulator Version 1.01, available from IBM with a no-cost license via the Internet at URL http://www.almaden.ibm.com/st/msim.

13. S. Benson and H. O'Neal, "Kinetic Data on Gas Phase Unimolecular Reactions," *NBS Reference Data Series,* **1970**, *21*, 189.

14. H. Ito, *J. Polym.Sci., Part A,* **1986**, *24*, 2971.

15. J. March , *Advanced Organic Chemistry,* Fourth Ed., Wiley-Interscience, New York, 1992; p. 250.

16. ref. 15, Chapters 8 and 10.

17. S. Benson, *Thermochemical Kinetics*, Second Ed., John Wiley & Sons, New York, 1976.

18. T. Long, S. Obendorf and F. Rodriguez, *Polym. Eng. Sci.,* **1992**, *32*, 1589.

19. H. Maskill, *The Physical Basis of Organic Chemistry*, Oxford University Press, New York, 1992.

20. G. Wallraff, W. Hinsberg, F. Houle, J. Opitz and D. Hopper, *Proc. Soc. Photo-Opt. Instr. Eng.,* **1995**, *2438*, 182.

21. W. Hinsberg, S. MacDonald, N. Clecak and C. Snyder,, *Proc. Soc. Photo-Opt. Instr. Eng.,* **1992**, *1672*, 24.

22. O. Nalamasu, M. Cheng, A. Timko, V. Pol, E. Reichmanis and L. Thompson, *J. Photopoly. Sci. Technol.,* **1991**, *4*, 299.

23. S. MacDonald, W. Hinsberg, R. Wendt, N. Clecak and G. Willson, *Chem. Matls.,* **1993**, *5*, 348.

24. see *Diffusion in Polymers,* J. Crank and G. Park, eds., Academic Press, New York, 1968.

25. A. Hult, S. MacDonald and G. Willson, *Macromolecules,* **1985**, *18*, 1804.

Chapter 26

The Effect of Macromolecular Architecture on the Thin Film Aqueous Base Dissolution of Phenolic Polymers for Microlithography

G. G. Barclay[1,3], M. King[1], A. Orellana[1], P. R. L. Malenfant[1], R. Sinta[1], E. Malmstrom[2], H. Ito[2], and C. J. Hawker[2,3]

[1]Shipley Company, 455 Forest Street, Marlborough, MA 01752–3092
[2]Center for Polymeric Interfaces and Macromolecular Assemblies, IBM Almanden Research Center, 650 Harry Road, San Jose, CA 95120–6099

The aqueous base development of phenolic thin films is the basis of modern microlithographic technology.[1] Thus, the dissolution behavior of phenolic homopolymers and copolymers in aqueous base plays a critical role in photoresist performance. As the minimum feature size required from photoresist compositions decreases below 0.20 μm, control of the macromolecular architecture of the matrix resin will play an increasingly important role. For example, in the case of novolac based photoresists the molecular weight distribution profoundly affects the dissolution behavior and lithographic performance of the resist.[2,3]

To date, little research has been carried out on the effect of the molecular weight, polydispersity and macromolecular architecture of linear phenolic polymers, such as poly(4-hydroxystyrene) (PHOST). However, Gabor et al.[4,5] reported that random and block methacrylate copolymers comprising siloxane monomers and t-butylmethacrylate show very different lithographic behavior. The block copolymers exhibited good imaging properties, while the random copolymers showed poor lithographic behavior. Further, copolymer systems used in advanced chemically amplified photoresists consist of monomeric units of very different polarity, for example, a polar phenolic unit and a nonpolar acid sensitive moiety used in deprotection chemistry.[6] To elucidate the effect of macromolecular architecture upon the dissolution of phenolic thin films in aqueous base, three types of well defined phenolic macromolecules were prepared and the dissolution behavior investigated:

1. **Homopolymers**: Poly(4-hydroxystyrene)s were prepared to investigate the effects of polydispersity and molecular weight on dissolution behavior.

[3]Corresponding authors.

2. Random Copolymers: Random copolymers incorporating various ratios of 4-hydroxystyrene and styrene were synthesized to study the effects of hydrophilic and hydrophobic groups upon aqueous base solubilization.

3. Block Copolymers: Block copolymers incorporating various ratios of 4-hydroxystyrene and styrene were prepared, to investigate the effect of macromolecular architecture upon thin film dissolution.

Recently tremendous interest has been generated by "living" free radical polymerizations.[7] This technique affords accurate control over molecular weight distribution and chain ends, and provides a simple synthetic technique for the preparation of well defined complex polymeric architectures. "Living" radical polymerization proceeds via the reversible capping of growing chain ends with 2,2,6,6-tetramethyl-1-piperidinyloxy (TEMPO), reducing free radical concentration and inhibiting premature chain termination. Poly(4-hydroxystyrene), and random and block copolymers incorporating various ratios of 4-hydroxystyrene and styrene were prepared using this *"living"* free radical technique.

Synthetic Procedures
Materials. 4-Acetoxystyrene (1) (Hoechst Celanese) and styrene (5) (Aldrich) were distilled under reduced pressure prior to use. 1-Phenyl-1-(2',2',6',6'-tetramethyl-1'-piperidinyloxy)ethane (2) was prepared as previously reported.[8]
Polymerization of 4-Acetoxystyrene (1). 4-Acetoxystyrene (1) (75.0 g, 0.463 mol) was placed in a 250 mL round bottom flask and purged with N_2. The appropriate molar quantity of 1-phenyl-1-(2',2',6',6'-tetramethyl-1'-piperidinyl-oxy)ethane (2) was then added to obtain the desired molecular weight. For example, to obtain a molecular weight of approximately 16,200 a.m.u. a molar ratio of 1:100, unimolecular initiator (4) to 4-acetoxystyrene (1), would be used. After addition of the initiator, the polymerization mixture was heated to 125-130°C, under N_2, and stirred for 48 hours. During the polymerization the polymer solidified in the reaction vessel. The reaction was then cooled to room temperature. The polymer was dissolved in acetone (225 mL) and isolated by precipitation into hexanes (2250 mL). The polymer was then filtered, washed with hexanes and dried in a vacuum oven overnight at 50°C. Typical isolated yields are between 80 to 95% of theory.
Deacetylation of Poly(4-acetoxystyrene) (3). To a slurry of poly(4-acetoxystyrene) (3) (50.0 g, 0.308 mol) in isopropyl alcohol or methanol at reflux (200 mL), under N_2, ammonium hydroxide (24.25 g, 0.692 mol), dissolved in water (36 mL), was added dropwise over 15 minutes. After addition, the reaction mixture was heated at reflux for 18 hours, during which time the polymer went into solution. The reaction was then cooled to room temperature, and the polymer isolated by precipitation into water (1500 mL), filtered, washed well with water, and dried in a vacuum oven overnight at 50°C. Typical isolated yields of poly(4-hydroxystyrene) (4) are between 80 to 90% of theory.

Using this novel polymerization technique, narrow polydispersity PHOST polymers have been synthesized (PD = 1.1 to 1.4), with a variety of molecular weights, ranging from M_n = 2,000 to M_n = 30,000.

Random Copolymerization. 4-Acetoxystyrene (1) (15.68 g, 0.097 mol) and styrene (5) (4.32g, 0.042 mol) were placed in a 100 mL round bottom flask and purged with N_2. 1-Phenyl-1-(2',2',6',6'-tetramethyl-1'-piperidinyloxy)ethane (2) (0.52g, 0.002 mol) was then added. After addition of the initiator, the polymerization mixture was heated to 125-130°C, under N_2, and stirred for 48 hours. During the polymerization the polymer solidified in the reaction vessel. The reaction was then cooled to room temperature. The polymer was dissolved in acetone (100 mL), and isolated by precipitation into hexanes (1000 mL). The poly(4-acetoxystyrene-co-styrene) (6) was then filtered, washed with hexanes and dried in a vacuum oven overnight at 50°C. Isolated yield 92% of theory. Mn = 9230, Mw = 10330, PD = 1.12 (Theoretical A.M.U = 10050).

A similar synthetic procedure was used to prepare the various random copolymers of 4-acetoxystyrene and styrene.

Deacetylation of Poly(4-acetoxystyrene-co-styrene) (6). To a slurry of poly(4-acetoxystyrene-co-styrene) (70:30) (6) (10.0 g, 0.069 mol) in methanol at reflux (50 mL), under N_2, ammonium hydroxide (5.33 g, 0.152 mol), dissolved in water (10 mL), was added dropwise over 15 minutes. After addition, the reaction mixture was heated at reflux for 18 hours, during which time the polymer went into solution. The reaction was then cooled to room temperature. The polymer was isolated by precipitation into water (500 mL), filtered, washed well with water, and dried in a vacuum oven overnight at 50°C. Isolated yield of poly(4-hydroxystyrene-co-styrene) (7) 85% of theory. Mn = 7278, Mw = 8297, PD = 1.14.

Block Copolymers. Well defined block copolymers of 4-hydroxystyrene and styrene were prepared by firstly, the synthesis of low molecular weight blocks of styrene capped with TEMPO. The molecular weights of these styrene blocks were controlled by the ratio of unimolecular initiator (2) relative to styrene monomer (5). For example, 1-phenyl-1-(2',2',6',6'-tetramethyl-1'-piperidinyloxy)ethane (2) (1.67g, 0.0064 mol) was added to styrene (5) (20.0g, 0.192 mol) and heated to 125-130°C, under N_2, for 48 hours. The reaction was then cooled to room temperature. The polymer was dissolved in tetrahydrofuran (100 mL) and isolated by precipitation into methanol (1000 mL). The TEMPO terminated polystyrene (8) was then filtered, washed with methanol and dried in a vacuum oven overnight at 50°C. Isolated yield 90% of theory. Mn = 2764, Mw = 3062, PD = 1.10 (Theoretical A.M.U = 3120).

These short TEMPO capped styrene blocks (8) were then used as "macroinitiators" for the further polymerization of 4-acetoxystyrene (1) using the same procedure as described above. The length of the acetoxystyrene block is controlled by the ratio of the "macroinitiator" (8) to acetoxystyrene monomer. The acetoxystyrene block was then converted to 4-hydroxystyrene by quantitative base hydrolysis of the acetoxy groups as previously described.

Characterization. Compositions were determined using a GE QE300 NMR spectrometer, [1]H (300 MHz) and [13]C (75.4 MHz) NMR, in acetone d_6, with tetramethylsilane as an internal standard. Molecular weights of the polymers were determined by gel permeation chromatography (GPC) using a Waters model 150C equipped with four Ultrastyragel columns in tetrahydrofuran at 40°C. The molecular weight values reported are relative to polystyrene standards.

Discussion.

1.The Synthesis of Well-defined Phenolic Macromolecules.

To determine the effect of PD and molecular weight on the dissolution behavior, a series of narrow PD PHOST were prepared (Scheme 1). One of the major advantages of "living" radical polymerizations is that the molecular weight of the polymer can be controlled. Furthermore, low molecular weight polymers (M_n < 20,000), which are usually used in microlithography, are easily attainable. Good control of PD was observed for the PHOST prepared in this manner. Over the range of molecular weights prepared (2,000 - 30,000) the molecular weight distribution was maintained below 1.5, Table 1.

A series of random and block copolymers containing various ratios of 4-hydroxystyrene and styrene were also prepared (Schemes 2 & 3). In all cases, both random and block, the monomer feed ratio was in agreement with the composition of the isolated polymers as determined by ^{13}C NMR. The molecular weights obtained were also in good agreement with the target experimental molecular weight (10,000 AMU). Complete deactylation of these acetoxy copolymers to form the polar hydroxyl functionality was determined by quantitative ^{13}C NMR. These model copolymers were then used to investigate the combined effects of macromolecular architecture and hydrophobicity upon thin film dissolution.

2. The Effects of Molecular Weight and Polydispersity (PD) on Dissolution Rate.

A number of narrow and broad polydispersity PHOST polymers were used to investigate the effect of M_n, M_w and PD, on the dissolution behavior of this class of phenolic polymer. Films of these PHOST polymers and blends were cast on 1" Si wafers and baked at 150°C for 1 minute to remove the casting solvent. The film thickness (approximately 1.0 μm) was measured on a Tencor Alphastep. The dissolution rates of these films were then measured by immersion in 0.21N aqueous tetramethylammonium hydroxide solution (MF321[®]) using a Single Point Development Rate Monitor.

The dissolution rate of the PHOST films as a function of the M_w for polymers with narrow and broad PD, is shown in Figure 1. It was observed that the dissolution rate exponentially and rapidly decreases with increasing molecular weight in M_w 10,000 g/nl range. Above approximately 15k, the effect of molecular weight on the dissolution rate is negligible. Rodriguez et al reported a steady decrease in the dissolution rate of PHOST with increasing M_w in a sodium hydroxide aqueous solution, an abrupt increase in the rate at M_w ~20,000, and a constant rate above M_w ~ 20,000.[9] As Figure 1 indicates, we only observe a monotonous molecular weight effect on the dissolution rate. The threshold-like molecular weight dependence of the dissolution rate is more pronounced in the

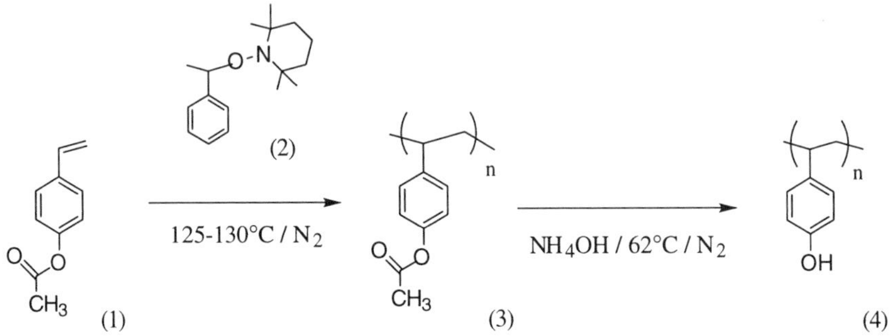

Scheme 1. Synthesis of Well-defined Poly(4-hydroxystyrene)

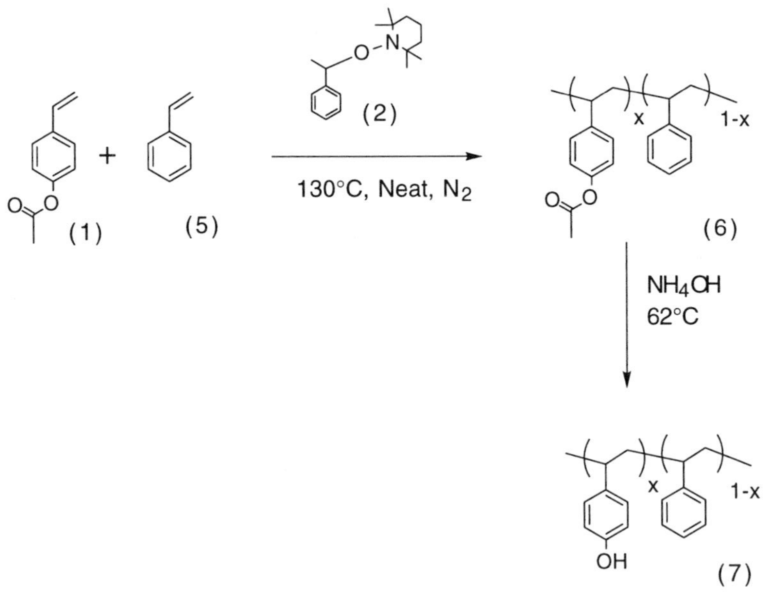

Scheme 2. Synthesis of Random Copolymers

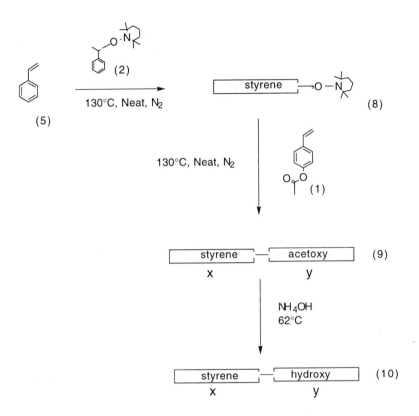

Scheme 3. Synthesis of Block Copolymers

Table 1. Molecular Weight Data on Poly(4-hydroxystyrene)s

M_n	M_w/M_n	T_g (°C)
2304	1.19	149
3874	1.18	172
6528	1.42	177
12726	1.38	185
24298	1.43	186

case of narrow PD polymers. The broad PD polymers exhibit a more gradual change of the dissolution rate when the molecular weight is changed. Furthermore, the narrow PD polymers tend to dissolve more slowly than the broad PD polymers, especially in the molecular weight range of M_n = 3,000 - 15,000. Two binary blends of narrow PD PHOST polymers are also shown in Figure 1. These blends consisted of a 1:1 blend of polymers giving an M_n = 5,000, M_w = 8,900, PD 1.76 and a 60:40 mixture which gave an M_n = 10,700, M_w = 17,200, PD = 1.6, respectively. The dissolution rates in 0.21 N TMAH were 81 and 38 nm/sec, respectively. Thus, the blends show dissolution rates that are an average of the rates of the individual components.

The dissolution rates of all the PHOST systems (narrow and broad PD polymers) with PD ranging from 1.2 to 5.5 are plotted against M_w in a log-log scale in Figure 2. This indicates that there is no straightforward relationship between the dissolution rate and M_w if the PD difference is not taken into consideration. In contrast, the similar plot of the dissolution rates of PHOST vs M_n can be nicely expressed by an exponential decay (linear log-log plot) even over a wide range of molecular weight distributions (Figure 3). Thus M_n is a more meaningful parameter for predicting the dissolution rates of PHOST. That is, the dissolution rate of PHOST can be estimated if M_n is known. This observation suggests that the dissolution of PHOST in aqueous base is controlled by lower molecular weight fractions which contrasts the novolac systems reported by Willson et al.[3]

3. The Effect of Macromolecular Architecture Upon Aqueous Base Dissolution.

A preliminary investigation into the effect of macromolecular architecture on the aqueous base dissolution behavior of these phenolic copolymers was undertaken. Well-defined random and block copolymers of 4-hydroxystyrene and styrene were prepared with a variety of molar compositions, ranging from 90:10 mol% (hydroxystyrene:styrene) to 55:45 mol%. The thin film aqueous base solubilization of these copolymers was then investigated. Thin films (1.0 μm thick) of these phenolic copolymers were cast on silicon wafers and baked at 110°C for 1 minute to remove the casting solvent (ethyl lactate). The dissolution rates were then measured by immersion in a variety of aqueous tetramethylammonium hydroxide (TMAH) solutions of increasing normality (0.14, 0.26, 0.30, 0.60 and 1.0 N).

Figure 4, shows the dissolution rates of a range of well-defined random and block copolymers of 4-hydroxystyrene and styrene in 0.26 N TMAH. As the styrene content is increased there is a dramatic reduction in the aqueous base solubility of both the random and block copolymers. The dissolution rate of these copolymers approaches 0 Å/sec for the 55:45 composition in 0.26 n TMAH. Interestingly, it was observed that over a composition range of approximately 20

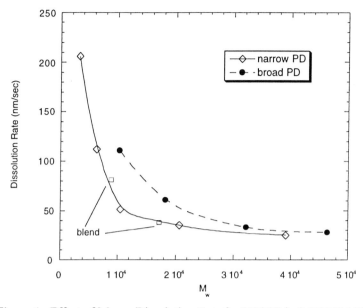

Figure 1. Effect of M_w on Dissolution Rate for PHOST in 0.21N TMAH

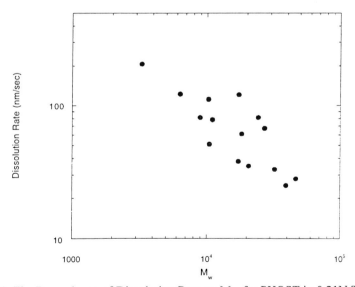

Figure 2. The Dependence of Dissolution Rate on M_w for PHOST in 0.21N TMAH. Narrow and Broad Polymers are plotted on a log to log scale.

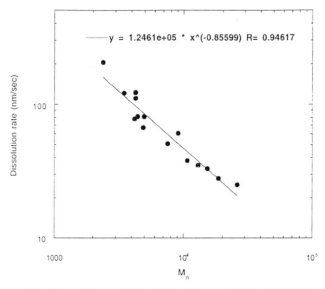

Figure 3. The Dependence of Dissolution Rate on M_n for PHOST in 0.21N TMAH. Narrow and Broad Polymers are plotted on a log to log scale.

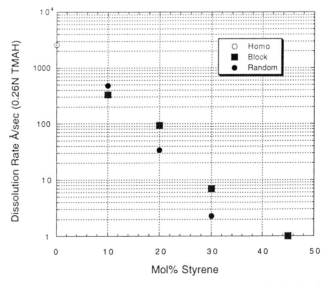

Figure 4. The Effect of Macromolecular Architecture on Dissolution in 0.26N TMAH

- 30 mol% styrene the block copolymers exhibited significantly faster dissolution than the corresponding random copolymers. At compositions containing less than 10 mol% or greater than 40 mol% styrene the aqueous base solubility of the random and block copolymers are similar.

To further investigate the improved aqueous base solubility of the block copolymers compared to the random copolymers, the dissolution behavior of the compositions containing the greatest mol% (45) of styrene were investigated as a function of TMAH normality (Figure 5). At normalities below 0.3 N, there is little difference in the dissolution rate between the random and block copolymers, however, at normalities greater than 0.3 N, there is a considerable difference in dissolution behavior between the random and block architectures. For example, the block 55/45 copolymer exhibited a 10x faster dissolution rate than the corresponding random copolymer in 0.6 N TMAH. The possible reasons for this dramatic and fundamental difference in the aqueous base dissolution behavior between random and block copolymers of 4-hydroxystyrene and styrene is under investigation. However, it is interesting to note that the block copolymers exhibit non-uniform dissolution behavior as observed by irregular development rate monitor (DRM) interferograms. This is in contrast to the random copolymers which exhibited normal, "smooth" dissolution behavior. One possible explanation for the abnormal dissolution behavior of the 55/45 block copolymer is microphase separation. These small domains of differing hydrophobicity result in the uneven solubilization of the polymer matrix at the developer front which causes diffuse reflection from rough surfaces at the interface. Small angle X-ray analysis is underway to confirm microphase separation in these higher styrene content block copolymers.

Conclusions

A series of well defined homopolymers and copolymers of 4-hydroxystyrene and styrene were prepared by *"living"* free radical polymerization. It was observed that as the molecular weight of poly(4-hydroxystyrene) is decreased, there is an exponential increase in dissolution rate. This study suggests that the dissolution behavior of PHOST in aqueous base is primarily governed by the lower molecular weight fractions. Further, the DR can be estimated if M_n is known. In the case of the 4-hydroxystyrene:styrene copolymers as the mol% of styrene is increased the dissolution rate in 0.26N TMAH dramatically decreases for a constant M_w. Interestingly a fundamental difference in the dissolution behavior of the block copolymers compared to the random copolymers was observed. That is, the "block" macromolecular architecture result in faster dissolution rates in aqueous base than equivalent random copolymers, for a constant M_w. Further, it was also found that at higher styrene contents the block copolymers show non-uniform dissolution behavior compared to the corresponding random copolymers.

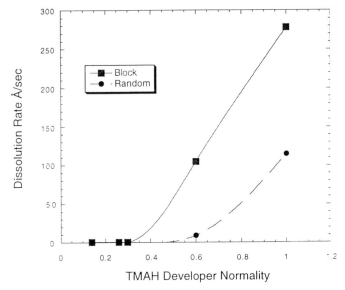

Figure 5. Effect of Developer Normality on Dissolution of 55:45
Poly(4-hydroxystyrene-co-styrene)

Acknowledgments
The authors gratefully acknowledge the financial support of the NSF Center for
Polymeric Interfaces and Macromolecular Assemblies, the IBM Corporation, and
Shipley Company.   Fellowship support (EM) from the Hans Werthen
Foundation, Sweden, is also acknowledged.

References
1. C. G. Willson. *Introduction to Microlithography (2nd Ed)*, L.F. Thompson, C.
G. Willson, M. J. Bowden, Eds., American Chemical Society, Washington, DC,
1994, p139.
2. R.D. Allen; K.J. Rex Chen; P.M. Gallagher-Wetmore, *Proc. SPIE*, **1995**, <u>2438</u>, 250.
3. P.C. Tsiartas; L.L. Simpson; A. Qin; C.G. Willson; R.D. Allen; V.J. Krukonis; P.M
Gallagher-Wetmore, *ibid*, 261.
4. A.H. Gabor, L.C. Pruette and C.K. Ober, *Chem. Mater.*, **1996**, <u>8</u>, 2282.
5. A.H. Gabor and C.K. Ober, *Chem. Mater.*, **1996**, <u>8</u>, 2272.
6. E. Reichmanis, F.M. Houlihan, O. Nalamasu, T.X. Neenan, *Chem. Mater.* **1991**,
<u>3</u>, 394.
7. M.K Georges; R.P.N. Veregin; P.M. Kazmaier; G.K. Hamer, *Macromolecules*,
1993, <u>26</u>, 2987. C.J Hawker, *J. Am. Chem. Soc.,* **1994**, <u>116</u>, 11314. K.
Matyjaszewski; S. Gaynor; J.S. Wang, *Macromolecules*, **1995**, <u>28</u>, 2093.
8. C.J. Hawker, G.G. Barclay, A. Orellana, J. Dao and W. Devonport,
Macromolecules, **1996**, <u>29(16)</u>, 5245.
9. T. Long; F. Rodriguez, *Proc. SPIE*, **1991**, <u>1466</u>, 188.

Chapter 27

Polymeric Organic–Inorganic Hybrid Nanocomposites

J. L. Hedrick[1,3], R. D. Miller[1], D. Yoon[1], H. J. Cha[1], H. R. Brown[1],
S. A. Srinivasan[1], R. Di Pietro[1], V. Flores[1], J. Hummer[2], R. Cook[2], E. Liniger[2],
E. Simonyi[2], and D. Klaus[2]

[1]IBM Almaden Research Center, 650 Harry Road, San Jose, CA 95120
[2]IBM T. J. Watson Research Center, Route 134 Kitchawan Road, Yorktown
Heights, NY 10589

Transparent, nanophase-separated inorganic-organic hybrid polymers
with high modulus and excellent crack resistance have been prepared
from reactively functionalized poly(amic ester) precursors of polyimide
and substituted, oligomeric silsesquioxanes. These polyimide/
poly(silsesquioxane) hybrid materials are stable to 400 °C and exhibit
an isotropic dielectric constant of ca. 2.9. Induced cracking and crack
propagation studies performed with the application of external stress
suggest a maximum critical film thickness in excess of 2.0 μm. These
hybrid materials appear to be significantly toughened by the chemical
incorporation of the polyimides relative to organically modified silicates
and spin-on-glasses without adversely affecting other important
properties of spin-on-glass.

As on-chip device densities increase and active device dimensions shrink, signal delays
increase due to capacitive coupling and crosstalk between the metal interconnects (1).
The situation is exacerbated by the need to keep conductor lines as short as possible in
order to minimize transmission delays, thus necessitating complex multilevel wiring
schemes for the chip. Since both capacitive delays and power consumption each
depend critically on the dielectric constant of the insulators (2,3), much attention has
focused recently on the replacement of the standard silicon oxide with new intermetal
dielectrics (IMD) having dielectric constants lower than conventional oxide
(k = 3.9–4.2) (2,4). This is not a simple matter given the complexities and demands of
current semiconductor integration processes which are such that replacement of oxide
as the IMD material with a material of significantly lower dielectric constant could be
considered to be one of the great materials challenges of the 90's. A replacement
dielectric must not only have a significantly lower and frequency independent dielectric
constant, but also must withstand temperatures of 400–450 °C associated with current
integration processes, provide a barrier to metal ion diffusion, adhere strongly to a

[3]Corresponding author.

variety of metals, ceramics and substrates, resist cracking under extreme conditions, be inert to common metallurgy and liners, be resistant to O_2-RIE etching and tolerant to chemical mechanical polishing (CMP) yet amenable to high resolution lithographic imaging, and show low water absorption, etc. In addition, an IMD material should be electrically isotropic, show minimal leakage currents and possess a high dielectric breakdown threshold. Although many organic polymers have dielectric constants below those of thermal and CVD oxide, their thermal stabilities and/or glass transition temperatures are often inadequate for current semiconductor processing. This has focused attention recently on inorganic-organic network materials such as spin-on-glasses (SOG), organically modified silicates, silsesquioxanes, etc., prepared by sol-gel condensation processes (4,5). These materials have many attributes. When properly cured, they have dielectric constants ranging from 2.7–3.2, are chemically unreactive and are exceptionally thermally stable. However, due to the loss of solvent and other small molecules through the polycondensation, shrinkage (6) occurs during curing which often leads to cracking, a situation which is exacerbated as the functionality number of the material and/or film thickness increases. For this reason, there have been numerous attempts to toughen silicates and organically modified derivatives through the incorporation of a variety of thermoplastic or network forming polymers (7). The advent of the sol-gel or spin-on-glass process for the preparation of high purity glass or ceramics has created the possibility to incorporate a polymeric component into the precursor solution (8–10). The chemistry of the sol-gel process involves the hydrolysis and condensation oligomerization of a group XIV metal alkoxide in solution (sol) and subsequent formation of a three-dimensional network. In such cases, the organic polymer must be thermally stable to allow proper curing of the silicate matrix. For use of such materials in IMD applications, the additive polymer must not only be thermally stable but should also have a relatively low dielectric constant.

Polyimide-inorganic hybrids are an emerging class of materials which may be designed to offer a range of properties depending on the relative composition of each component, the size scale of phase separation, and the chemistry and interactions between the organic and inorganic components (11–20). The initial attempts at hybridization involved mixing of a poly(amic-acid) solution with tetraethylorthosilicate (TEOS) in an aprotic dipolar solvent (11). In these systems, gross macroscopic phase separation was observed. Partial control of the phase separation and resulting morphology was accomplished by introducing inorganic functionality in the polyimide so as to chemically incorporate the organic component into the TEOS-based network (12–14). This was accomplished for low inorganic compositions by binding the metal alkoxide precursor to the carboxylic sites of the poly(amic-acid) (15,16). Upon imidization, the metal alkoxide is released and condenses, producing dispersed inorganic particles in the polyimide matrix. The rigidity of the polyimide prevents coarsening of the inorganic component. The resulting films were clear and small particles were reported.

Alternatively, Mascia and Kionl reported the use of glycidylpropyltrimethoxysilane as a means to compatibilize TEOS with poly(amic-acids) (17–20). The use of the coupling agent served to significantly reduce the size scale of phase separation, which occurred via a spinodal decomposition process. Other key factors in structure control included the molecular weight of the poly(amic-acid), and the reaction time for the coupling agent and catalyst. Other polyimide-silica hybrid materials were prepared by hydrolysis and cocondensation of TEOS with polyimides containing the *triethoxysilane functionality*. Relatively tough, free standing films were obtained for silica compositions as high as 70 wt.%. The degree of phase separation was controlled by the composition of TEOS in the mixture and the number of inorganic functionalities. In general, the size of the phases increased with increasing silica content and decreased with triethoxysilane content (*18*).

The synthetic procedure for the preparation of polyimide-silica hybrids involves mixing varying quantities of TEOS and water with the poly(amic-acid) solution. The resulting mixtures are cast and cured to effect imidization and concurrent network formation. For this, water is essential, and the stability of the poly(amic-acid) to hydrolysis is of concern. To this end, we have surveyed an alternative route to polyimide-silica hybrids based on a poly(amic ethyl ester) precursor to the polyimide (*20*). In this route, pyromellitic dianhydride (PMDA) is opened by ethanol to yield a meta, para mixture of half acid esters which can be separated by fractional recrystallization and converted to the respective acid chlorides. Polymerization with a diamine yields the target poly(amic alkyl esters). These precursors are hydrolytically stable allowing isolation, characterization and copolymerization in a wide variety of solvents and solvent mixtures. We have primarily used the poly(amic ethyl ester) precursor since imidization occurs at a substantially higher temperature than the poly(amic-acid) analog. In polyimide-silica hybrids derived from poly(amic-acid) solutions, the morphology is strongly influenced by the hydrogen bonding between the organic and inorganic components. If this interaction is lost prior to vitrification (i.e., imidization), significant coarsening during phase separation is observed. Chujo and coworkers (*21*) and others (*22*) have reported that nano-level phase separation is obtained for organic-inorganic hybrids only when there is inorganic functionality on the organic component and there is a strong interaction (i.e., hydrogen bonding) between the components. The onset of imidization of the poly(amic ethyl ester) is 250 °C and a cure temperature of 350 °C is required for quantitative imidization. In this article, we will discuss the preparation of triethoxysilyl functional poly(amic ethyl ester) oligomers, and the chemical modification of inorganic precursors as a route to microphase separated inorganic-organic hybrids.

Experimental

Materials. The N-methyl-2-pyrrolidone (NMP), dimethylpropylene urea (DMPU), and pyridine were purchased from Aldrich and used without further purification. The aminophenyltrimethoxysilane was purchased from Gelest Inc. and used without further purification. The bis(p-aminophenyl) phenyl trifluoromethylmethane was prepared according to a literature procedure (*23*).

Polymer Synthesis

Trimethoxysilyl Functionalized Poly(Amic Ethyl Ester) Oligomers ($<M_n>$ = 10,000 g/mol) (Scheme 1) (*24*). To a three-neck flask equipped with an overhead stirrer, nitrogen inlet and addition funnel was charged 3.2456 g (9.48 mmol) of 3FDA, 0.2218 g (1.04 mmol) of aminophenyltrimethoxysilane, 2 g (25 mmol) of pyridine and 50 mL of distilled NMP. The reaction mixture was maintained under a positive nitrogen pressure and cooled to 0 °C. The PMDA diethyl ester diacyl chloride (3.4716 g, 10 mmol) was dissolved in ~ 100 mL of methylene chloride, quantitatively transferred to the addition funnel and added dropwise to the cold, stirred reaction mixture. After the addition was complete, the polymerization was allowed to proceed overnight at room temperature. The poly(amic ethyl ester) oligomer was isolated by precipitation under high shear conditions (Waring blender) in methanol, filtered and dried in a vacuum oven at 60 °C. An oligomer ($<M_n>$ = 20,000 g/mol) was synthesized likewise employing the appropriate stoichiometry.

Characterization

The inherent viscosities were obtained with 0.5 g/dL solutions in NMP at 30 °C employing a Cannon Ubbelohde viscometer and were calculated from an average of five different runs. ^{1}H NMR spectra were obtained on a Bruker AC 250 MHz NMR spectrometer in either $CDCL_3$ or deuterated DMSO and are reported in ppm (δ) downfield from TMS. In the case of the trimethoxysilyl functionalized oligomers, the number average molecular weights were estimated from the ratio of the methyl protons of the ester functionality to the methoxy protons of the trimethoxysilyl end-group. FTIR analyses were carried out on a Nicolet FTIR on films prepared from a solution of poly(amic ethyl ester) in THF on NaCl plates. Dynamic TGA was performed on a Perkin-Elmer TGA-7 in air at a heating rate of 10 °C/min. Dynamic DSC was carried out on a DuPont 1090 instrument at 10 °C/min.

Measure of Fracture Properties. The hardness and modulus of the films were determined using an instrumented nano-indenter. A Berkovich diamond-pyramid triangular indenter was loaded onto the surface of 1 μm thick films to a peak load of 2 mN and then unloaded. The monitored load and displacement of the indenter exhibited a hysteresis and residual displacement on loading typical of brittle materials. Fracture properties were characterized using a controlled-flaw stressing technique in which the films were exposed to a reactive environment under applied stress. Strips 10 mm wide and 30 mm long were cleaved from the coated silicon wafers and a 5N indentation made in the center of the strips. The Vickers diamond-pyramid square indenter used produced a contact impression with cracks approximately 30 μm long emanating from the corners that extended through the film to the silicon substrate. The indented strips were then immersed in water either as indented or in a small four point bend fixture that applied 70 MPa tension to the outer fiber of the silicon substrate resulting in approximately 6 to 9 MPa additional stress on the film. Control strips that contained no indentations were also monitored or were exposed to air

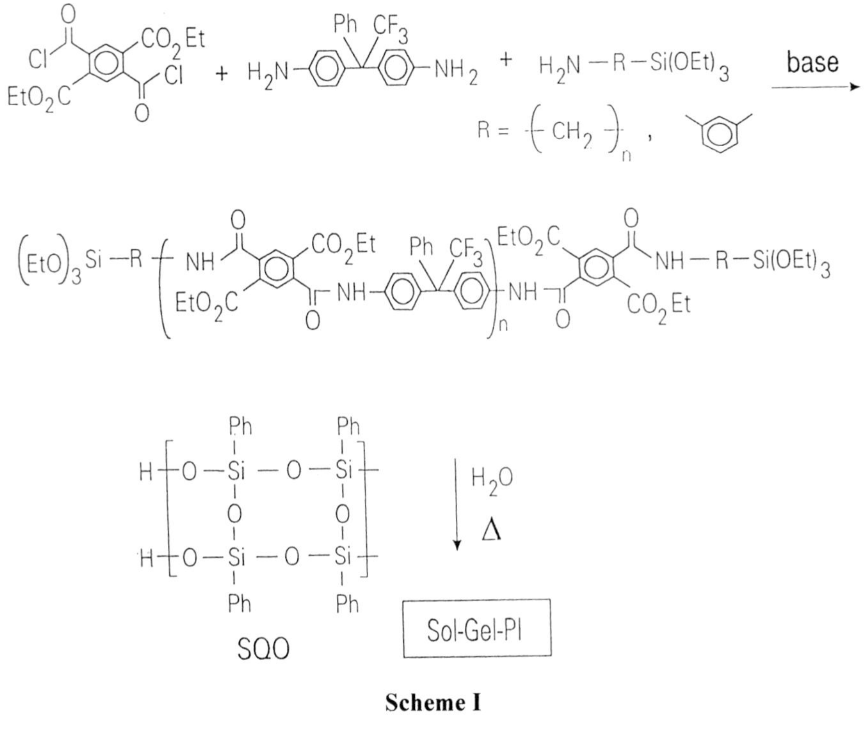

SQO

Sol-Gel-PI

Scheme I

(approximately 50% relative humidity). The behavior of the cracks was monitored using optical microscopy.

Results and Discussion

For the organosilicate matrix we have chosen silsesquioxane derivatives which are easily prepared from the hydrolysis of $RSiX_3$ derivatives (R = alkyl, aryl, X = halogen, alkoxy). These materials can be prepared in a variety of molecular weights depending on the condensation conditions (*25*). The structure of silsesquioxanes has been described variously as either ladder or cyclorandom polymers (see Figure 1). The terminating end groups are predominately –SiOH and the oligomers are prone to chain extension and crosslinking either by heating or in the presence of acids or bases. To date, we have examined a number of silsesquioxane derivatives whose polymer properties are shown in Figure 1. The organic substituents are either methyl or phenyl and the ratios can easily be determined by ^{1}H NMR.

We report here on our hybridization studies using Spl-4 (Figure 1) which is the predominately phenylated material. For simplicity, we designate this material as phenyl silsesquioxane or PSSQ. PSSQ is oligomeric with a degree of polymerization around 5 based on the number average molecular weight and is soluble in a variety of common organic solvents (THF, $CHCl_3$, γ-butyrolactone, NMP, etc.). The chain ends are predominately SiOH as determined by IR and ^{29}Si NMR analysis. The GPC data suggests a material which is ~ 40% chain ends based on a ladder polymer structure. A similar number was obtained by integration of the ^{29}Si peaks from –66 to –73 ppm ($PhSi(OR)$– $(OSi)_2$, R=H or alkyl)) relative to the upfield signals at –75 to –83 ppm ($PhSi$– $(OSi)_3$). Films of PSSQ can be spun from γ-butyrolactone, NMP, etc., and cured in situ. The overall weight loss from the solid PSSQ resin measured by TGA upon heating to 400 °C and holding at this temperature for 6 hrs is about 10% (much of this loss is solvent, reactants and low molecular weight by-products from the synthesis). The rate of weight loss at 400 °C after curing as measured by isothermal TGA, is about 0.2%/hr., confirming that PSSQ is quite thermally stable.

For the polyimide, we have chosen the polymeric material formally derived from pyromellitic anhydride and bis(p-aminophenyl) phenyltrifluoromethylmethane (*23,24*). We selected the polyamic ester route for the preparation of the functionalized precursor polymer, since these materials are isolable and are more stable than the corresponding poly(amic acids). The required diethylester acid chloride was prepared from reaction of pyromellitic anhydride with ethanol to first yield the half acid esters as a meta/para mixture. The isomers were partially separated by fractional crystallization (meta/para = 90/10) and the respective acid chlorides generated with thionyl chloride. The functionalized precursor polymer was generated as shown in Scheme I by reaction of the diethyl ester diacid chloride with the 3F-diamine (3FDA) in the presence of a monofunctional terminator such as trimethoxysilylaniline (95/5 para/meta). The molecular weights were controlled by stoichiometry according to the Carothers equation (*26*). The molecular weights M_n of the oligomers prepared were 8,900 and 17,000 gmol (Table I). A representative ^{1}H NMR spectra of the oligomer is shown in

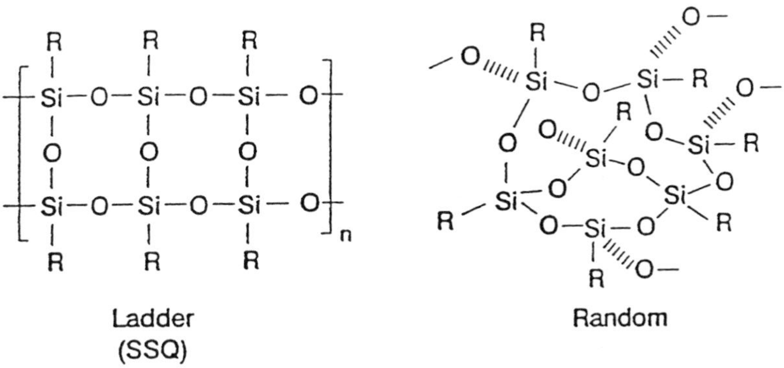

Figure 1. (a) Proposed silsesquioxane structures. (b) Polymer properties of various silsesquioxane oligomers.

Material	^{1}H NMR		M_W [a]	M_n [a]	wt. loss (%) at 425 °C (860m)
	% Ph	% Me			
Spl - 1	30	70	1,121	765	9.2
Spl - 2	45	55	1,127	796	11.5
Spl - 3	69	31	2,300	1,001	11.7
Spl - 4	95	5	3,455	1,207	11.9

Figure 2. The singlet at δ 3.16 corresponds to the methyl protons of the trimethoxysilyl group of the PI-Si(OCH$_3$)$_3$ terminated oligomers. The number average molecular weight of the trimethoxysilyl functionalized oligomers was estimated by the ratio of the methyl proton signal of the ethyl ester centered at δ 1.7 to the trimethoxysilyl protons at δ 3.16. The calculation was based on the assumption that there were two such end-groups per chain. Trends in inherent viscosities in the oligomers indicate fairly good molecular weight control and, as expected, the inherent viscosity values increased with increasing molecular weight.

Table I. Functionalized Precursor Polymer

Material	Poly(amic ester)		
	Target M_n	M_n ^{1}H-NMR	M_n NIMP
PI-Si(OMe)$_3$	10,000 20,000	8,900 17,000	0.17 0.24

Homogeneous solutions of the phenyl-substituted PSSQ and poly(amic ethyl ester) were prepared in either NMP or N,N-dimethylpropylene urea (DMPU), case and cured to 410 °C which effects both imidization and condensation. The resulting films were defect free and could be prepared with film thicknesses as high as 10 μm, depending on the polyimide composition. Interestingly, the temperature range for imidization is between 250 and 350 °C, which is nearly identical to that for the polysilsesquioxane condensation chemistry and the loss of solvent. In contrast, hybrids from the methyl-substituted SSQ cast from NMP or DMPU produced poor films (i.e., significant dewetting). In addition, these solutions containing the methyl-substituted SSQ tended to gel in hours. Alternatively, a hydroxy-ethyl NMP and NMP solvent mixture gave high film quality and better solution stability for the high-methyl hybrids. The characteristics of the hybrid systems prepared are shown in Table I. The composition of polyimide in the composite was intentionally maintained low and ranged from 5 to 30 wt%. The thermal stability of the composites was not compromised by the incorporation of the polyimide and was comparable to that of the cured SSQ derivatives.

The preliminary mechanical characteristics of the PMDA-3F/PSSQ hybrids cast from DMPU and cured to 410 °C together with those of fused silica, used as a PECVD silica model, are shown in Table II. The modulus of the cured PSSQ is lower than that of the fused silica, an effect expected from the organic substituents on the PSSQ. The addition and curing of the poly(amic alkyl ester) results in a hybrid with a modulus of 3 GPa. A summary of the fracture behavior of the hybrids in comparison with that of bulk silica after 24 h of exposure is given in Table II. In this table, the term "unstable" refers to spontaneous global fragmentation of the film, "stable extension" refers to the growth of the indentation cracks to a new finite length and "unstable extension" refers to the growth of cracks completely across the specimen. Although qualitative, these observations clearly suggest the addition of polyimide enhances the resistance of the hybrids to cracking whether spontaneously initiated in the film or induced by deliberate

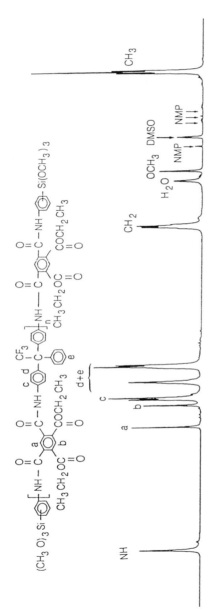

Figure 2. ¹H-NMR spectra of poly(amic ethyl ester) containing trimethoxysilyl end groups.

indentation. The maximum critical thickness for suppression of film fragmentation under the conditions appears to be at least 2 microns for each sample studied. However, the sample containing 23 wt% polyimide appears to offer greater resistance to continued crack propagation of introduced cracks over that of the silica model.

Table II. Mechanical Characteristics of Hybrids

Material	Thickness	Treatment Conditions				
		Air	Water	Indent/Water	Indent/Water /Stress	Modulus,[a] E(GP$_a$)
Fused Silica	Bulk	Stable	Stable	Very little stable extension	Very little stable extension	7.2
PSQ-20 %PI	3 μm	Unstable (day)	Unstable (hours)	Unstable extension	Unstable extension	9.2
	2 μm	Stable	Stable	No extension	Large stable extension	
	1 μm	Stable	Stable	No extension	Large stable extension	
PSQ-23 %PI	3 μm	Unstable (day)	Unstable (day)	Unstable extension	Unstable extension	
	2 μm	Stable	Stable	No extension	No extension	
	1 μm	Stable	Stable	No extension	No extension	6.3

[a]Modulus of neat PSQ resin is 326 Pa.

Over the compositional ranges and film thicknesses studied (0.5 to 5.0 phase-separated microns), the hybrids prepared from 3FDA/PMDA polyimide produced clear, transparent films when the films were cast from DMPU. Conversely, films cast from NMP were translucent, suggesting the presence of phase-separated structures on the order of the wavelength of visible light. TEM micrographs of the samples containing 23 wt.% polyimide and cast from either NMP (Bp = 206 °C) or DMPU (Bp = 284 °C) are shown in Figure 3. The sample cast from NMP has polyimide "rich" regions which are on the order of 250 nm. In contrast, the regions of significant electron density contrast are smaller than 100 nm for the samples case from DMPU. The extent of phase separation appears to be kinetically rather than thermodynamically controlled. The higher boiling DMPU presumably keeps the composite homogeneous until vitrification limits the extent of phase separation.

Summary

We have described the preparation of hybrid polyimide-silsesquioxane materials where the polyimide becomes chemically incorporated into the organosilicate matrix through reactive chain ends. A variety of thin film morphologies can be obtained depending on the composition, solvent, processing conditions, etc. Films of this type seem to be significantly toughened relative to pure silsesquioxanes and resist crack propagation

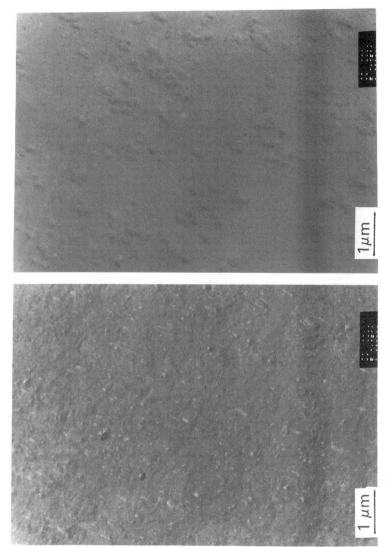

Figure 3. TEM micrographs of polyimide/PSSQ hybrids cast from NMP (left) and DMPU (right)

382

under aggressive testing procedures. The electrical properties of these films seem acceptable for IMD applications and the dielectric constant is less than 3.0. The data presented suggest that these toughened inorganic-organic hybrid polymeric materials have a significant potential for multilayer IMD applications.

Literature Cited

1. Tummala, R. R.; Keyes, R. W.; Grobman, W. D.; Kapur, S. In *Microelectronics Packaging Handbook*; Tummala , R. R., Rymaszewski, E. J., Eds.; Van Nostrand Reinhold: 1989; Chapter 9.
2. Murarka, S. P. *Solid State Technology* **1996** (March), 83.
3. Jang, S.-P.; Havemann, R. H.; Chang, M. C. *Mater. Res. Soc. Symp. Proc.* 1994, *337*, 25.
4. Singer, P. *Semiconductor International* **1996** (May)**,** 88.
5. (a) Chee, J. Y.; Draye, J. S.; Gupta, S.; Hopkins, R.; Wiesner, J. Proc. VIMC Conference **1993**, 128; (b) Hacker, N. P.; Drage, J. S.; Katsanes, R.; Sebahar, P. Proc. VIMC Conference **1995**, 138.
6. (a) Scherer, G. W. *J. Non-Cryst. Solids* **1986**, *87*, 199; (b) Brinker, C. J.; Scherer, G. W.; Roth, E. P. *J. Non-Cryst. Solids* **1985**, *72*, 345; (c) Scherer, G. W.; Brinker, C. J.; Roth, E. P. *J. Non-Cryst. Solids* **1985**, *72*, 369; (d) Strawbridge, I. In *Chemistry of Glasses*; 2nd edition, Paul, A., Ed.: Chapman and Hall, 1990; pp. 57, 85; (e) Schmidt, H.; Seiferling, B. *Mater. Res. Soc. Symp. Proc.* **1986**, *73*, 731.
7. (a) Noell, J. W. W.; Wilkes, G. L.; Mohanty, D. K.; McGrath, J. E. *J. Appl. Polym. Sci.* **1990**, *40*, 1177; (b) Wang, B.; Huang, H. H.; Wilkes, G. L.; Liptak, S.; McGrath, J. E. *Polym. Mater. Sci. & Eng.* **1990**, *62*, 892; (c) Novak, B. M. et al. *J. Am. Chem. Soc.* **1991**, *113*, 2756; (d) Mascia, L. *Trends in Polymer Science* **1995**, *3(2)*, 61; (e) Morikawa, A.; Iyoku, Y.; Kakimoto, M.; Imai, Y. *J. Mater. Sci.* **1992**, *2(7)*, 679; (f) Wang, B.; Wilkes, G. L.; Smith, C. D.; McGrath, J. E. *Polym. Commun.* **1991**, 32(13), 400; (g) Wang, B.; Wilkes, G. L.; Hedrick, J. C.; Liptak, B. C.; McGrath, J. E. *Macromolecules* **1991**, *24*, 3449; (h) Nandi, M.; Conklin, J. A.; Salvati, L., Jr.; Sen, R. *Chem. Mater.* **1991**, *3*, 201.
8. Schmidt, H. *J. Non-Crystalline Solids* **1988**, *100*, 51.
9. Brinker, D. J.; Scherer, G. W. *Sol-Gel Science, The Physics and Chemistry of Sol-Gel Processing;* Academic: San Diego, CA, 1990.
10. *Better Ceramics Through Chemistry, Mat. Res. Soc. Symp. Proc.*; Brinker, C. J., Clark, D. E., Ulrich, D. R., Eds.; North-Holland: New York, 1984; vol. 32.
11. Morikawa, A.; Kyoku, Y.; Kakimoto, M.; Imai, Y. *Polym. J.* **1991**, *28*, 107.
12. Morikawa, A.; Kyoku, Y.; Kakimoto, M.; Imai, Y. *J. Maater. Chem.* **1992**, *1*, 679.
13. Johnen, N.; Beecroft, L. L.; Ober, C. K. In *Step Growth Polymers for High Performance Materials: New Synthetic Methods*; Hedrick, J. L., Labadie, J. W., Eds.; ACS Symp. Series 624; p. 392.
14. Iyoku, Y.; Kakimoto, M.; Imai, Y. *High Performance Polym.* **1994**, *6*, 95.
15. Nandi, M.; Conklin, J. A.; Salvati, L., Jr.; Sen, A. *Chem. Mater.* **1991**, *3*, 201.
16. Nandi, M.; Conklin, J. A.; Salvati, L., Jr.; Sen, A. *Chem. Mater.* **1990**, *2*, 772.

17. Mascia, L.; Kioul, A. *J. Mater. Sci.* **1994**, *13*, 641.
18. Kioul, A.; Mascia, L. *J. Non-Cryst. Solids* **1994**, *174*, 169.
19. Mascia, L. *Trends in Polym. Sci.* **1995**, *3*, 61.
20. Volksen, W.; Yoon, D. Y.; Hedrick, J. L.; Hofer, D. *Mater. Res. Soc. Symp. Proc.* **1991**, *227*, 23.
21. Chujo, Y.; Saegusa, T. *Adv. in Polym. Sci.* **1992**, *100*, 11.
22. Landry, D. J. T.; Coltrain, B. K.; Wesson, J. A.; Lippert, J. L.; Zumbulyadis, N. *Polymer* **1992**, *33(7)*, 1496.
23. Rogers, M. E.; Brink, M. H.; McGrath, J. E. *Polymer* **1993**, *34*, 849.
24. Srinivasan, S.; Hedrick, J. L.; Di Pietro, R.; Miller, R. D. *Polymer*, submitted 1996.
25. (a) Brown, J. F., Jr.; Vogt, L. H., Jr.; Katchman, A.; Eustance, J. W.; Kiser, K. M. *J. Am. Chem. Soc.* **1960**, *82*, 6194; (b) Baney, R. H.; Itoh, M.; Sakakibara, A.; Suzuki, T. *Chem. Rev.* **1995**, *95*, 1409; (c) Loy, D. A.; Shea, K. *Chem. Rev.* **1995**, *95*, 1431.
26. Stevens, M. P. In *Polymer Chemistry: An Introduction*; 2nd edition; Oxford University Press: 1990; Chapter 10.

Author Index

Subject Index

387

T